# REINFORCED CONCRETE DESIGN TO BS8110

*Simply explained*

# REINFORCED CONCRETE DESIGN TO BS8110

## *Simply explained*

A. H. ALLEN
*MA, BSc, CEng, FICE, FIStructE*
*Formerly Head of Design Department*
*Training Division*
*Cement and Concrete Association*

*London*
*New York*
E. & F.N. SPON

*First published in 1988 by*
*E. & F. N. Spon Ltd*
*11 New Fetter Lane, London EC4P 4EE*
*Published in the USA by*
*E. & F. N. Spon*
*29 West 35th Street, New York NY 10001*



*Phototypeset in 10/12pt Photina by*
*Cotswold Typesetting Ltd, Gloucester*
*Printed in Great Britain by*
*J. W. Arrowsmith Ltd, Bristol*

ISBN 0 419 14550 8



British Library Cataloguing in Publication Data

Allen, A.H., 1924–
Reinforced concrete design to BS8110.
1. Reinforced concrete structures. Design
I. Title
624.1'8341

ISBN 0–419–14550–8

Library of Congress Cataloging in Publication Data

Allen, A.H. (Arthur Horace)
Reinforced concrete design to BS8110: simply explain/A.H. Allen.
p. cm.
Includes index.
ISBN 0–419–14550–8 (pbk.)
1. Reinforced concrete construction—Standards—Great Britain.
I. Title.
TA683.24.A36 1988
624.1'8341—dc 19 88–9702
CIP

# CONTENTS

# PREFACE

At the request of many practising design engineers the author has updated his previous book *Reinforced Concrete Design to CP110 – Simply Explained* (Cement and Concrete Association, London, 1974). With the knowledge gained from writing that book and lecturing on design courses presented by the Cement and Concrete Association he was invited to join the drafting panel to revise Section 3, *Reinforced Concrete*, of CP110. Dr Matthews has said in the Foreword of BS8110 that there are no major changes in principle, but the text has been largely rewritten with alterations in the order and arrangement of topics. This is very true, but in the alterations new material has also been introduced.

One of the inherent problems in updating a Code of Practice is that the members of the Committee are familiar with the existing Code. They are aware of the ideas and principles involved when it was written and assume that engineers using the revised version will also be familiar with the background. This is not always the case, and in the first chapter of this new book the author has repeated some of the material from the previous book to ensure that young engineers do not forget the basic principles of limit state design.

A criticism of the earlier Code of Practice CP114 was that it appeared to concentrate on the strength criterion, even though permissible stresses at working conditions were involved. A feature of the new design method was that not only did it involve strength (ultimate limit state), but also deflection and cracking (serviceability limit states) were taken more seriously.

With the passage of time the emphasis on serviceability appears to be diminishing. Minor calculations still have to be carried out and in Part 2 of the Code there is a great deal of information; but will the majority of engineers read Part 2? For many of what are referred to as 'normal' building structures they will not need to do this, and some of the basic understanding of what they are doing in Part 1 will be overlooked.

As with the previous Code, a Handbook has been produced which gives background information, but mainly in the form of references*. Many design engineers are still seeking guidance in using the Code itself and it is the hope of the author that this book will be found to be as useful as the previous one. Design examples and design aids are given and these should prove to be of use to new and experienced engineers.

For those engineers familiar with CP110 the terminology should not be a problem, but they will have to get used to looking in a different place for the information they need and, in several cases, a slightly different approach. The change is obviously not as great as when engineers had to convert from CP114 to CP110.

The notation is essentially the same as that of CP110, but in the revision it was decided that the notation relevant to each topic should be included at the beginning of that topic. This means that symbols are often repeated and the reader has to make certain that the same symbol means the same thing every time.

**Handbook to British Standard BS8110: 1985 Structural use of Concrete*, Rowe, R. E. *et al.* (1987).

It is expected that as reference to clauses is made the reader will have at hand a copy of the Code, thus avoiding the constant repetition of many clauses.

Although this book has been prepared from the author's own lecture notes he would like to acknowledge contributions made by his colleagues in the Cement and Concrete Association. These contributions have been in the form of comments, (helpful) criticism and discussions. He thanks, in alphabetical order, Andrew Beeby, John Clarke, Ray Rogers and Tony Threlfall, and also Susan Munday who typed it all.

# NOTATION

BS8110 now provides a list of symbols at the beginning of each section defining the symbols used in that section rather than a general list at the start of the Code. An attempt has been made here to give a general list of symbols relating to reinforced concrete. In a number of cases the list appears to contain ambiguities, but as this book is meant to be read in conjunction with the Code, the reader should find that no ambiguities actually occur in use.

| | |
|---|---|
| $A_c$ | Area of concrete |
| $A_{cc}$ | Area of concrete in compression |
| $A_h$ | Area of steel required to resist horizontal shear |
| $A_s$ | Area of tension reinforcement |
| $A_{sb}$ | Area of bent-up bars |
| $A_{sc}$ | Area of compression reinforcement, or, in columns, the area of reinforcement |
| $A'_s$ | Area of compression reinforcement |
| $A_{s,prov}$ | Area of tension reinforcement provided at midspan (at support for a cantilever) |
| $A'_{s,prov}$ | Area of compression reinforcement provided |
| $A_{s,req}$ | Area of tension reinforcement required at midspan to resist the moment due to design ultimate loads (at support for a cantilever) |
| $A_{st}$ | Area of transverse steel in a flange |
| $A_{sv}$ | Area of shear reinforcement, or area of two legs of a link |
| $a$ | Deflection |
| $a'$ | Distance from the compression face to the point at which the crack width is being calculated |
| $a_b$ | Centre-to-centre distance between bars (or groups of bars) perpendicular to the plane of bend |
| $a_{cr}$ | Distance from the crack considered to the surface of the nearest longitudinal bar |
| $a_f$ | Angle of internal friction between the faces of the joint |
| $a_u$ | Deflection of column at ultimate limit state |
| $a_{uav}$ | Average deflection of all columns at a given level at ultimate limit state |
| $a_v$ | Length of that part of a member traversed by shear failure plane |
| $b$ | Width (breadth) or effective width of section |
| $b'$ | Effective section dimension of a column perpendicular to the $y$ axis |
| $b_c$ | Breadth of the compression face of a beam measured midway between restraints (or the breadth of the compression face of a cantilever) |
| $b_e$ | Breadth of effective moment transfer strip (of flat slab) |
| $b_t$ | Width of section at the centroid of tension steel |
| $b_v$ | Width (breadth) of section used to calculate the shear stress |
| $b_w$ | Breadth or effective breadth of the rib of a beam |
| $c$ | Width of column |

| | |
|---|---|
| $c_{min}$ | Minimum cover to the tension steel |
| $c_x$, $c_y$ | Plan dimensions of column, parallel to longer and shorter side of base respectively |
| $d$ | Effective depth of section or, for sections entirely in compression, distance from most highly stressed face of section to the centroid of the layer of reinforcement furthest from that face |
| $d'$ | Depth to the compression reinforcement |
| $d_h$ | Depth of the head (of a column) |
| $d_n$ | Depth to the centroid of the compression zone |
| $E_c$ | Static modulus of elasticity of concrete |
| $E_{cq}$ | Dynamic modulus of elasticity of concrete |
| $E_{c,t}$ | Static modulus of elasticity of concrete at age $t$ |
| $E_{eff}$ | Effective (static) modulus of elasticity of concrete |
| $E_n$ | Nominal earth load |
| $E_s$ | Modulus of elasticity of reinforcement |
| $E_t$ | Modulus of elasticity of concrete at the age of loading $t$ |
| $E_u$ | Modulus of elasticity of concrete at age of unloading |
| $E_0$ | Initial modulus of elasticity at zero stress |
| $e$ | Eccentricity, or the base of Napierian logarithms |
| $e_a$ | Additional eccentricity due to deflections |
| $e_x$ | Resultant eccentricity of load at right angles to the plane of the wall |
| $e_{x,1}$ | Resultant eccentricity calculated at the top of a wall |
| $e_{x,2}$ | Resultant eccentricity calculated at the bottom of a wall |
| $F$ | Total design ultimate load on a beam or strip of slab |
| $F_b$ | Design force in a bar used in the calculation of anchorage bond stresses |
| $F_{bt}$ | Tensile force due to ultimate loads in a bar or group of bars in contact at the start of a bend |
| $F_s$ | Force in a bar or group of bars |
| $F_t$ | Basic force used in defining tie forces |
| $f$ | Stress |
| $f_{bs}$ | Bond stress |
| $f_{bu}$ | Design ultimate anchorage bond stress |
| $f_c$ | Maximum compressive stress in the concrete under service loads |
| $f_{cu}$ | Characteristic strength of concrete |
| $f_s$ | Estimated design service stress in the tension reinforcement |
| $f_t$ | Maximum design principal tensile stress |
| $f_y$ | Characteristic strength of reinforcement |
| $f_{yv}$ | Characteristic strength of shear or link reinforcement |
| $G$ | Shear modulus |
| $G_k$ | Characteristic dead load |
| $H$ | Storey height |
| $h$ | Overall depth of the cross-section measured in the plane under consideration |
| $h'$ | Effective section dimension in a direction perpendicular to the $x$ axis |
| $h_{agg}$ | Maximum size of the coarse aggregate |
| $h_c$ | Effective diameter of a column or column head |
| $h_f$ | Depth (thickness) of flange |
| $h_{max}$ | Larger dimension of a rectangular section |
| $h_{min}$ | Smaller dimension of a rectangular section |
| $I$ | Second moment of area of the section |
| $K$ | Coefficient, as appropriate |

| | |
|---|---|
| $L$ | Span of member or, in the case of a cantilever, length |
| $l$ | Span or effective span of member, or anchorage length |
| $l_a$ | Clear horizontal distance between supporting members |
| $l_{b,1}$ | Breadth of supporting member at one end or 1.8 m, whichever is the smaller |
| $l_{b,2}$ | Breadth of supporting member at the other end or 1.8 m, whichever is the smaller |
| $l_c$ | Dimension related to columns (variously defined) |
| $l_e$ | Effective height of a column or wall |
| $l_{ex}$, $l_{ey}$ | Effective height in respect of the major or minor axis respectively |
| $l_h$ | Effective dimension of a head (of column) |
| $l_o$ | Clear height of column or wall between end restraints |
| $l_r$ | Distance between centres of columns, frames or walls supporting any two adjacent floor spans |
| $l_s$ | Floor-to-ceiling height |
| $l_x$, $l_y$ | Length of sides of a slab panel or base |
| $l_z$ | Distance between point of zero moment |
| $l_1$ | Panel length parallel to span, measured from centres of columns |
| $l_2$ | Panel width, measured from centres of columns |
| $M$ | Design ultimate resistance moment |
| $M_{add}$ | Additional design ultimate moment induced by deflection of beam |
| $M_i$ | Initial design ultimate moment in a column before allowance for additional design moments |
| $M_t$ | Design moment transferred between slab and column |
| $M_{t,max}$ | Maximum design moment transferred between slab and column |
| $M_u$ | Design moment of resistance of the section |
| $M_x$, $M_y$ | Design ultimate moments about the $x$ and $y$ axis respectively |
| $M_{x'}$, $M_{y'}$ | Effective uniaxial design ultimate moments about the $x$ and $y$ axis respectively |
| $M_1$ | Smaller initial end moment due to design ultimate loads |
| $M_2$ | Larger initial end moment due to design ultimate loads |
| $m_{sx}$, $m_{sy}$ | Maximum design ultimate moments either over supports or at midspan on strips of unit width and span $l_x$ or $l_y$ respectively |
| $N$ | Design axial force |
| $N_{bal}$ | Design axial load capacity of a balanced section |
| $N_d$ | Number of discontinuous edges ($0 \leqslant N \leqslant 4$) |
| $N_{uz}$ | Design ultimate capacity of a section when subjected to axial load only |
| $n$ | Design ultimate load per unit area, or number of columns resisting sidesway at a given level or storey (in clause 3.8.1.1) |
| $n_o$ | Number of storeys in a structure |
| $n_w$ | Design ultimate axial load |
| $Q_k$ | Characteristic imposed load |
| $R$ | Restraint factor (against early thermal contraction cracking) |
| $r$ | Internal radius of bend |
| $r_{ps}$ | Radius of curvature |
| $1/r_b$ | Curvature at midspan or, for cantilevers, at the support section |
| $1/r_{cs}$ | Shrinkage curvature |
| $1/r_x$ | Curvature at $x$ |
| $S_s$ | First moment of area of reinforcement about the centroid of the cracked or gross section |
| $s_b$ | Spacing of bent-up bars |

| | |
|---|---|
| $s_v$ | Spacing of links along member |
| $T$ | Torsional moment due to ultimate loads |
| $t_e$ | Effective thickness of a slab for fire resistance assessment |
| $t_f$ | Thickness of non-combustible finish (for fire resistance) |
| $u$ | Length (or effective length) of the outer perimeter of the zone considered |
| $u_o$ | Effective length of the perimeter which touches a loaded area |
| $V$ | Shear force due to design ultimate loads, or design ultimate value of a concentrated load |
| $V_b$ | Design shear resistance of bent-up bars |
| $V_c$ | Design ultimate shear resistance of the concrete |
| $V_{co}$ | Design ultimate shear resistance of a section uncracked in flexure |
| $V_{cr}$ | Design ultimate shear resistance of a section cracked in flexure |
| $V_{eff}$ | Design effective shear force in a flat slab |
| $V_t$ | Design shear force transferred to column |
| $v$ | Design shear stress |
| $v_c$ | Design shear stress in the concrete |
| $v_{c'}$ | Design concrete shear stress corrected to allow for axial forces |
| $v_{max}$ | Maximum design shear stress |
| $v_{sx}$, $v_{sy}$ | Design end shear on strips of unit width and span $l_x$ or $l_y$ respectively and considered to act over the middle three-quarters of the edge |
| $v_t$ | Torsional shear stress |
| $v_{t,min}$ | Minimum torsional shear stress, above which reinforcement is required |
| $v_{tu}$ | Maximum combined shear stress (shear plus torsion) |
| $W_k$ | Characteristic wind load |
| $x$ | Neutral axis depth, or dimension of a shear perimeter parallel to the axis of bending |
| $x_1$ | Smaller centre-to-centre dimension of a rectangular link |
| $y_o$ | Half the side of the end block |
| $y_{po}$ | Half the side of the loaded area |
| $y_1$ | Larger centre-to-centre dimension of a rectangular link |
| $z$ | Lever arm |
| $\alpha$ | Coefficient of expansion, or angle between shear reinforcement and the plane of beam or slab |
| $\alpha_{c,1}$, $\alpha_{c,2}$ | Ratio of the sum of the column stiffness to the sum of the beam stiffness at the lower or upper end of a column respectively |
| $\alpha_{c,min}$ | Lesser of $\alpha_{c,1}$ and $\alpha_{c,2}$ |
| $\alpha_e$ | Modular ratio ($E_s/E_{eff}$) |
| $\alpha_{sx}$, $\alpha_{sy}$ | Bending moment coefficients for slabs spanning in two directions at right angles, simply supported on four sides |
| $\beta$ | Coefficient, variously defined, as appropriate |
| $\gamma_f$ | Partial safety factor for load |
| $\gamma_m$ | Partial safety factor for strength of materials |
| $\Delta t$ | Difference in temperature |
| $\varepsilon$ | Strain |
| $\varepsilon_{cc}$ | Final (30 year) creep strain in concrete |
| $\varepsilon_{cs}$ | Free shrinkage strain |
| $\varepsilon_{c,1}$ | Strain in concrete at maximum stress |
| $\varepsilon_m$ | Average strain at the level where the cracking is being considered |
| $\varepsilon_r$ | Thermal strain assumed to be accommodated by cracks |
| $\varepsilon_{sh}$ | Shrinkage of plain concrete |

| | |
|---|---|
| $\varepsilon_1$ | Strain at the level considered, calculated ignoring the stiffening effect of the concrete in the tension zone |
| $\mu$ | Coefficient of friction |
| $\xi$ | Proportion of solid material per unit width of slab |
| $\rho$ | Area of steel relative to area of concrete |
| $\phi$ | Creep coefficient, or diameter |
| $\phi_e$ | Effective bar size |

# LIMIT STATE PRINCIPLES 1

Codes of Practice have developed considerably since the first attempt in 1934. The way in which they have done so is outlined in Table 1.1.

From Table 1.1 it can be seen that working stresses have gradually increased and the load factor or factor of safety has decreased. This has arisen mainly from the satisfactory performance of structures and the general improvement in construction standards.

Quality control of concrete took a large jump forward in the 1965 edition of CP114, when statistical control was introduced and the allowable compressive stress in bending was increased if there was a design mix. Substantial progress had also been made in the philosophical approach to structural design, mainly due to the work of the international committees. One of these, the Comité Européen du Béton (CEB), published in 1963 its *Recommendations for an International Code of Practice for Reinforced Concrete*, generally known as the Blue Book, and later, in conjunction with the Fédération Internationale de la Précontrainte (FIP), a complementary report dealing with prestressed concrete. Further to these there was published in 1970 the *International Recommendations for the Design and Construction of Concrete Structures* giving the Principles and Recommendations and generally known as the Red Book.

When the drafting committees for CP114 and 115 were reconvened in 1964, they agreed to adopt the CEB report as a guide in the preparation of the new British Codes with the proviso that the new recommendations should not change unduly the

**Table 1.1** Development of Codes of Practice since 1934

| *Code* | *Steel stress (working load)* | *Load factor* | *Deflection* | *Cracking* | *Comments* |
|---|---|---|---|---|---|
| 1934 DSIR | $0.45f_y$ (140N/mm$^2$) | 2.2 | Nothing | Nothing | Concrete – nominal proportions. Beams – straight line theory |
| 1948 CP114 | $0.5f_y$ (189N/mm$^2$) | 2.0 | Warning | Nothing | Ditto |
| 1957 CP114 | $0.5f_y$ (210N/mm$^2$) | 2.0 | Warning + span/depth | Nothing | Concrete – nominal or strength |
| 1965 CP116 } CP114 } | $0.55f_y$ (230N/mm$^2$) | 1.8 | Warning + expanded span/depth | Warning | Concrete – statistical control for quality |
| 1972 CP110 | $0.58f_y$ (267N/mm$^2$)* (without redistribution) | 1.6*–1.8 | Span/effective depth ratios | Bar spacing rules | Ditto |

*It should be noted that these values are included as an indication to show the trend; specific values are not given in the Code.

proportions of structures compared with those designed to the recommendations of the current codes. The main consequence of this decision was that limit state design was accepted as the basis for the preparation of the new drafts. Later, these two committees in conjunction with the drafting committee for CP116 agreed to the unification of the three codes into a single document, which would rationalize design and coordinate detailed interpretation, for concrete construction. The Code for Water Retaining Structures was not included and although that Code (now BS8007) relies very heavily on the building structures document it has retained its independence.

In drafting CP110 it was decided to go right back to square one and establish the engineer's intentions and problems. The purpose of design may, perhaps oversimply, be stated as the provision of a structure complying with the client's and the user's requirements. In design appropriate attention must be paid to overall economy, the safety, serviceability and aesthetics of the structure. In most cases the design process entails finding the cheapest solution capable of satisfying the appropriate safety, serviceability and aesthetic considerations.

The design of a structure for a specific function is usually a two-stage process, involving first the selection of an appropriate type or form of structure and secondly the detailed design of the various parts of the chosen structure. In selecting the type or form of structure the question of the relative costs of different types of structures and of different methods of construction of the same structure will be of great importance. In this selection the designer must rely to a large extent on his experience, judgement and intuition. A preliminary study of several types of structure may be necessary.

Having selected the type of structure the designer then has to proceed with the detailed design of the chosen one, always bearing in mind the factors of safety considerations and cost. In most cases the aesthetic requirements will have been substantially met in the selection of the type of structure and will now be completely satisfied by the specification of surface finishes, colour, etc. Fundamentally, then, the design process consists of finding and detailing the most economical structure consistent with the safety and serviceability requirements.

In design the following points have to be taken into consideration:

1. variations in materials in the structure and in test specimens
2. variations in loading
3. constructional inaccuracies
4. accuracy of design calculations
5. safety and serviceability.

For (1) we know that the cube test is a reliable guide as regards quality of concrete from the mixer but does not guarantee that the concrete in the structure is the same. If we get consistent cube results of the required strength this means that the potential of the concrete in the structure is higher. This is why we took a higher proportion of the cube strength as a permissible stress when we have quality control, i.e. a design mix. There is, however, still no guarantee that the concrete in the structure is of the same consistent strength and properties, as has been shown from tests that have been performed. The same applies to reinforcement, as tests are carried out on small samples which may or may not be truly representative of the whole. For (2) we must enquire how near the truth is the loading given in BS6399, Part 1. Constructional inaccuracies (3) are probably accidental. For (4) designers can and do make mistakes in calculations but very often in analysis they assume a structure will behave in a certain way or that certain conditions exist. Item (5) is dealt with quite arbitrarily in previous codes – if the structure does not collapse it is deemed to be satisfactory.

So, having the purpose of design which, as previously stated, consists of finding and dealing with the most economical structure associated with safety and serviceability requirements, and conscious that variability exists between construction materials and the construction process itself, if we now list the various criteria required to define the serviceability or usefulness of any structure we should be able to state a design philosophy to cope with these in a rational manner.

The various criteria required to define the serviceability or usefulness of any structure can be described under the following headlines. The effects listed may lead to the structure being considered 'unfit for use'.

1. *Collapse*: failure of one or more critical sections; overturning or buckling.
2. *Deflection*: the deflection of the structure or any part of the structure adversely affects the appearance or efficiency of the structure.
3. *Cracking*: cracking of the concrete which may adversely affect the appearance or efficiency of the structure.
4. *Vibration*: vibration from forces due to wind or machinery may cause discomfort or alarm, damage the structure or interfere with its proper function.
5. *Durability*: porosity of concrete.
6. *Fatigue*: where loading is predominantly cyclic in character the effects have to be considered.
7. *Fire resistance*: insufficient resistance to fire leading to 1, 2 and 3 above.

When any structure is rendered unfit for use for its designed function by one or more of the above causes, it is said to have entered a *limit state.* The Code defines the limit states as:

1. *Ultimate limit state*: the ultimate limit state is preferred to collapse.
2. *Serviceability limit states*: deflection, cracking, vibration, durability, fatigue, fire resistance and lightning.

The purpose of design, then, is to ensure that the structure being designed will not become unfit for the use for which it is required, i.e. that it will not reach a limit state. The essential basis of the design method, therefore, is to consider each limit state and to provide a suitable margin of safety. To obtain values for this margin of safety it was proposed that probability considerations should be used and the design process should aim at providing acceptable probabilities so that the structure would not become unfit for use throughout its specified life.

Accepting the fact that the strengths of constructional materials vary, as do also the loads on the structure, two partial safety factors will now be used. One will be for materials and is designated $\gamma_m$; the other, for loading, is termed $\gamma_f$. These factors will vary for the various limit states and different materials. As new knowledge on either materials or loading becomes available the factors can be amended quite easily without the complicated procedures to amend one overall factor used in previous Codes.

## 1.1 Characteristic strength of materials

For both concrete and reinforcement the Code uses the term 'characteristic strength' instead of 28-day works cube strength and yield stress, although it is still related to these. The characteristic strength for all materials has the notation $f_k$ and is defined as the value of the cube strength of concrete ($f_{cu}$), the yield or proof stress of reinforcement

($f_y$), below which 5% of all possible test results would be expected to fall. The value therefore is

$$f_k = f_m - 1.64s$$

where $f_m$ is the mean strength of actual test results determined in accordance with a standard procedure, $s$ is the standard deviation, and 1.64 is the value of the constant required to comply with 5% of the test results falling below the characteristic strength, as indicated in Fig. 1.1.

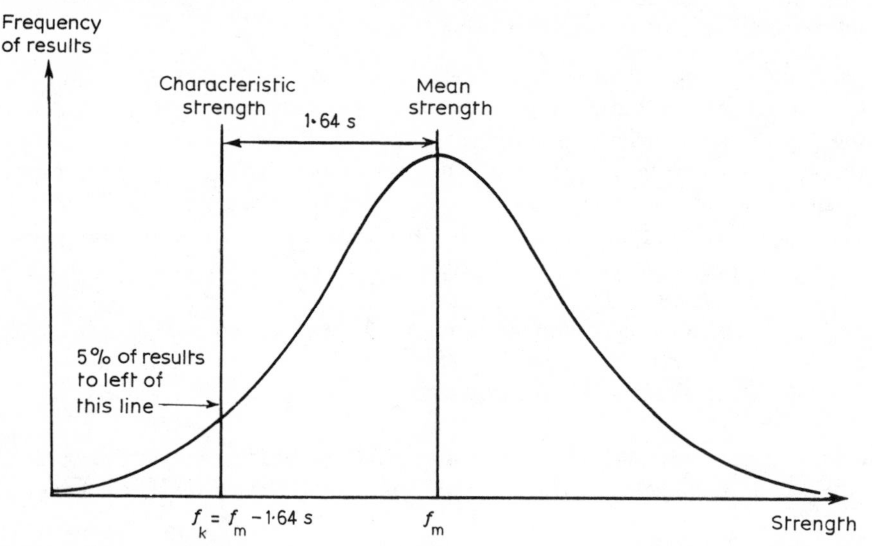


FIG. 1.1 Characteristic strength.

## 1.2 Concrete

This is dealt with in Section 6 of the Code, but BS5328: *Methods for Specifying Concrete Including Ready-Mixed Concrete* gives a much fuller treatment of the subject. Durability is given more importance in BS8110 than it was in CP110 and we shall therefore concentrate on how this requirement affects the designer, as given in Section 3.

The strength of concrete for design purposes will be based on tests made on cubes at an age of 28 days unless there is satisfactory evidence that a particular testing regime is capable of predicting the 28-day strength at an earlier age. These 28-day characteristic strengths determine the grade of the concrete and it is important to select the correct grade appropriate for use. The concrete has to provide the durability for the environmental conditions as well as adequate strength for the loading requirements.

For example, clause 3.1.7.2 says that for reinforced concrete the lowest grade should be C25 for concrete made with normal-weight aggregates. Reference to Table 3.4 of the Code, however, does not reveal a grade lower than C30. It is only by reading clause 3.3.5.2 that it can be found that under certain specific conditions a Grade C30 can be classed as Grade C25. As compliance with these conditions is not easy to achieve this will not be expanded upon (a concrete technologist is best consulted). We shall deal with the normal approach.

In selecting an appropriate grade of concrete, the designer has to determine the environment and exposure conditions to which the members of the structure will be subjected. These are given in Table 3.2 of the Code, and it is probable that there will be more than one condition of exposure. Moving then to Table 3.4 of the Code will give the lowest grade of concrete to meet the durability requirements for nominal cover to all reinforcement. For a severe exposure the lowest grade is C40 and a nominal cover of 40 mm is required. The same grade of concrete can be used for a mild exposure when the nominal cover is 20 mm.

The subject of cover to reinforcement will be discussed more fully later, but at this stage it is important for the designer to realize that a grade of concrete is selected for durability as well as for strength. Having chosen a grade, the concrete supplier will then have to add the value of 1.64 times the standard deviation to obtain the target mean strength.

In general the minimum cement content given in Table 3.4 of the Code will be exceeded but the maximum free water/cement ratio should not be exceeded.

It should also be noted that the age allowance for concrete increasing in strength with age has now been deleted.

## 1.3 Reinforcement

The reinforcement should comply with BS4449, BS4461 or BS4483, all of which specify the tests for compliance to obtain the characteristic strength. Section 7 of the Code deals with specification and workmanship. The designation of the reinforcement with its specified characteristic strength is shown in Table 1.2.

**Table 1.2** Designation of reinforcement

| *Designation* | *Nominal sizes* | *Specified characteristic strength* ($f_y$) ($N/mm^2$) |
|---|---|---|
| Hot-rolled mild steel | All sizes | 250 |
| High-yield steel (hot rolled or cold worked) | All sizes | 460 |

From the table it can be seen that the characteristic strength of high-yield bars is independent of whether they are hot-rolled or cold-rolled worked. A subdivision is made later in the Code to determine the bond characteristics which depend on the surface shape of the bar.

## 1.4 Characteristic loads

For loading we use the 'characteristic' load ($F_k$) as the basis. Ideally this should be determined from the mean load and its standard deviation from the mean, and using the same probability as for the materials we should say that $F_k = F_m + 1.64s$. The characteristic load would be that value of loading such that not more than 5% of the spectrum of loading throughout the life of the structure will lie above the value of the characteristic load (see Fig. 1.2).

Although several surveys have been carried out, we are not yet able to give a statistical interpretation and reasoning to them. The surveys do imply that, in general, the overall imposed loads to be considered in design on floors are well above the loading that occurs most of the time, although in individual areas the loading may well be above that for which the floors have been designed. For characteristic loads we shall use those given and defined in BS6399: Part 1 for dead and imposed loads and CP3:

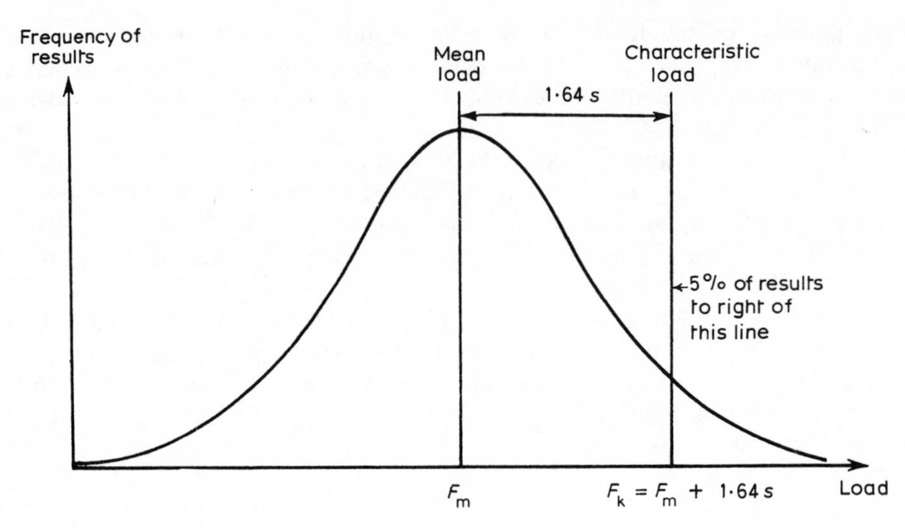


FIG. 1.2 Characteristic load.

Chapter V: Part 2 for wind loads, although the loading conditions during erection and construction should be considered in design and should not be such that the subsequent compliance of the structure with the limit state requirements is impaired.

The characteristic dead, imposed and wind loads have the notation $G_k$, $Q_k$, $W_k$ respectively, where the upper-case letters denote the total load on a span. Lower-case letters denote uniform load per square metre, although in design examples for beams the lower-case letters have been used for a uniformly distributed load, so that $G_k = g_k l$.

## 1.5 Design strengths of materials

We obtain the design strengths of the materials by dividing the characteristic strengths by the partial safety factor $\gamma_m$, i.e. design strength $= f_k/\gamma_m$.

$\gamma_m$ takes account of possible differences between the material in the actual structure and the strength derived from test specimens. In concrete, this would cover such items as insufficient compaction, differences in curing, etc. For reinforcement it would cover such items as the difference between assumed and actual cross-sectional areas caused by rolling tolerances, corrosion, etc. The values of $\gamma_m$ for each material will be different for the different limit states by virtue of the different probabilities that can be accepted.

Table 1.3 sets out these values, and it should be noted that ultimate limit state values only are given in Part 1 of the Code. For serviceability, clause 2.4.6.1 refers to 3.2 in Part 2 of the Code, but the values given in the table below are taken from the Handbook (see Preface).

**Table 1.3** Values of $\gamma_m$ for concrete and steel at different limit states

| *Limit state* | *Values of $\gamma_m$* | |
|---|---|---|
| | *Concrete* | *Steel* |
| Ultimate | 1.5 | 1.15 |
| Deflection | 1.0 | 1.0 |
| Cracking | 1.3 | 1.0 |

For both materials, the factor for the ultimate limit state is higher than the others because not only must the probability of failure be decreased but failure should also be localized. The $\gamma_m$ factor therefore also contains an allowance for this; as a compressive failure in concrete is sudden and without warning the factor for concrete is higher than for reinforcement.



Deflection is related to the whole member and the factor for both materials is 1.0. For cracking only parts of the member are affected and a factor between 1.0 and 1.5 for concrete has been selected, but kept at 1.0 for reinforcement. When one is analysing any cross-section within the structure the properties of the materials should be assumed to be those associated with their design strengths appropriate to the limit state being considered.

The short-term design stress–strain curve for concrete is shown in Fig. 1.3. By putting in the relevant value of $\gamma_m$ depending on the limit state being considered we can obtain the appropriate design stress–strain curve. Two things here are worthy of note.

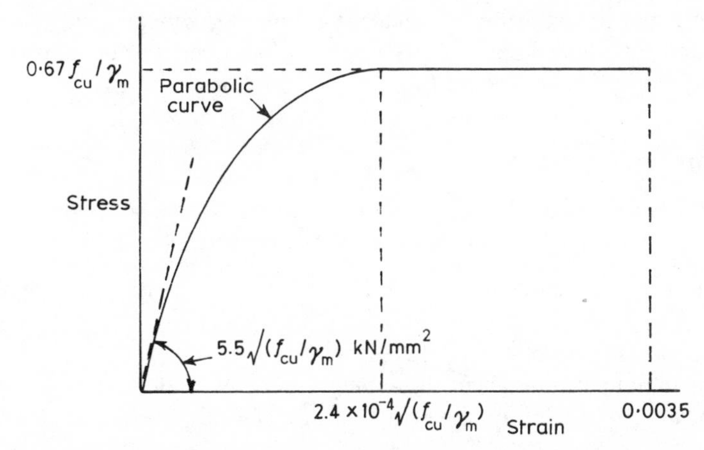


FIG. 1.3 Short-term design stress–strain relation for normal weight concrete ($f_{cu}$ in N/mm$^2$).

First, design strength has been defined as characteristic strength divided by $\gamma_m$, and yet the maximum stress value is given as $0.67f_{cu}/\gamma_m$. The reason for this is that the characteristic strength has been derived from tests on cubes. It is well established from tests that the maximum compressive stress at failure in a member of the same concrete as a cube has a value in the region of $0.8f_{cu}$. This is a peak value and as an additional safety factor against compressive failure this value has been reduced to $0.67f_{cu}$, which agrees with the design methods using ultimate load. If one were using cylinders in determining the characteristic strength the factor would be of the order of 0.85, as the cylinder strength is nearer the actual behaviour and is approximately 0.8 × cube strength.

Secondly, it is suggested in the Code that for the serviceability limit states the short-term elastic modulus may be taken from a table in the Code, i.e. a linear stress–strain relationship is assumed with a specified value for $E_c$ depending on $f_{cu}$.

The initial tangent modulus for serviceability limit states will seldom be used unless a rigorous analysis is being carried out. Where sustained loading is being considered, however, allowance for shrinkage and creep should be made. For the serviceability limit states Poisson's ratio may be taken as 0.2.

For reinforcement the short-term stress–strain relationship is shown in Fig. 1.4. This differs quite considerably from CP110 in that the relationship is now bilinear for

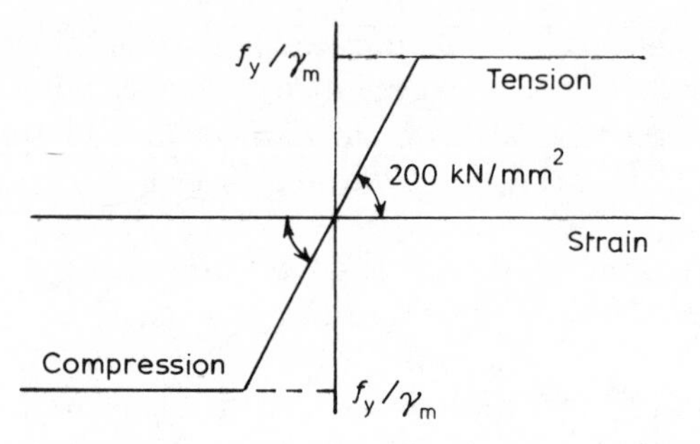


FIG. 1.4 Short-term design stress–strain relation for reinforcement ($f_y$ in N/mm$^2$).

reinforcement and also the maximum design stress in compression is the same as in tension. The elastic modulus remains at 200 kN/mm$^2$.

It is important to point out at this stage that in the majority of design calculations this is the last time that the partial safety factors for materials will be referred to. They are built in to formulae and design charts so that a designer will usually refer to the characteristic strengths, i.e. $f_{cu}$ and $f_y$.

## 1.6 Design loads

We obtain the design load by multiplying the characteristic load by the other partial safety factor $\gamma_f$; this factor $\gamma_f$ is introduced to take account of:

1. possible unusual increases in the load beyond those in deriving the characteristic load
2. inaccurate assessment of effects of loading
3. variations in dimensional accuracy achieved in construction
4. the importance of the limit state being considered.

$\gamma_f$ varies for different limit states and also for different combinations of loading. As with the design strengths of materials, Part 1 of the Code gives numerical values at ultimate limit state but the reader is referred to Part 2 to assess values at serviceability limit states. The effect of the load is now classed as adverse or beneficial. Values of $\gamma_f$ for ultimate limit state are given in Table 1.4.

**Table 1.4** Values of $\gamma_f$ at ultimate limit state

| | *Load type* | | | | | |
|---|---|---|---|---|---|---|
| | *Dead* | | *Imposed* | | *Earth and water pressure* | *Wind* |
| *Load combination* | *Adverse* | *Beneficial* | *Adverse* | *Beneficial* | | |
| 1. Dead and imposed (and earth and water pressure) | 1.4 | 1.0 | 1.6 | 0 | 1.4 | — |
| 2. Dead and wind (and earth and water pressure) | 1.4 | 1.0 | — | — | 1.4 | 1.4 |
| 3. Dead and imposed (and earth and water pressure) | 1.2 | 1.2 | 1.2 | 1.2 | 1.2 | 1.2 |

The arrangement of loads should be such as to cause the most severe effects, i.e. the most severe stresses. The 'adverse' partial factor is applied to any loads that tend to produce a more critical design condition. The 'beneficial' factor is applied to loads that tend to produce a less critical condition.

So, in a normal building structure with dead and imposed loads, the maximum design load on a span from load combination (1) is $1.4G_k + 1.6Q_k$. The minimum design load is $1.0G_k$.

Under load combination (2), which will generally be a stability condition, the most critical case may arise when moments due to $1.4G_k$ on some parts of the structure are additive to the wind moments (using $1.4W_k$) and moments due to $1.0G_k$ on other parts of the structure form the restoring moment.

For load combination (3) a factor of 1.2 is used throughout the structure, with no variations for loaded and unloaded spans.

However, when considering load combinations (2) and (3) the horizontal wind load should not be less than the notional horizontal load as given in clause 3.1.4.2 and will be dealt with when considering robustness.

In addition to the above factors we may have to consider the effects of excessive loads. In this case the $\gamma_f$ factor should be taken as 1.05 on the defined loads and only those loads likely to be acting simultaneously need be considered. Also, if a structure can sustain localized damage, to maintain continued stability the $\gamma_f$ factor is again 1.05 and is applied to all loads which are likely to occur before temporary or permanent remedial measures are taken. In general, the effects of creep, shrinkage and temperature will not be considered for ultimate limit state.

For serviceability limit states, clause 3.3 in Part 2 says that the loading assumed in these calculations will depend on whether the aim is to produce a best estimate of the likely behaviour of the structure or to comply with a serviceability limit state requirement. It does go on to say, however, that for limit state calculations the characteristic values will be used generally. This implies a partial factor of 1.0.

The design loads for the serviceability limit states apply when estimating the immediate deflections of a structure but in most cases it will be necessary to estimate the additional time-dependent deflections due to creep, shrinkage and temperature. It will also be necessary to assess how much of the live load is permanent and how much is transitory.

## EXAMPLE 1.1

A five-storey building of the cross-section shown has the following characteristic loads on the frame:

Roof: Dead 22 kN/m
Imposed 7 kN/m
Parapet 1 m high – point load 12 kN

Floors: Dead 20 kN/m
Imposed 25 kN/m
Cladding – point load 15 kN

Wind: 7 kN/m.

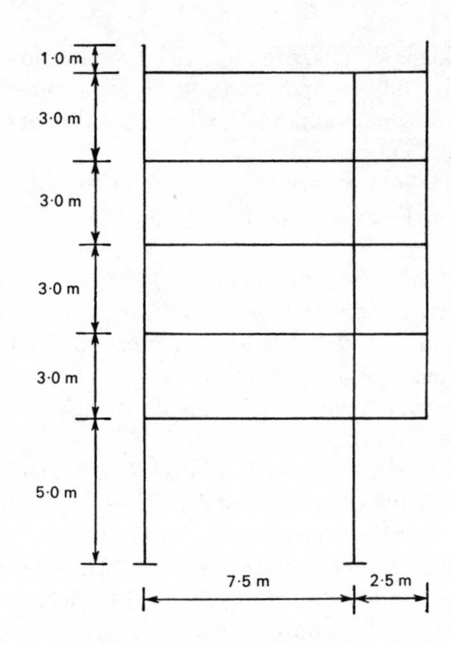


Calculate the maximum ultimate design load for the left-hand column and check if tension can occur (ignore the self load of the columns).

The structure can be simplified as follows:

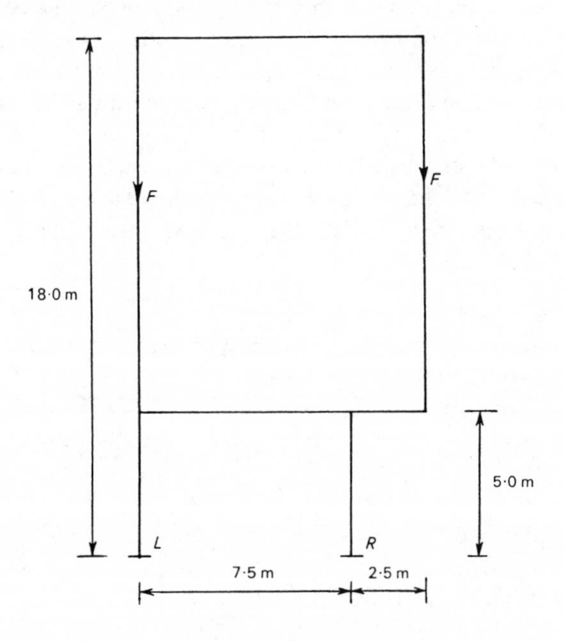


Characteristic loads:
Dead load ($g_k$) $= 22 + 4 \times 20 = 102$ kN/m
Imposed load ($q_k$) $= 7 + 4 \times 25 = 107$ kN/m
Wind load ($w_k$) $= 7$ kN/m
Point load $F_k$ $= 12 + 4 \times 15 = 72$ kN/m

## Load combination (1)

Design loads:

UD load – maximum $= 1.4\, g_k + 1.6\, q_k$
$= 1.4 \times 102 + 1.6 \times 107$
$= 142.8 + 171.2 = 314$ kN/m

minimum $= 1.0\, g_k = 1.0 \times 102 = 102$ kN/m.

Point load:

– maximum $= 1.4\, F_k = 1.4 \times 72 = 100.8$ kN (say 101 kN)

minimum $= 1.0\, F_k = 1.0 \times 72 = 72$ kN.

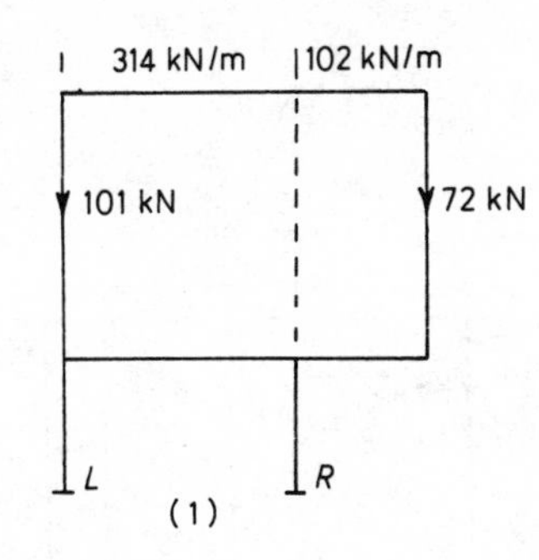


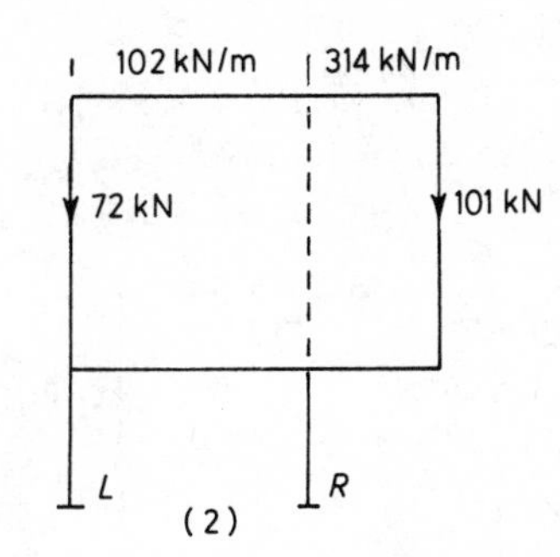


Moments about $R$

(1) $7.5L = 101 \times 7.5 + 314 \times 7.5^2/2 - 72 \times 2.5 - 102 \times 2.5^2/2$
$= 9090$
$L = 1212$ kN.

(2) $7.5L = 72 \times 7.5 + 102 \times 7.5^2/2 - 101 \times 2.5 - 314 \times 2.5^2/2$
$= 2175$
$L = 290$ kN.

## Load combination (2)

Design loads:

UD load – maximum $= 1.4\, g_k = 143$ kN/m

minimum $= 1.0\, g_k = 102$ kN/m

Point load – maximum $= 1.4\, F_k = 101$ kN

minimum $= 1.0\, F_k = 72$ kN

Wind load $= 1.4 W_k = 9.8$ kN/m

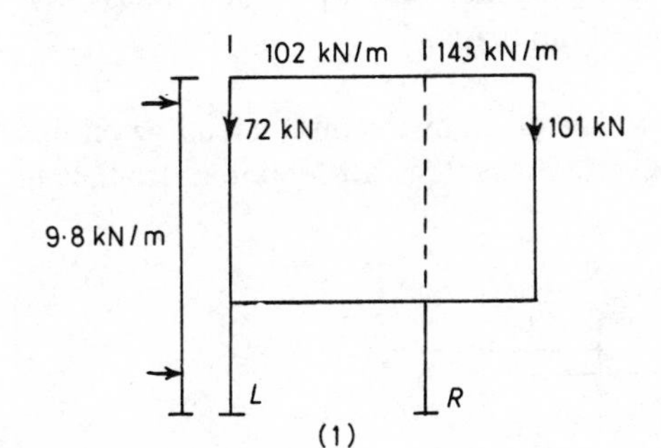


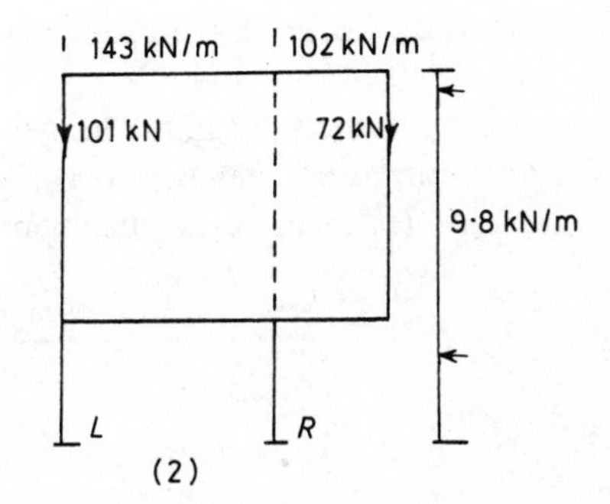

Moments about $R$

(1) $7.5L = 72\times7.5 + 102\times7.5^2/2 - 143\times2.5^2/2 - 101\times2.5 - 9.8\times18^2/2$
$= 1122$
$L = 150$ kN.

(2) $7.5L = 101\times7.5 + 143\times7.5^2/2 + 9.8\times18^2/2 - 102\times2.5^2/2 - 72\times2.5$
$= 5868$
$L = 782$ kN.

## Load combination (3)

Design loads:
UD load $= 1.2\ g_k + 1.2\ q_k = 1.2\times102 + 1.2\times107 = 251$ kN/m
Point load $= 1.2\ F_k = 1.2\times72 = 86$ kN
Wind load $= 1.2\ w_k = 1.2\times7 = 8.4$ kN/m.

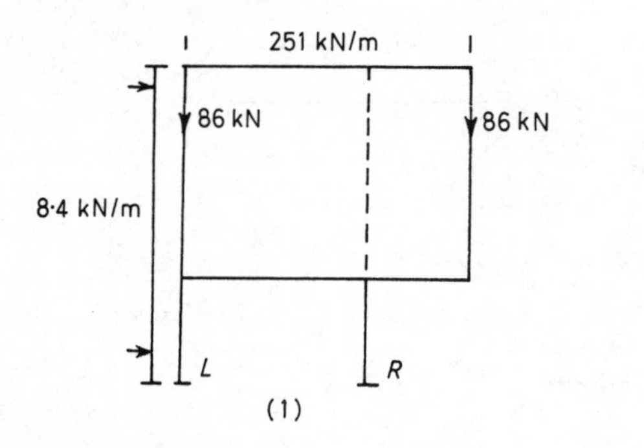


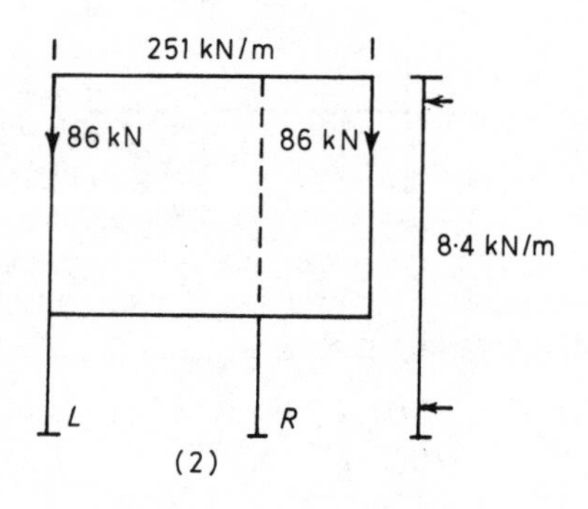


Moments about $R$

(1) $7.5L = 86\times7.5 + 251\times10\times2.5 - 86\times2.5 - 8.4\times18^2/2$
$= 5344$
$L = 713$ kN.

(2) $7.5L = 86\times7.5 + 251\times10\times2.5 + 8.4\times18^2/2 - 86\times2.5$
$= 8066$
$L = 1075$ kN.

Maximum load on column = 1212 kN (load combination 1)
Minimum load on column = 150 kN (load combination 2).

## EXAMPLE 1.2

Taking a typical floor beam in the previous example and assuming vertical loads only, which loading patterns will give the maximum moments?

Maximum span moment in the longer span will be given by maximum loading on this span and minimum loading on the cantilever. The values are obtained from load combination (1) in the previous example.

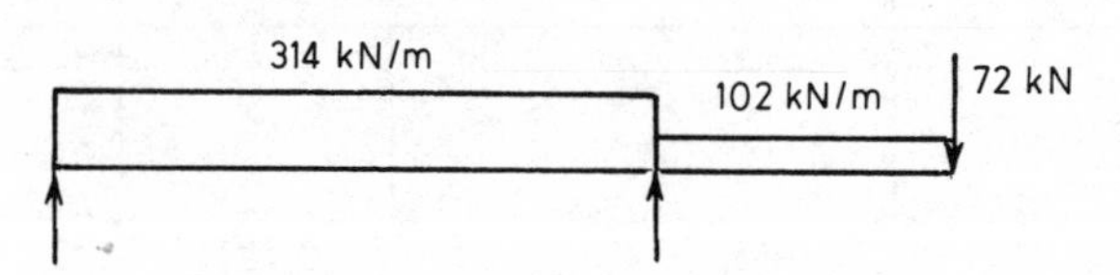

Maximum support moment will be when main span has minimum load and cantilever has maximum load.

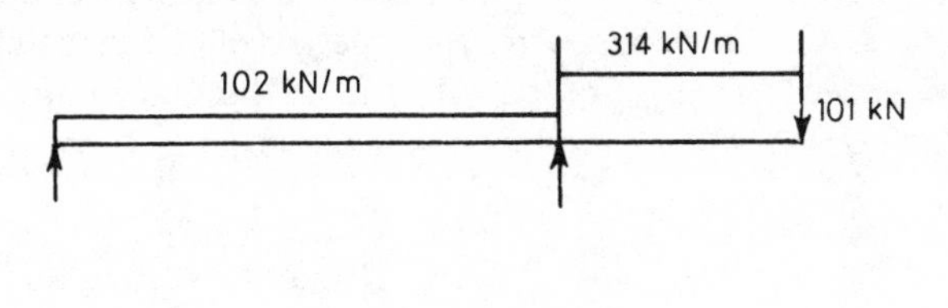


## 1.7 Limit state requirements

All relevant limit states have to be considered in design, but in reinforced concrete structures the three most important ones for design calculations will be ultimate limit state and the serviceability limit states of deflection and cracking. Durability and fire resistance will be complied with by grade of concrete, cement content, cover to reinforcement, etc., all of which will be decided before calculations begin. The usual approach will be to design on the basis of the most critical limit state and then check that the remaining limit states will not be reached. In general, therefore, we shall design for the ultimate limit state, i.e. analyse the structure and analyse the critical sections. After that we shall check that deflection is satisfactory and check that cracking is satisfactory. This does not mean that three sets of calculations will have to be carried out, although in very special circumstances this may be so. In some circumstances we may find that the critical limit state is deflection, in which case we would design for this limit state and check the other two.

The criteria with which we have to comply for the various limit states are outlined below.

### 1.7.1 Ultimate limit state

The strength of the structure should be sufficient to withstand the design loads. The layout of the structure should be such as to ensure a robust and stable design.

### 1.7.2 Serviceability limit states

(a) DEFLECTION

In general this will be met by complying with the span/effective depth ratios obtained from Section 3 in Part 1. However, Section 3 in Part 2 does give limits which give a designer the basis on which the deemed-to-satisfy rules have been obtained. These are as follows:

*Appearance*

The final deflection (including the effects of creep, shrinkage and temperature) measured below the as-cast level of the supports of horizontal members should not in general, exceed span/250. The Code talks about a 'sag', but the above description is retained from CP110.

*Damage*

Unless partitions, cladding and finishes have been specifically detailed to allow for anticipated deflections, the movement after the erection of partitions or application of finishes should not exceed the lesser of span/350 and 20 mm for non-brittle materials and the lesser of span/500 and 20 mm for brittle materials.

*Horizontal deflections*

For horizontal deflections affecting non-structural elements it is suggested that the lateral deflection in any one storey should not exceed $H/500$, where $H$ is the storey height. For excessive accelerations under wind loads reference should be made to specialist literature.

(b) CRACKING

The assessed surface width of cracks should not exceed 0.3 mm for appearance and corrosion, but where loss of performance such as watertightness is affected the limit could be less.

A fuller explanation of the criteria for the various limit states will be given in the chapters dealing with the individual limit states. To comply with the criteria we can use calculations, model analysis and testing or experimental development of analytical procedures. The most usual way will be by calculations, and the Code gives procedures for doing these. It states that the methods of analysis used in assessing compliance with the requirements of the various limit states should be based on as accurate a representation of the behaviour of the structure as is practicable but the methods and assumptions which are given in Section 2 are generally adequate. In certain cases a more fundamental approach may be an advantage.

In analysing the structure or part of the structure to determine force distributions, the properties of the materials may be assumed to be those associated with their characteristic strength, irrespective of the limit states being considered. Thus, if one has different grades of concrete, $E$ would vary with this, but not for each limit state. When one is analysing any cross-section within the structure, the properties of the materials should be those associated with their design strengths.

When analysing the structure for any limit state we use an elastic analysis in most cases to determine the force distributions within the structure. In this case the relative stiffnesses may be based on any one of the following, but it must be the same throughout:

1. the concrete section – the entire concrete cross-section, ignoring the reinforcement
2. the gross section – the entire concrete cross-section including the reinforcement on the basis of modular ratio
3. the transformed section – the compression area of the concrete cross-section combined with the reinforcement on the basis of modular ratio.

In the initial stages of design (1) is the easiest to use as generally a cross-section will have been chosen, and throughout the design examples that follow the concrete section will be used, unless stated otherwise.

In a framed structure with monolithic beam and slab construction there are various ways of selecting the concrete section. One may use a rectangular section throughout, a Tee section throughout, or a combination of rectangular and Tee section. Computer programs are available for all three but again we shall use a rectangular section unless stated otherwise.

At ultimate limit state only, a limited amount of redistribution of the calculated forces may be made provided the members concerned possess adequate ductility.

When considering slabs, yield line theory or other appropriate plastic theory may be used for the ultimate limit state.

# 2 ROBUSTNESS

## 2.1 Compliance with the Building Regulations 1985

Irrespective of which Code of Practice is being used it must be remembered that such documents are still only recommendations. The real power lies in the Building Regulations.

Regulation A1 of Part A of Schedule 1 to the Building Regulations 1985 states that the building shall be so constructed that the combined dead, imposed and wind loads are sustained and transmitted to the ground (a) safely, and (b) without causing such deflection or deformation of any part of the building, or such movement of the ground, as will impair the stability of any part of another building.

For reinforced concrete structures, BS8110: Parts 1, 2 and 3: 1985 are deemed-to-satisfy documents. It is interesting to note that CP110: 1972 and CP114: 1969 were also approved Codes, but CP110 was deleted with the publication of BS8110, and CP114 was deleted 2 years after the publication of BS8110.

There is, however, a further Regulation A3 dealing with 'Disproportionate collapse'. It applies only to buildings having five or more storeys and to public buildings where the structure incorporates a clear span exceeding 9 m between supports. The requirement here is that in the event of an accident the structure will not be damaged to an extent disproportionate to the cause of the damage. BS8110: 1985 is again a deemed-to-satisfy document. CP110: 1972 contained stability clauses which met this requirement, but also contained recommendations for structures of less than five storeys. Although it referred to stability it was not talking about overturning but the ability of the building as a whole to remain stable (i.e. not collapse) even though some parts may have collapsed. In other words, the structural failure of an element would affect only a local part of the building, not the whole building.

## 2.2 Requirements

In revising CP110 the Committee considered that the structural integrity of structures should be given more importance and the term robustness is now used. This is obviously an ultimate limit state and although an acceptable probability of a structure collapsing under defined loads is treated by formal calculations, the possibility of collapse being initiated by foreseeable, though indefinable and perhaps remote, effects should be considered in design. In particular, situations should be avoided where damage to small areas of a structure or failure of single elements may lead to collapse of major parts of the structure.

Clause 2.2.2.2 of Part 1 lists the precautions which should generally prevent unreasonable susceptibility to the effects of misuse or accidents. A summary of these is as follows:

1. All buildings should be capable of resisting a minimum horizontal force (the Code uses the word notional).
2. All buildings are provided with effective horizontal ties.

3. For buildings of five or more storeys, the layout should be checked to identify key elements. A key element is such that its failure would cause the collapse of more than a limited area close to it. BS8110 does not attempt to define a limited area, but Regulation A3 suggests an area within a storey of 70 m$^2$ or 15% of the area of the storey, whichever is the lesser. If key elements exist the layout should be modified if at all possible. If this cannot be done then these key elements should be designed in accordance with 2.6 of Part 2.
4. Again for buildings of five or more storeys, any vertical load-bearing element (other than a key element) should be detailed so that its loss will not cause considerable damage. Vertical ties will generally achieve this, but where this cannot be done the element should be considered to be removed and the surrounding members designed to bridge the gap in accordance with 2.6 of Part 2.

Figure 3.1 of Part 1 gives a very useful flow chart which summarizes the design procedure to ensure robustness. With *in situ* monolithic construction, by the very nature of the detailing we do tend to get horizontal and vertical tying actions, but with some details this is not as effective as in others.

The subject of key elements has not always been considered in design and the Code now makes it quite clear that an engineer should be responsible for the overall stability (robustness) of a structure. The same engineer should ensure the compatibility of the design and detail of **all** components, even those not made by him/herself – clause 2.2.2.1.

## 2.3 Recommendations

### 2.3.1 Notional horizontal load

To ensure that all structures have a reasonable minimum strength a notional horizontal force is to be applied to the structure. For load combinations (2) and (3) a wind load will be applied automatically, but this should not be less than the notional horizontal load. As with wind forces it will be applied at each floor or roof level (loads at the nodes) and is equal to 1.5% of the characteristic dead load between mid-height of the storey below and either mid-height of the storey above or the roof surface. So we have to compare 1.4 $W_k$ or 1.2 $W_k$ with 0.015 $G_k$ and take the larger value (see Fig. 2.1).

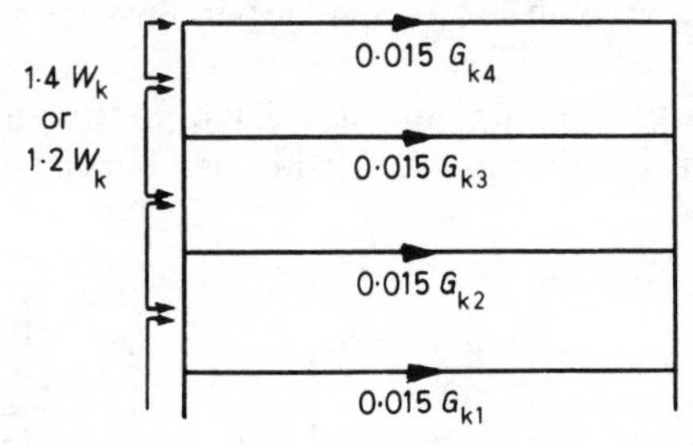


FIG. 2.1 Horizontal loads.

### 2.3.2 Design of ties

The horizontal forces to be resisted by ties are all related to $F_t$, where $F_t$ is determined from the number of storeys (including basements). The philosophy behind this is that the consequences of collapse are generally more serious for high buildings. Furthermore, simply because of their greater size, the probability of misuse or the occurrence of exceptional accidental loads is greater, and the objective is to ensure that the risk is approximately the same at all heights.

$$F_t = (20 + 4\ n_0)$$

where $n_0$ is the number of storeys, or

$$F_t = 60$$

and we use the lesser value. So $F_t$ varies from 24 for a single storey, up to 60 for a building of ten or more storeys.

In proportioning the ties, it may be assumed that no other forces are acting and the reinforcement is acting at its characteristic strength. As reinforcement provided for other purposes may be regarded as forming part or the whole of the ties, it will be found that for many structures the reinforcement provided for the usual dead, imposed and wind loads will, with minor modifications, fulfil these tie requirements. So the normal procedure will be to design the structure for the usual loads and then carry out a check for the tie forces. The types of tie to be provided are given in clause 3.12.3.1 as (a) peripheral ties, (b) internal ties, (c) horizontal ties to columns and walls, and (d) for buildings of five or more storeys, vertical ties. These will now be dealt with separately.

### (a) PERIPHERAL TIE – clause 3.12.3.5

At each floor and roof level an effectively continuous tie should be provided capable of resisting a tensile force of $F_t$ kN, located within 1.2 m of the edge of the building or in the perimeter wall. The force thus varies from 24 kN for a single-storey building up to 60 kN for ten storeys or more. For high-yield reinforcement with a characteristic strength of 460 N/mm$^2$ this works out at 2.2 mm$^2$ (approximately) per kN of force. For the maximum force of 60 kN this is 131 mm$^2$, slightly more than one 12 mm bar.

If we have a perimeter beam spanning between columns the obvious position for this tie is in the beam, as (a) we shall have reinforcement there anyway, and (b) tying the beams and columns together is the most effective resistance. However, as we shall see when dealing with internal ties, the internal tie has to be anchored to the peripheral tie and the existing reinforcement in the beam may not be in the right place.

If we have cantilever slabs, supporting external cladding, projecting in front of the columns and these are more than 1.2 m then the peripheral tie must go in the slab – see Fig. 2.2.

Even when we have a peripheral beam it is still within the rules to put the peripheral ties in the slab inside the beams and use existing slab bars as part or whole of the tie.

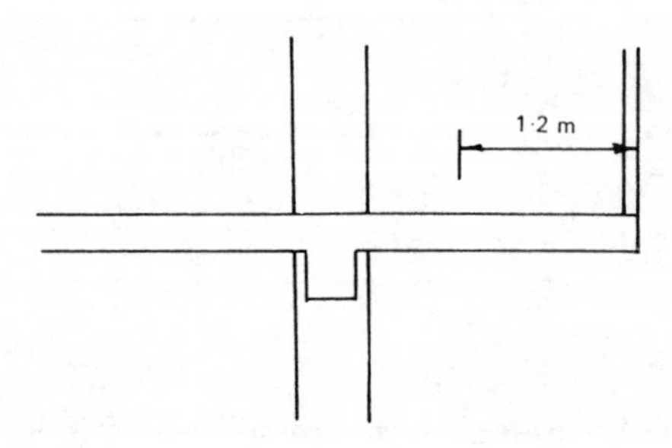


FIG. 2.2 Position for peripheral tie.

These are to be provided at each floor and roof level in two directions approximately at right angles. They should be effectively continuous throughout their length and be anchored to the peripheral tie at both ends, unless continuing as horizontal ties to columns or walls. The tensile force, in kN per metre width, is to be the greater of

$$\frac{(g_k + q_k)}{7.5} \frac{l_r}{5} F_t$$

and

$$1.0\ F_t$$

where $(g_k + q_k)$ is the sum of the average characteristic dead and imposed floor loads (in $kN/m^2$), and $l_r$ is the greater of the distances between the centres of columns, frames or walls supporting any two adjacent floor spans in the direction of the tie under consideration.

Where we have beam and slab construction such that there are secondary beams framing into main beams, and main beams framing into columns, the distance $l_r$ is the distance between the main beams, not the secondary beams. So, if we have three consecutive frames (i.e. main beams) at 8 m, 10 m and 6 m intervals, $l_r$ would be 10 m for all three spans. With $l_r = 10$ m and $(g_k + q_k) = 7.5\ kN/m^2$ the force per metre width is $2.0\ F_t$.

The bars providing these ties may be distributed evenly in the slabs or may be grouped at or in beams, walls or other appropriate positions, but at spacings generally not greater than $1.5\ l_r$.

Assuming the bars are distributed evenly in a floor slab, we have to consider (i) continuity, and (ii) anchoring effectively to peripheral tie.

For (i) in continuous slabs, this means lapping some bottom steel at supports, either by extending existing bars or the addition of splice bars (see Fig. 2.3).

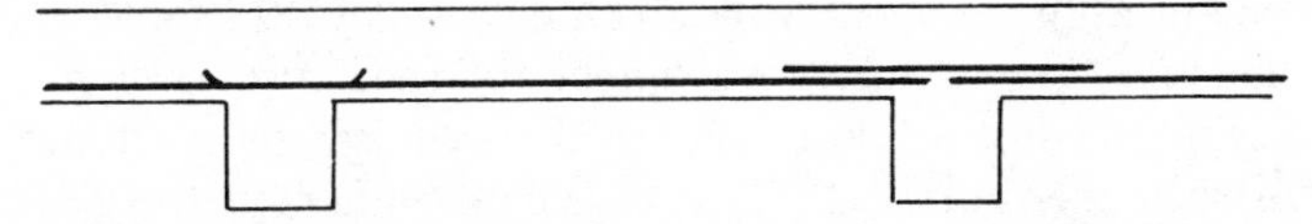

FIG. 2.3 Continuity requirement for slab.

The length of lap required is not quite clear. The reinforcement is to be at its characteristic strength, which would give an enhancement factor of 1.15 (i.e. 1/0.87). The Handbook, however, in talking about anchorage suggests that the bond stresses could also be characteristic values, and as these stresses include a partial factor of 1.4 there would be a reduction factor of 1.4. The net effect would, therefore, be a length of 1.15/1.4 times the normal design length. This does not really agree with the comment in clause 2.4.4.2 when talking about values of $\gamma_m$ for localized damage. It is suggested that the normal design anchorage length, and hence lap length, is used. The length of the lap will depend on the area provided compared with the area required, but will be not less than the minimum lap of 15 times the bar size or 300 mm, whichever is the greater.

It is also recommended that the lapping bars are in contact. It is not permitted to go from bottom bars to top bars as continuously effective, without using links as demonstrated in Section 5, Precast Concrete.

For (ii) the Code says that the tie bars may be considered anchored to the peripheral tie if they extend either (a) $12\phi$ or an equivalent anchorage beyond all the bars forming the peripheral tie; or (b) an effective anchorage length beyond the centre line of the bars forming the peripheral tie.

If the peripheral tie consists of bars in the slab then either of these two requirements can be met quite easily. If the pripheral tie consists of bars in an edge beam, then the bottom bars in the slab shown in Fig. 2.3 will not be at the same level as the peripheral tie bars (Fig. 2.4). It is suggested that either an additional bar be used for the peripheral tie or the internal tie bars be extended and anchored around the top bar in the beam.

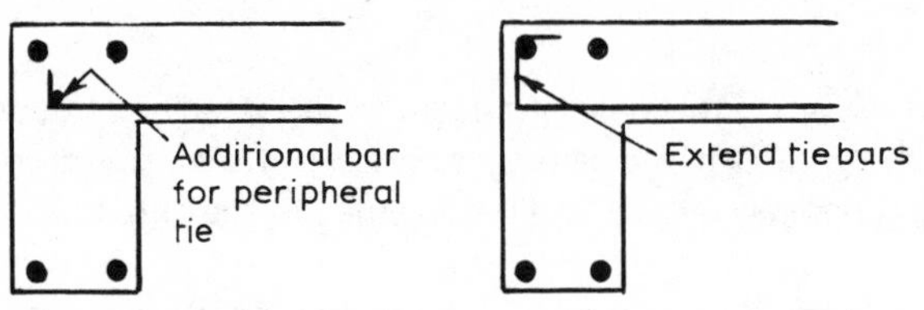


FIG. 2.4 Anchoring of ties in slabs.

As stated earlier, these horizontal tie forces required can be grouped together in beams if there are any. As will be seen in the next section it will probably be an advantage to do so where beams are available. In this case it can be the top or bottom beam bars which are made effectively continuous.

For precast concrete floor units the Code, in clause 5.1.8.2 and the Handbook, in the commentary to clause 5.3.4, give examples on how continuity can be achieved but, as with *in situ* construction, anchorage to the peripheral tie is not quite so straightforward.

### (c) COLUMN AND WALL TIE

All external load-bearing members such as columns and walls should be anchored or tied horizontally into the structure at each floor and roof level. This force is to be the greater of: (i) $2F_t$ kN or $(l_s/2.5)F_t$ kN, whichever is the lesser, for a column, or for each metre length if we have a wall ($l_s$ is the floor to ceiling height in metres); or (ii) 2% of the total ultimate vertical load in the column or wall at that level.

Thus if the clear height is less then 5.0 m the force will be $(l_s/2.5)F_t$, from (i), which we then compare with (ii). For corner columns we need this force in each of two directions approximately at right angles.

The Code does not say that these ties should be anchored to the peripheral tie.Indeed, the implication from (b) is that if the internal ties are grouped together in the beams and continue as column ties they need not be.

These column ties must be properly anchored for the force involved and it is suggested that these are terminated in a hook or bend. If they go round other bars at right angles and are continued for $8\phi$ beyond the bend they can be considered as fully anchored. Otherwise a full anchorage length beyond the face of the column is required. With high stresses in the bars this may need a radius greater than the standard radius. This can be modified by supplying more reinforcement than is needed and so reducing the stress.

The full anchorage length from the face will be the same length as at the full design stress (cf lap length under (b)). A useful detail at this junction can be seen in *Standard Reinforced Concrete Details* published by the Concrete Society (1973), using loose U-bars.

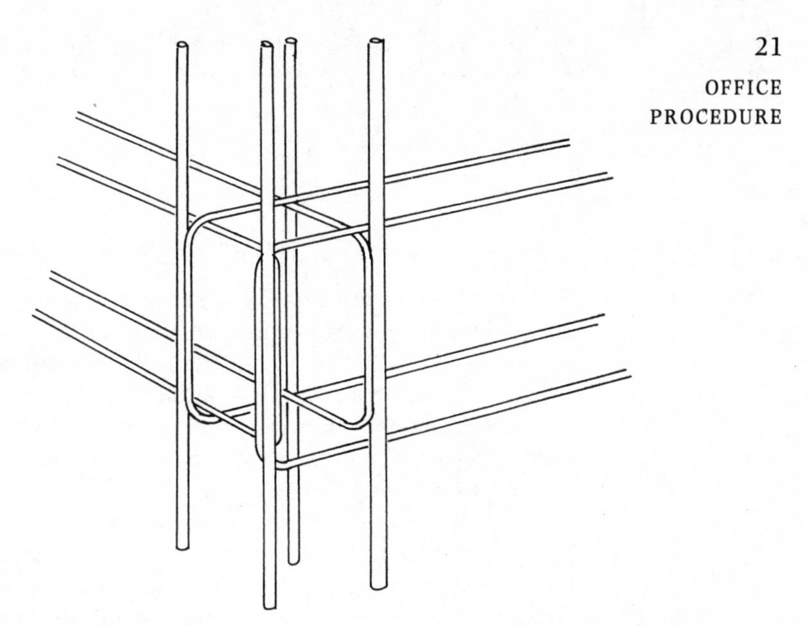

FIG. 2.5 Detail at corner column.

At a corner column this detail would be required in both directions and if the peripheral tie is incorporated in the perimeter beam it is considered that the tie would be effectively uninterrupted. This is illustrated diagrammatically in Fig. 2.5.

(d) VERTICAL TIES

If we have five storeys or more we have to provide an effective vertical tie in all columns and walls, and this is to be continuous from the foundations to the roof level. Although effectively continuous, the reinforcement provided is required only to resist a tensile force equal to the maximum design ultimate load (dead and imposed) received from any one storey.

For *in situ* construction this will generally be met for all buildings, including those of less than five storeys, by the normal design and detailing rules. For precast structures this may not be so straightforward, but clause 5.3.4 of the Code gives recommendations on how this can be achieved.

## 2.4 Office procedure

The method of providing ties described previously will be the normal method of ensuring robustness. There may, however, be cases where there are key elements which have to be designed or where effective ties cannot be provided. In such cases, Section 2.6 of Part 2 of the Code gives guidance on how they should be treated.

Although every case will have to be treated on its merits it does seem that design offices will have to decide on some standard details or principles for complying with the tying requirements. It may be that tables or charts could be prepared so that a detailer could refer to these to see the reinforcement required and then check against some standard details to see (a) if this reinforcement is already provided and (b) if it is properly anchored. Checks should be carried out as the design and detailing proceeds as this may affect the layout of reinforcement required.

# 3 ANALYSIS OF STRUCTURES

The critical limit state for reinforced concrete structures will generally be ultimate, so we shall start the design procedure by assuming this.

First we have to find the distribution of moments and forces by analysing the structure. Quite obviously we can analyse a structure as a complete structural frame and where computer programs are available this will be done. In other cases hand methods or small computers will be used and in this case it will generally be more convenient to analyse parts of a complete frame.

Irrespective of the methods used, an assessment of section sizes will have to be made. So-called rules of thumb are available for depths and widths of beams, and these have generally been developed from previous experience. It is not the intention here to elaborate on these rules, but bearing in mind that limit state design has not drastically changed member sizes obtained from other design methods, it is suggested that any existing rules the reader may have should be retained. For example, with continuous slabs spanning in one direction a span/depth ratio of 30 will still generally be satisfactory, provided the durability and fire resistance requirements are satisfied. For continuous beams a span/depth ratio of 12 to 15 is an acceptable starting point.

Having assumed some section sizes we can now analyse the structure. If a computer is available the complete frame can be analysed using the relevant load combinations. As regards the arrangement of loads or load patterning within each load combination the number of load cases that could be considered will increase very rapidly with the number of floors and number of bays in the frame. The designer must bear in mind that he is trying to find the worst effect, not every effect. For example, the maximum moment in a continuous beam over a support occurs when the adjacent spans have maximum design load. This will also give the maximum load on the column. For the maximum moment in a span, that span will have the maximum design load and the adjacent spans will have minimum design load. This will also give the maximum moment in the column, but with a smaller load. So, for column design the maximum load will have the same load pattern on each floor, but for the maximum moments and loads associated with them the load pattern will probably have to be different on different floors. The designer will have to make his own decision; there is no quick answer.

The Code does, however, give methods whereby the frame can be divided into smaller units which can be analysed separately and the results combined as required, e.g. for the design of columns. Which clause is used depends on whether or not the frame is providing lateral stability to the structure. If the frame is not providing lateral stability then it will be subjected to vertical loads only and clause 3.2.1.2 is appropriate. If the frame is providing lateral stability then it will be subjected to vertical *and* lateral loads and clause 3.2.1.3 is appropriate. The first case will be referred to as a braced frame and will not sway. The second case will be referred to as an unbraced frame, and will sway.

## 3.1 Braced frame, i.e. frame not providing lateral stability

In this case we are concerned with vertical loads only and we shall consider a series of subframes. Each subframe consists of the beams at one level together with the columns above and below, assumed to be fixed at their ends remote from those beams, unless the assumption of a pinned end is more reasonable. This would be the case of a foundation detail unable to develop moment restraint.

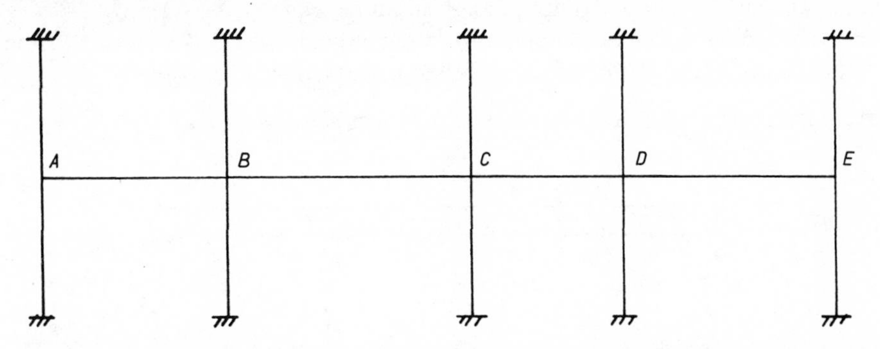


For an individual beam we can consider a simplified subframe consisting only of that beam, the columns attached to the ends of the beam and the beams on either side, if any. The column and beam ends remote from the beam under consideration should be taken as fixed unless the assumption of a pinned end is more reasonable. The stiffnesses of the outer beams will be taken as half their actual stiffnesses if the ends are taken as fixed. So for an interior beam we have

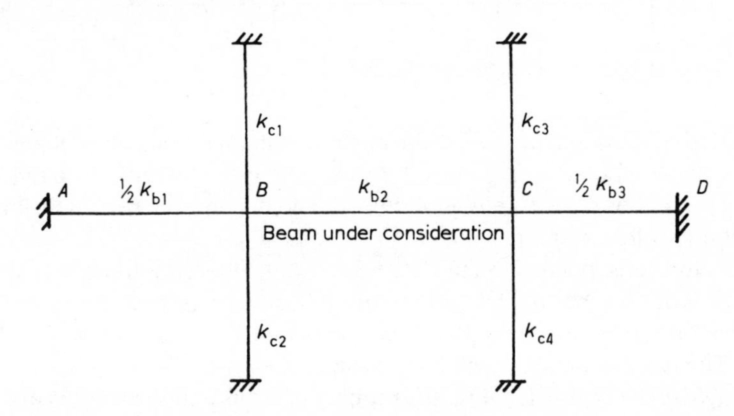


It should be noted in passing that the moments in the columns may also be found from this simplified subframe provided that the span under consideration is longer than both the adjacent spans.

For an exterior beam we have

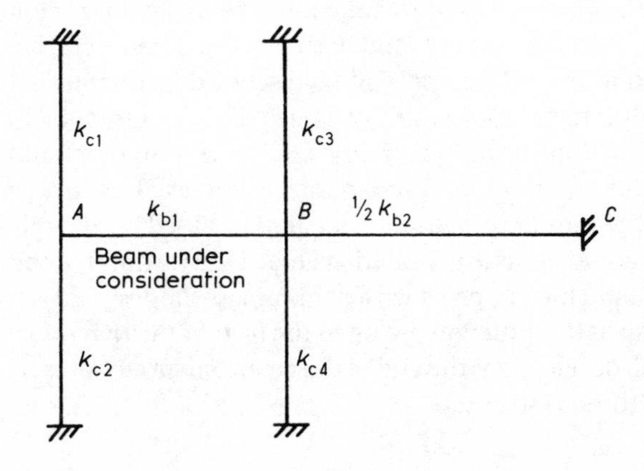

For the loading patterns we consider load combinations (1) only (i.e. dead and imposed) and we take

1. all spans loaded with maximum design ultimate load ($1.4\ G_k + 1.6\ Q_k$)
2. alternate spans loaded with maximum design ultimate load ($1.4\ G_k + 1.6\ Q_k$) and all other spans loaded with minimum design ultimate load ($1.0\ G_k$).

Typical load patterns for a three-bay subframe are shown in Fig. 3.1.

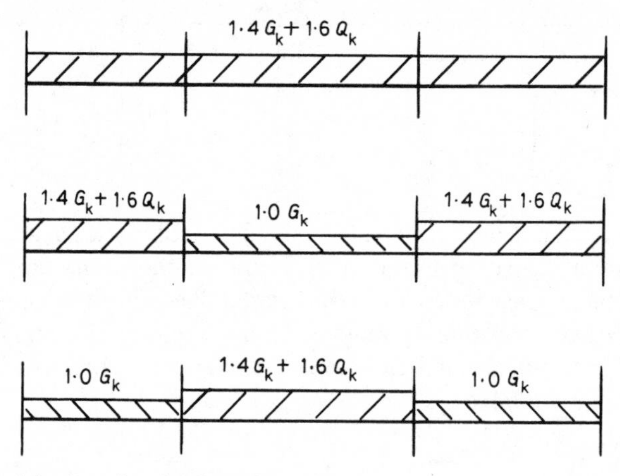


FIG. 3.1 Vertical loads only – load combination (1).

Although the frame will not sway, the answers obtained from a full frame will differ from those obtained from a series of subframes due to the fact that in a full frame all joints will be allowed to rotate, not just the joints in the run of beams under consideration, as in a subframe.

There is a further simplification in clause 3.2.1.2.4, where continuous beams can be analysed by ignoring the columns and treating the supports as simple; i.e. the beams are capable of free rotation about the supports and there is no restraint from the columns. The loading patterns will be the same as above.

Irrespective of the type of loading, the number of spans and the lengths of the spans, not more than three separate analyses will be necessary to prepare a bending moment envelope.

Provided a proper analysis is carried out (i.e. by preparing a bending moment envelope, not using coefficients), be it full frame, series of subframes or simplifications, redistribution of moments can be carried out within the limits to be discussed shortly.

Full frame and subframe will automatically give the column moments. Simplifications ignoring the columns will not, and in this case moments can be found by a single moment distribution procedure and will be discussed in Chapter 11. Any change in beam moments due to column restraint is generally ignored, including that at the external beam junction, although many engineers put in nominal or anti-crack reinforcement either by eye or by standard office practice. In many cases this has been satisfactory, but the author feels that a calculation based on the relative column and beam stiffnesses should be done to find the value of the restraint at the end of the beam. Reinforcement should then be provided for this unless the designer is prepared to accept the risk of cracking at this junction owing to the fact that with monolithic construction some restraint will develop, but the end of the beam will try to rotate as if it were simply supported, i.e. with no restraint.

We now have to consider lateral loads as well as vertical loads. Where we have an unbraced frame it will sway even under vertical loads where we have asymmetrical loading. Anyone who has taken this into account when doing a hand analysis will appreciate how tedious and time-consuming this can be. The obvious answer is to use a computer program, but to state in a Code of Practice that a sway frame should be analysed by a computer would not be practical, particularly as many frames of this type have been analysed, designed and constructed with the effects of sway ignored. It is the magnitude of these effects that does sometimes cause concern, particularly when one is talking in terms of percentage variation. In actual numerical values the effect is not so great, bearing in mind that some of the assumptions made can also give variations.

Single-bay frames and two-bay frames appear to give the largest variations, but for three or more bays, particularly where the spans are approximately equal, the answers ignoring sway should be within acceptable limits of accuracy of the analysis being carried out. The Code thus warns a designer that the effects of sway should be considered, but that for frames of three or more equal bays a simple approach (i.e. by hand if required) can be adopted by using a series of subframes as before.

As there are lateral and vertical loadings to consider we have to find which loading gives the worst effects. We look at load combination (1) with its load factors and load combination (3) with its different load factors.

For load combination (1) we can find the moments and forces as for 'vertical loads' above.

For load combination (3) we can consider a series of subframes consisting of a run of continuous beams with the columns above and below assumed to be fixed at their ends remote from the beams (exactly the same as before) and all the beams are loaded with $1.2\ G_k + 1.2\ Q_k$. This takes care of the vertical loads. To these moments and forces we then add those obtained from considering the whole frame loaded with a lateral loading of $1.2\ W_k$, ignoring vertical loads, and for the frame we assume points of contraflexure at midspan of beams and mid-height of columns. Such arrangements are shown in Fig. 3.2.

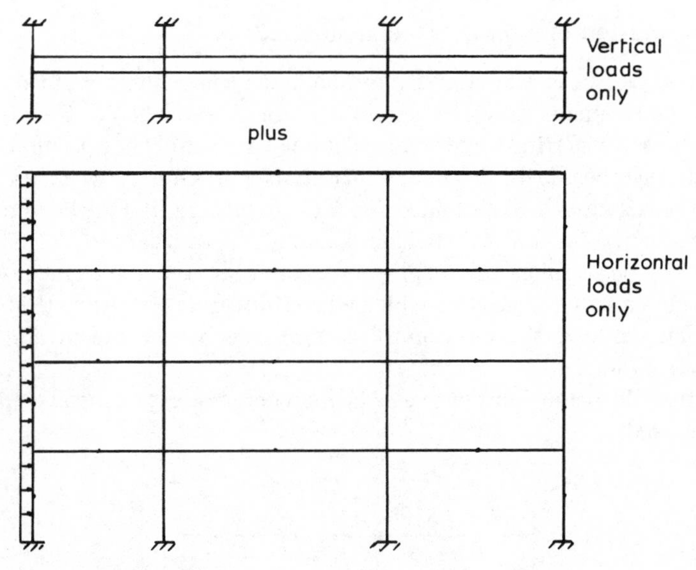


FIG. 3.2 Load combination (3).

In this case we have two loading combinations to analyse and it will only be with experience that the designer will be able to decide whether it is necessary to analyse both.

In certain cases it will be necessary to consider the effects of load combination (2), i.e. 1.0 $G_k$ with 1.4 $W_k$, particularly with tall narrow buildings where tension may develop in the columns. This was illustrated in Chapter 1.

In analysing an unbraced frame it is not recommended that the simplifications used for braced frames should be used, i.e. simplified subframe and continuous beam simplification.

Redistribution of moments obtained by the methods of analysis given above can be done within the same limits as for a braced frame except for structures over four storeys in height, as will be mentioned later.

The definitions given in clause 3.4.1 have a bearing on the spans and section sizes when setting up the mathematical model (i.e. line diagram) for the analysis. If one has a monolithic beam and slab construction, the decision has to be made as to whether to use a Tee section, a rectangular section or a combination when considering the beam section. In strength-of-sections calculations it is usual to use the Tee section, but in analysis opinion varies. For a Tee or Ell section the Code gives the effective flange width in clause 3.4.1.5.

For a Tee section, the effective flange width is the web width plus $l_z/5$, and for an Ell section it is the web width plus $l_z/10$. The distance $l_z$ is defined as the distance between points of zero moment, and for a continuous beam can be taken as 0.7 times the effective span. For a single span simply supported, $l_z$ will be the effective span. For the end span of a continuous run of beams it could presumably be taken as $0.85l$, but it is proposed to use the value of $0.7l$. For a continuous beam the effective flange width becomes $b_w + 0.14l$ for a Tee section, and $b_w + 0.07l$ for an Ell section.

The slenderness limits are quite liberal and as the effective depth is one of the dimensions involved, this has been modified so that if a beam is made far deeper than necessary (for aesthetic reasons), the slenderness limits are not exceeded and the member penalized.

## 3.3 Redistribution of moments – clause 3.2.2

Having carried out an elastic analysis and obtained the bending moments one may redistribute the moments provided certain conditions are satisfied. Redistribution of moments means transferring some of the calculated moment at one position to another position in the member. So the first condition is that equilibrium between internal forces and external loads must be maintained, i.e. the overall height of the bending moment remains the same for any particular loading. 'Redistribution' usually means 'reduction', so if the calculated moment at a support is reduced, then it means providing a resistance moment at that position which will be incapable of resisting the total elastic moment it can get. So at this position the member will become plastic and yield, and there will be rotation.

Consider now the behaviour of a loaded beam where there is a central point load and the ends are fixed:

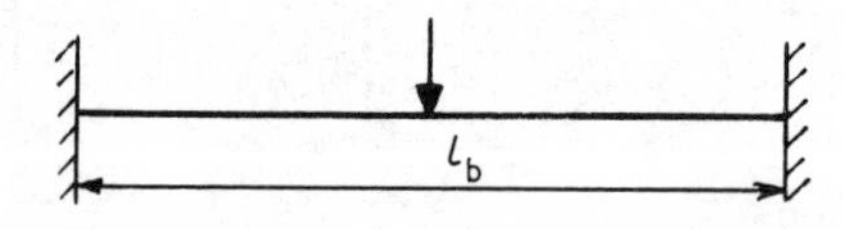

1. As the load increases the beam behaves elastically until the plastic moment at one or more critical sections is reached, i.e. where the steel has yielded. In this case it will be assumed that the resistance moment provided at the supports is half that at midspan and has a value $Ql_b/8$. So for a particular load $Q$, the moment at the supports and midspan is $Ql_b/8$ and the steel now yields at the supports.

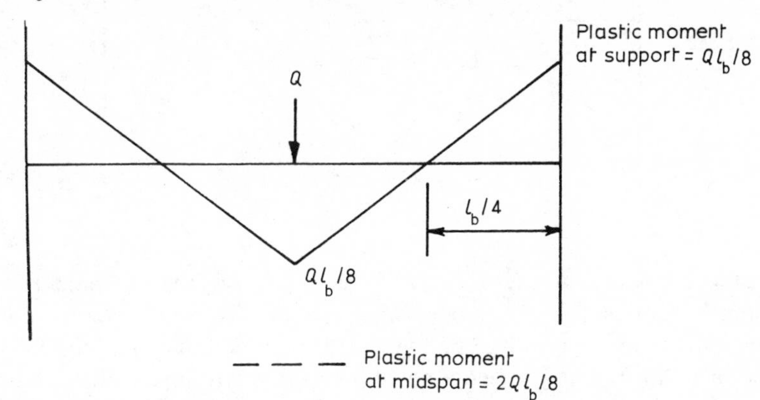


2. Further loading will cause these hinges to rotate, but the moments at the hinges do not change. The extra moment required to balance the load is carried by other parts of the member.
3. This will continue until the midspan moment reaches the plastic moment of that section, when the structure fails. The beam has, in effect, behaved as a simply supported beam with a moment capacity at midspan of $2Ql_b/8 - Ql_b/8 = Ql_b/8$. This will take a central point load of $Q/2$, so the gross central point load is $3Q/2$.

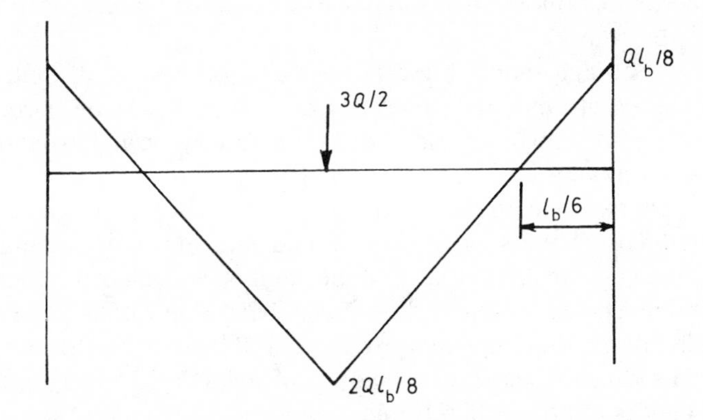


4. From the design point of view, this same bending moment diagram can be obtained by calculating the elastic bending moment under the failure loading of $3Q/2$ then reducing the support moments while increasing the midspan moment by the corresponding amount to maintain equilibrium.

This operation is what is meant by redistribution of moments, and although it can be a useful design tool it must not be done indiscriminately. There are two major factors to be taken into consideration. First, how will the modified design at the ultimate limit state affect the behaviour at the serviceability limit states? Secondly, how may one ensure that sufficient rotation occurs at the section where the elastic moment has been reduced, prior to failure?

For the first of these considerations, condition 3 of clause 3.2.2.1 states that the

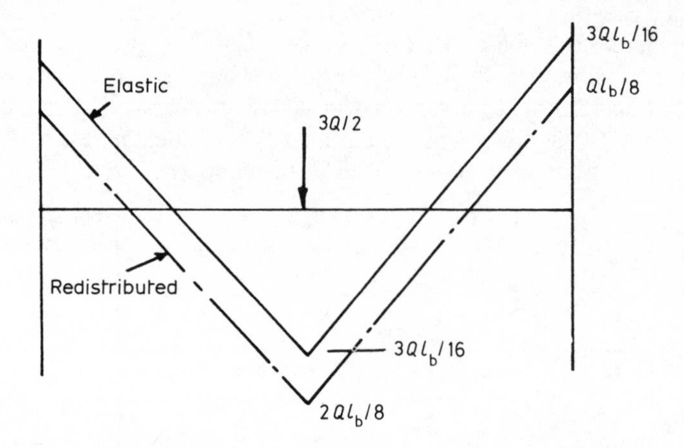


'resistance moment at any section should be at least 70% of moment at that section obtained from an elastic maximum moments diagram covering all appropriate combinations of design ultimate load'. This means that when one has obtained an elastic bending moment envelope and drawn a new envelope with ordinates 70% of the original, any diagram or part of a diagram arrived at after redistribution must never fall within this new boundary.

The reasoning behind this is that redistribution is not permitted at serviceability limit state, and the modified ultimate bending moment envelope must not come within the service bending moment envelope. There are two methods of ensuring this. One method is to prepare a service envelope, and the other is to assume that the service envelope is a percentage of the ultimate envelope. Although preparing a service envelope is more correct, the percentage method is very close and a value of 70% has been arrived at, which is an approximation for the ratio of service load to ultimate load if dead and imposed loads are equal.

By using the percentage approach it can be seen that the points of contraflexure for the service and ultimate moment envelopes remain in the same position. This is important from reinforcement considerations, as can be seen by referring to the diagrams above for a point load.

The first diagram can be regarded as the service moment diagram and the point of contraflexure is $l/4$ from the support. The second diagram is the ultimate moment diagram after theoretical redistribution and the point of contraflexure has now moved to $l/6$ from the support. Reinforcing to this smaller dimension would not be satisfactory for service conditions. As the design moment must be at least 70% of the elastic analysis moment it means that the maximum amount of reduction is 30% of the analysis moment. This applies at any section being considered.

An example of redistribution using the 70% line is as follows. Suppose we have a bending moment envelope thus:

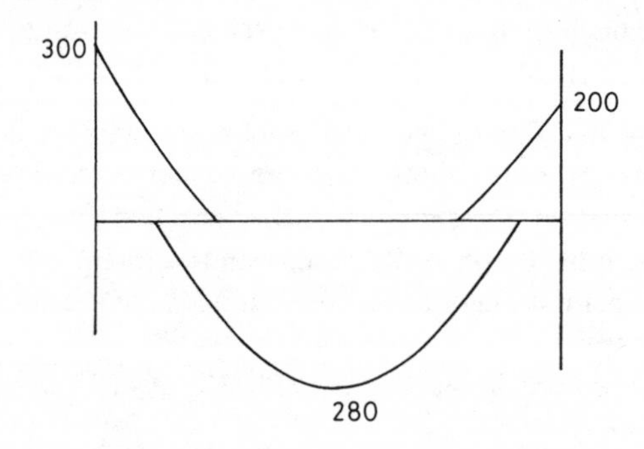

obtained from (a):

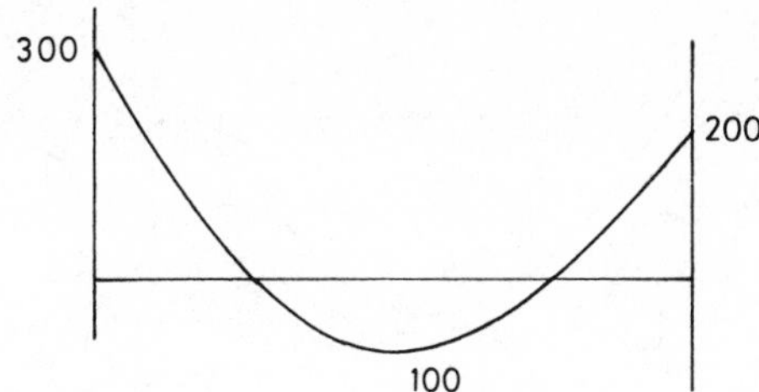


and (b):

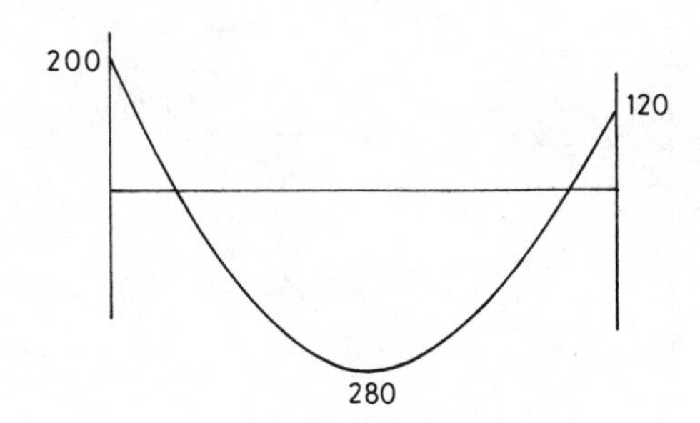


The maximum support moments are 300 kN m and 200 kN m, with the span moment of 280 kN m. The support moments could be reduced to 210 kN m and 140 kN m respectively, which means the span moment in (a) would be increased to 175 kN m:

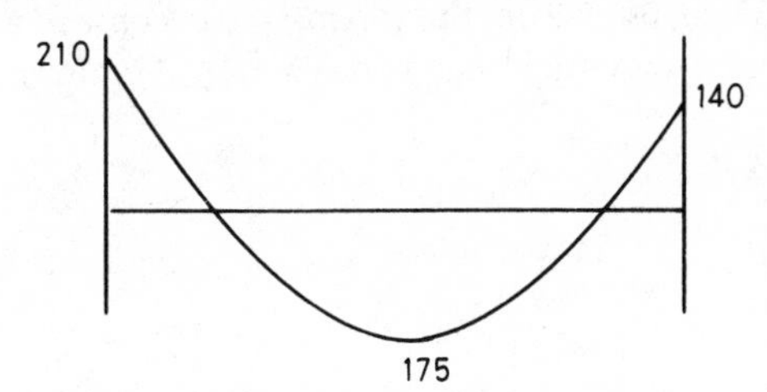


For the span moment in (b), this could be reduced to 196 kN m, but this would mean the support moments would be increased to 284 kN m and 204 kN m respectively, i.e. almost back to where we started.

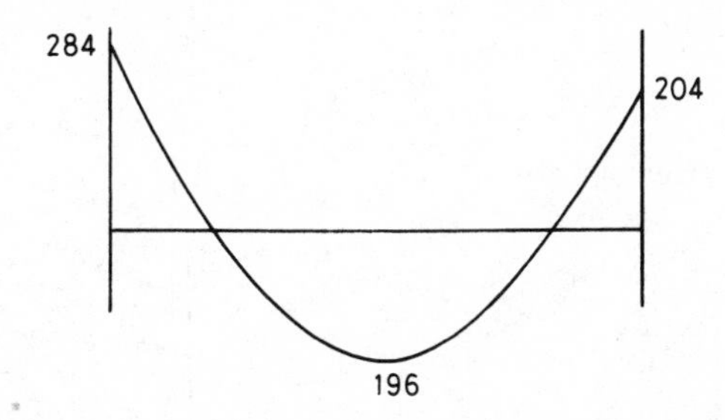


So it is not possible to take full advantage of the maximum allowable redistribution at all sections simultaneously. By increasing the support moments in (b) to 210 kN m and 140 kN m (i.e. those obtained by reducing the maximum support moments) we can reduce the span moment to 265 kN m – a modest reduction of 15 kN m.

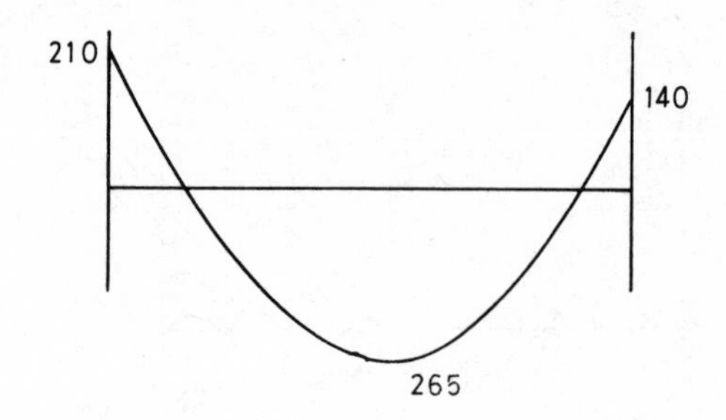


The redistributed envelope would be

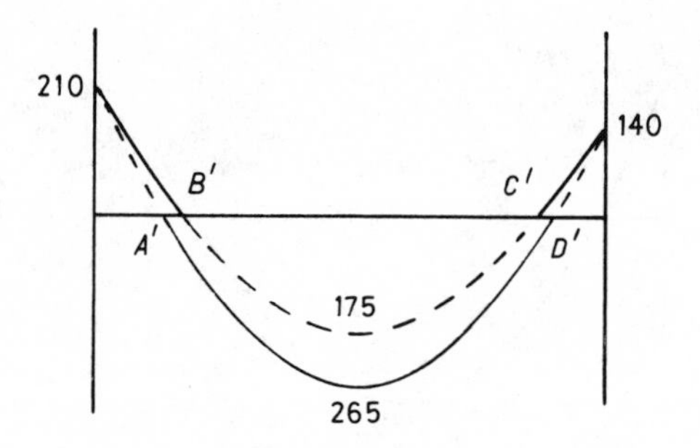


If we now consider the original elastic bending moment envelope and superimpose the 70% line, it is fairly obvious that the points of contraflexure for the hogging part of the diagram, i.e. points $B$ and $C$, have moved towards the supports, becoming $B'$ and $C'$, and points $A$ and $D$ have moved away from the supports to become $A'$ and $D'$. All four points contravene the 70% line rule which means the points of contraflexure remain where they were.

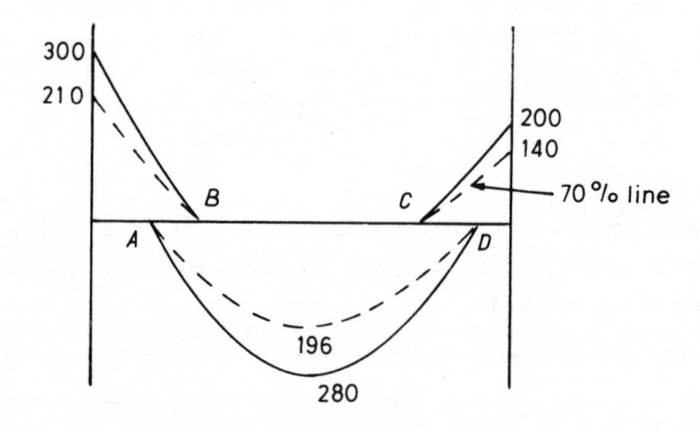


We finish up with an envelope thus:

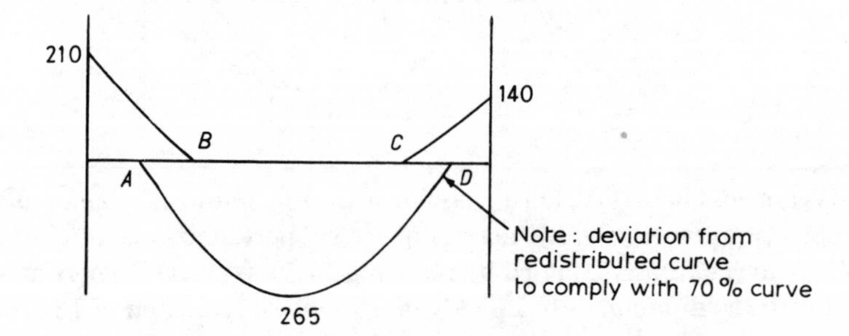

There are many variations on this theme. One of these is to make the support moments the same at both ends of the beam. This means increasing the right-hand support moment, and there is no limit by which a moment can be increased. Another is to try for equal span and support moments. When adjusting diagrams, however, it is most useful to have a 70% moments diagram available, as it then becomes very obvious whether one is encroaching on this area.

The second consideration as stated above is that if some reduction in moment is made at a section there will be rotation and the section design must cater for this. The amount of rotation which any section can undergo depends on how under-reinforced it is. If the reinforcement reaches its yield stress at the same time as the concrete reaches its ultimate strain, little rotation can take place. If the reinforcement reaches its yield stress well before the concrete fails then considerable rotation can take place. The depth of the neutral axis at failure gives a good idea of the amount of rotation capacity. With a large neutral-axis depth the concrete will fail before the reinforcement yields – whereas with a small neutral-axis depth the reinforcement will yield first.

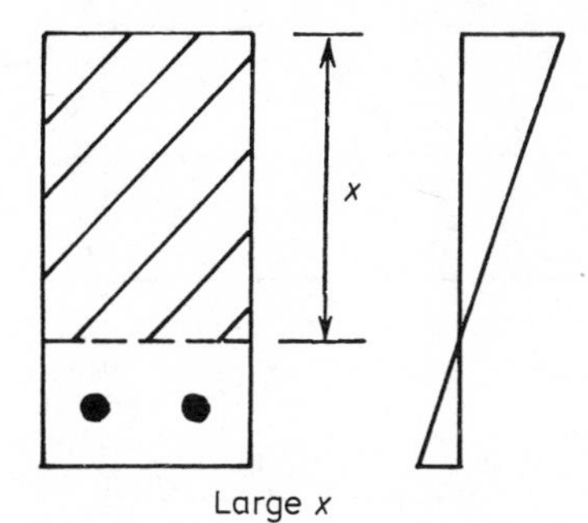


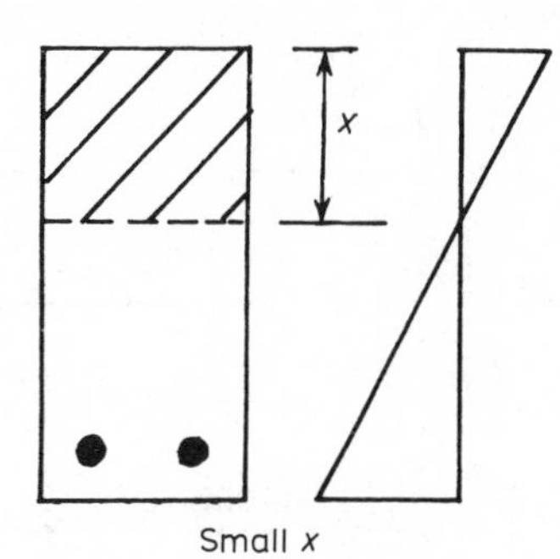


Condition 2 in 3.2.2.1, therefore, is written in terms of $x$. Where the maximum moments have been reduced, neutral-axis depth, $x$, should be checked and this should not be greater than $(\beta_b - 0.4)d$, where $d$ is the effective depth and $\beta_b$ is the ratio

$$\frac{\text{moment at the section after redistribution}}{\text{moment at the section before redistribution}} \leqslant 1.$$

Note that this requirement only applies in regions of maximum moments.

With 10% reduction, i.e. $\beta_b = 0.9$, the depth of the neutral axis is restricted to 0.5 times the effective depth. For the maximum amount of 30% reduction, i.e. $\beta_b = 0.7$, the neutral-axis depth is decreased to 0.3 times the effective depth.

What this means, in practice, is that if redistribution has been done, one must find the limit on the neutral-axis depth, and this, from the design charts or equations, will determine the reinforcement required to ensure that this limit is not exceeded. This will be dealt with more fully in Chapter 5.

This condition concerning the neutral-axis depth will rule out the possibility of reduction in moments in a column unless the axial load is very small. We shall, therefore, have the odd position in framed structures that if we redistribute the beam moments at the junction with a column we cannot adjust the column moments and so we shall not get a balance of moments at the junction. Where a structural frame provides the stability for a building we are restricted to a 10% reduction in moments if the frame is more than four storeys in height.

## EXAMPLE 3.1 Vertical loads only

Three-span continuous beam, with equal spans of 10.0 m. Beam and slab construction thus:

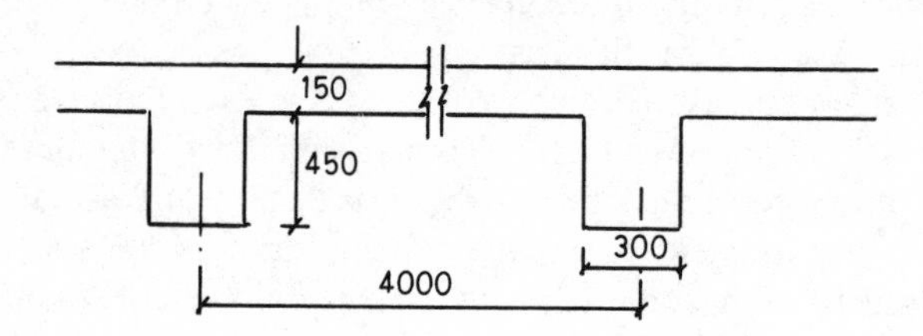


A line diagram in the direction of the span of the beams is shown below. External columns are 300 mm square and internal columns are 400 mm square. The beams to be considered are at first floor level.

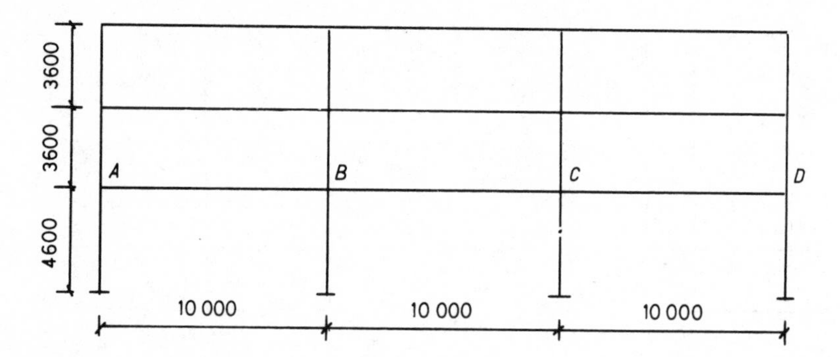


We assume:

1. all supports classed as simple
2. complete subframe taking columns into account
3. simplified subframe.

The imposed load for occupancy is 5.0 kN/m$^2$ and we shall take 1.4 kN/m$^2$ for finishes.

Characteristic dead loads:

Slab $\quad 0.150 \times 24 \times 4 = 14.4$ kN/m
Finishes $\quad 1.4 \times 4 = 5.6$ kN/m
Self $\quad 0.45 \times 0.3 \times 24 = 3.2$ kN/m

Total $= 23.2$ kN/m

Characteristic imposed load:

Slab $\quad 5.0 \times 4 = 20$ kN/m

Design loads – ultimate limit state, load combination (1):

Minimum $= 23.2 \times 1 = 23.2$ kN/m
Maximum $= 23.2 \times 1.4 + 20 \times 1.6 = 64.5$ kN/m
Minimum free moment $= 23.2 \times 10^2/8 = 290$ kN m
Maximum free moment $= 64.5 \times 10^2/8 = 806$ kN m

For obtaining the values in all of the following sections a small computer program has been used, but they could be quite easily obtained by hand methods such as moment distribution. The values of the moments have been rounded off to the nearest whole number.



## Section 1 All simple supports

Loading cases are:

1. all spans loaded with maximum design load
2. *AB, CD* maximum design load
   *BC* minimum design load
3. *AB, CD* minimum design load
   *BC* maximum design load.

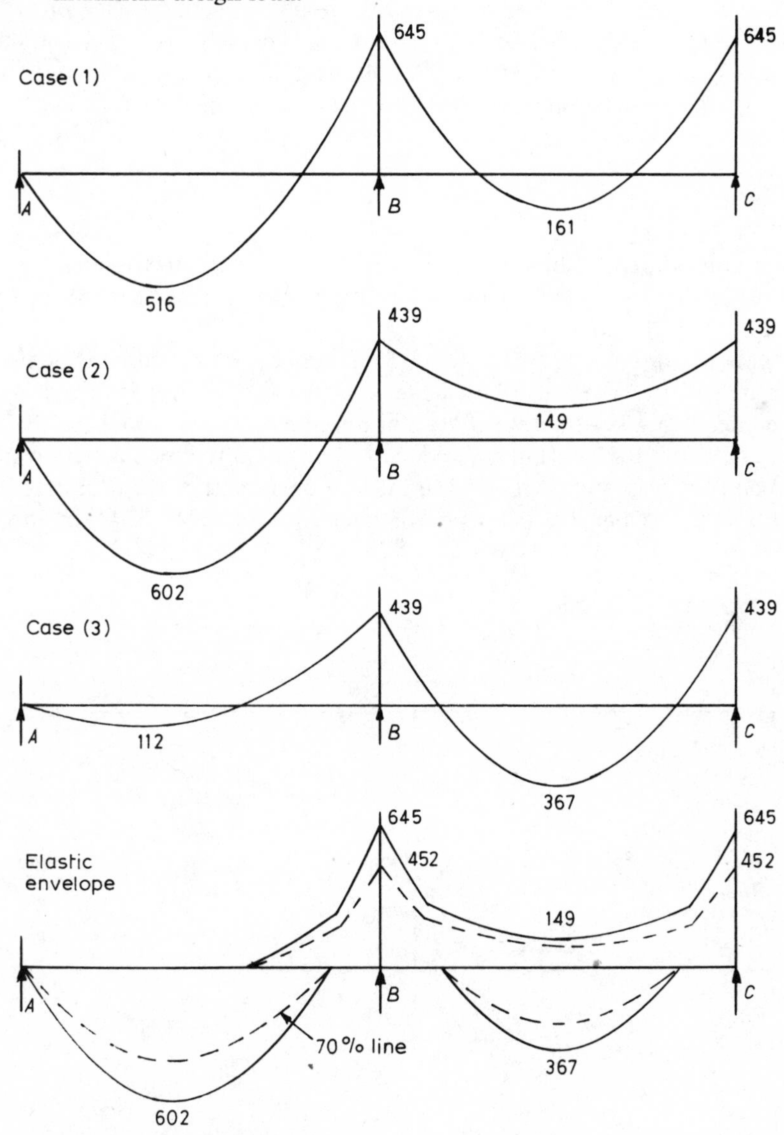


FIG. 3.3 Simple supports – no redistribution.

The bending moment diagrams for the separate cases are shown in Fig. 3.3 together with the bending moment envelope. On the envelope diagram the 70% envelope has also been drawn. As the diagrams are symmetrical only two spans have been drawn.

As the criteria for uniformly-loaded continuous beams with approximately equal spans are met we can compare these moments with those obtained from Table 3.6 of the Code.

Near middle of end span: $0.09 \times 645 \times 10 = 581$ kN m
At supports *B* and *C*: $0.11 \times 645 \times 10 = 710$ kN m
At middle of span *BC*: $0.08 \times 645 \times 10 = 516$ kN m

From a comparison it can be seen that the support moments and span *BC* moment are greater if taken from Table 3.6. This is because these moments are less for a three-span beam than for a four- or five-span beam and Table 3.6 has to cover all cases.

It will also be seen that analysis gives a hogging moment in span *BC*, whereas Table 3.6 does not. As will be seen later the simplified detailing rules cope with this. It must also be remembered that in using this table we cannot do any redistribution. With the bending moment diagrams we can, and this will now be done.

## Redistribution

The maximum moment at the supports is 645 kN m and this can be reduced by 30% to 452 kN m. For load case (1) the moment in span *AB* will now increase to 586 kN m and in span *BC* to 354 kN m.

If we now examine load cases (2) and (3) it can be seen that the moments at the supports are less than the redistributed moment calculated above. By increasing these elastic moments to 452 kN m, we could reduce the span moment in *AB* to 596 kN m (load case (2)) and to 354 kN m in *BC* (load case (3)), but it does not seem worth the effort. When the reinforcement has been determined for the redistribution support moment it will probably give a resistance moment in excess of 452 kN m anyway.

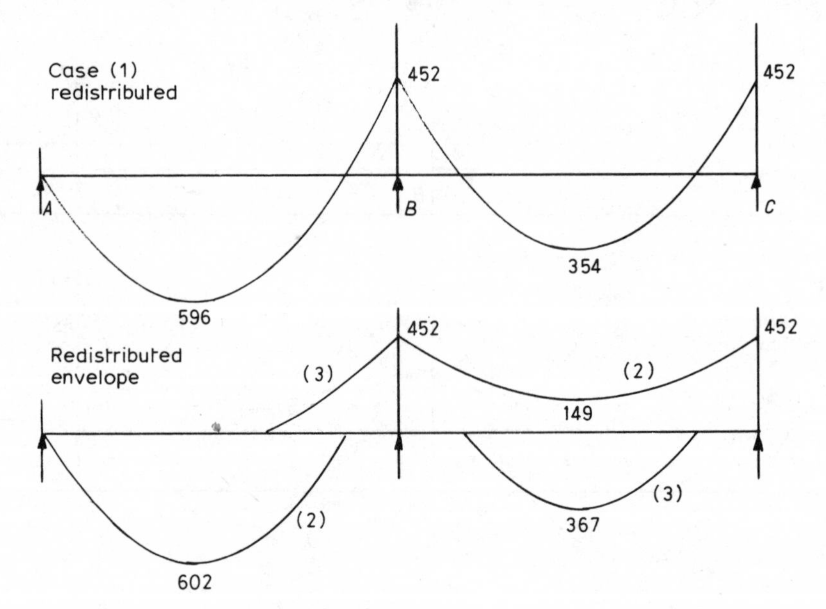


FIG. 3.4 Simple supports – after redistribution.

So a reasonable rule to follow is: if the maximum support moment is reduced and the elastic moments from other load cases are less, then leave the load cases alone.

The modified diagram for load case (1) and the redistributed envelope are shown in Fig. 3.4. If this redistributed envelope is superimposed on the elastic envelope it will be found that it is always outside the 70% envelope.

## Section 2 Complete subframe

In this analysis we take the columns into account and it will be assumed that the bases provide nominal restraint only. The design loads and loading cases are identical to those of section 1. The bending moment diagrams and envelope are shown in Fig. 3.5. As can be seen the columns have quite an effect. One of the big differences is that the moments at the support are different on opposite sides of the column. To make a

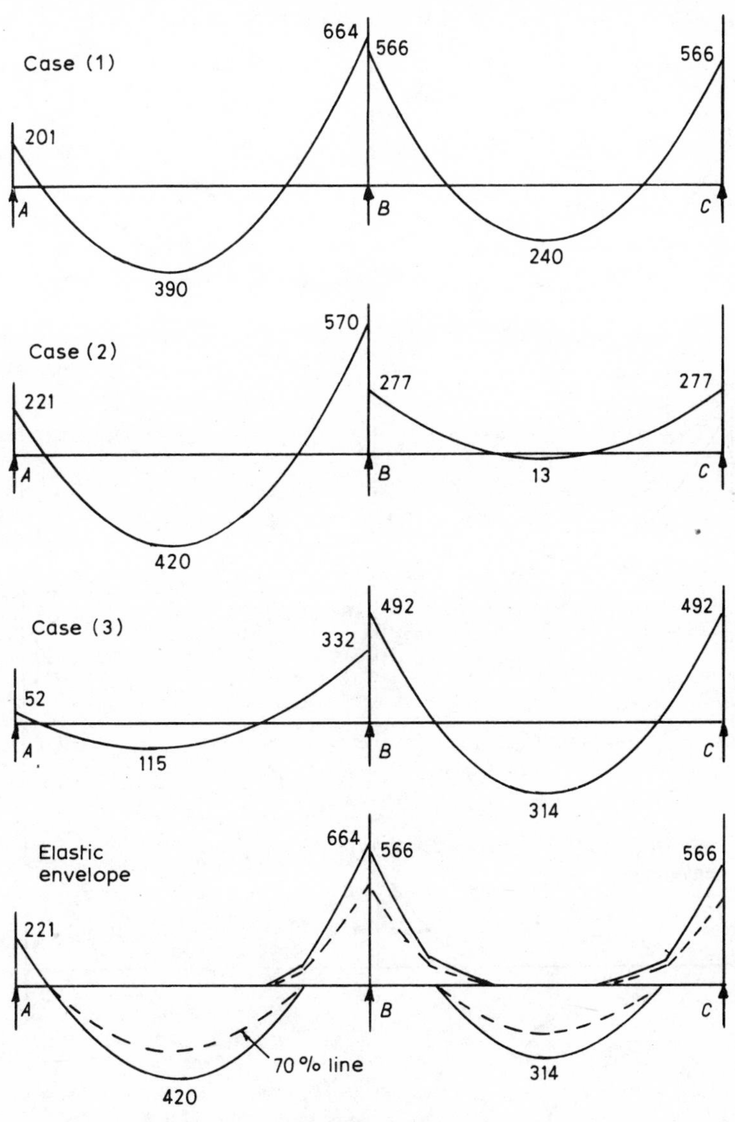


FIG. 3.5 Complete subframe – no redistribution.

comparison with the results in section 1, redistribution will be carried out away from the supports.

The maximum moments at interior supports occur under load case (1). In span *AB* the moment of 664 kN m will be reduced by 30% to 465 kN m. In span *BC* the moment of 566 kN m will be reduced to the same value of 465 kN m, a reduction of 18%. Having established this support moment of 465 kN m we shall now look at the other load cases. For load case (2) in span *AB* reduce the moment at *B* of 570 kN m to 465 kN m, but leave span *BC* as it is. In load case (3), reduce the support moment in span *BC*, but leave span *AB* as it is.

For the support at *A* we will use the moment of 201 kN m which means reducing the moment at *A* in load case (2) only. The revised diagrams are shown in Fig. 3.6. The envelope is also given, but note that in the regions of the interior support this has been modified to coincide with the 70% elastic envelope.

Remember that although the moments in the beams at the beam column junction have been modified, the moments in the columns remain unaffected.

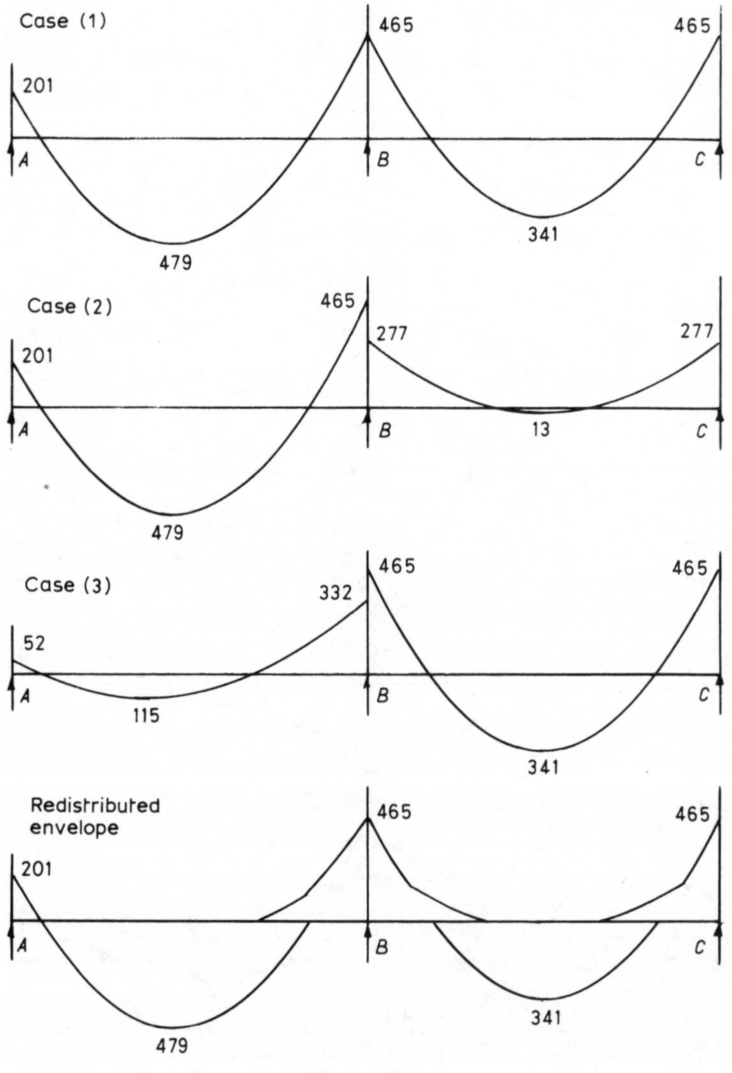


FIG. 3.6 Complete subframe – with redistribution.

## Section 3 Simplified subframe

For a three-span beam, the use of a simplified subframe does not save any time – it would in fact take longer. It is shown here to demonstrate the principles.

Span *BC*

The subframe we have to consider is shown below with the stiffnesses indicated to illustrate the point that the stiffnesses of spans *AB* and *CD* are taken as half their actual stiffness.

Using the same loads and load cases as in section 2 we can obtain a combined bending moment diagram for span *BC* as shown in Fig. 3.7. From this it can be seen that the moments are similar to those obtained in section 2, but the support moments are slightly less.

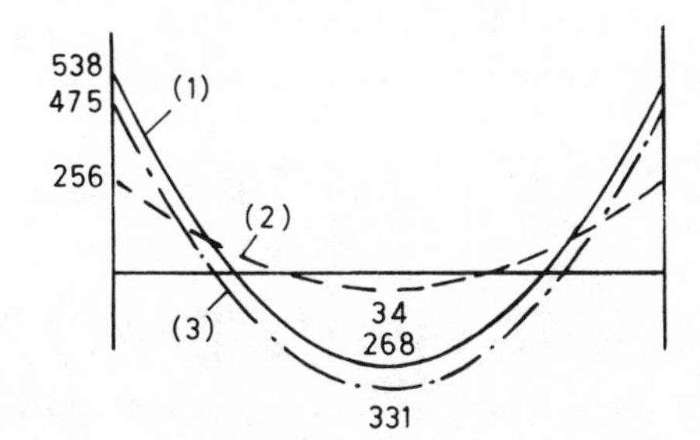


FIG. 3.7 Interior span – no redistribution.

Span *AB*

The subframe is as shown below. Note that the stiffness of *BC* is taken as half its actual stiffness.

For this the load cases will be:

1. both spans maximum load
2. span *AB* maximum, span *BC* minimum
3. span *AB* minimum, span *BC* maximum.

Again we can obtain a combined bending moment diagram which, as can be seen in Fig. 3.8, gives moments identical to those of section 2.

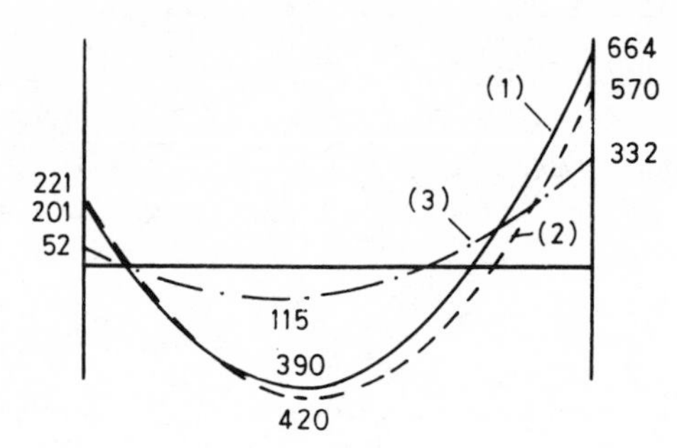


FIG. 3.8 Exterior span.

## Conclusions

The results from sections 2 and 3 are very similar, but section 3 does not really give an overall picture.

Many designs of a framed structure have been carried out using section 1, i.e. ignoring the columns, and finding the column moments from the ratio of the stiffnesses and the maximum out-of-balance moments, but not altering the beam moments. This is conservative but safe. This is dealt with more fully in Chapter 11. The effect of the exterior column, however, will be dealt with here as this can be considered in the beam design. The moment in the column at *A* can be taken as

above *A*: $$\left(\frac{0.19}{0.19+0.11+0.54/2}\right)\times 538 = 179 \text{ kN m}$$

below *A*: $$\left(\frac{0.11}{0.19+0.11+0.54/2}\right)\times 538 = 104 \text{ kN m}$$

where 538 kN m is the fixed end moment for span *AB* with the maximum load. So the moment in the beam at *A* would be $179+104=283$ kN m, which is rather larger than the moment obtained from section 2 or section 3.

The analysis carried out so far has been for a braced frame and any of the methods discussed will give a satisfactory answer. For an unbraced frame where lateral loads have to be considered the procedures carried out in section 2 of Example 3.1 should be performed (i.e. a subframe) and the results then compared with those obtained from load combination (3) (i.e. a 1.2 load factor on all loads) to see which gives the maximum moments. This will be done in Example 3.2.

## EXAMPLE 3.2 Lateral and vertical loads

The same member sizes as in Example 3.1 will be used although it is probable that some column sizes may be increased if the lateral stability depends on the frame. It will be assumed that the building is 12.0 m high and the characteristic wind load is 1.0 kN/m$^2$, i.e. 8.0 kN/m height per frame.

We now have to consider the design loads under load combination (3). To make a straight comparison with Example 3.1 consider the first floor.

Vertical loads $(23.2+20.0)\times 1.2 = 51.8$ kN/m
Horizontal loads $8.0\times 1.2 = 9.6$ kN/m.

For the vertical loads a subframe will be used as in section 2 of Example 3.1. For the horizontal loads we shall consider the whole frame with the wind forces acting at the external beam column connection, with the internal columns taking twice the horizontal force taken by the external columns.

## Vertical loads

There is only one case to consider now. From the free bending moment in any span of $51.8 \times 10^2/8 = 648$ kN m the bending moment diagram is as shown in Fig. 3.9 (note that the effects of sidesway have again been neglected).

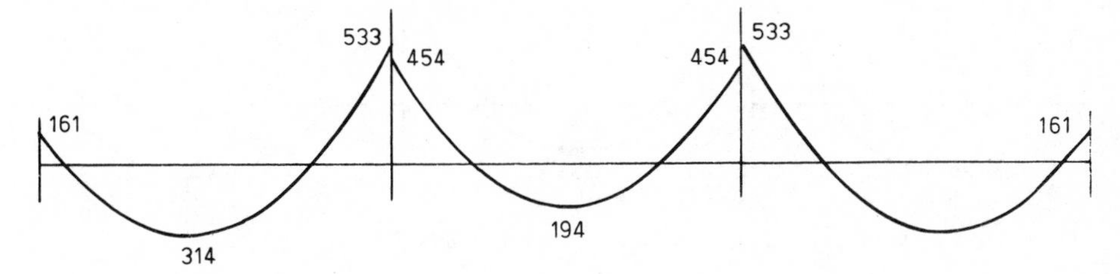


FIG. 3.9 Complete subframe – vertical loads for load combination (3).

## Horizontal loads

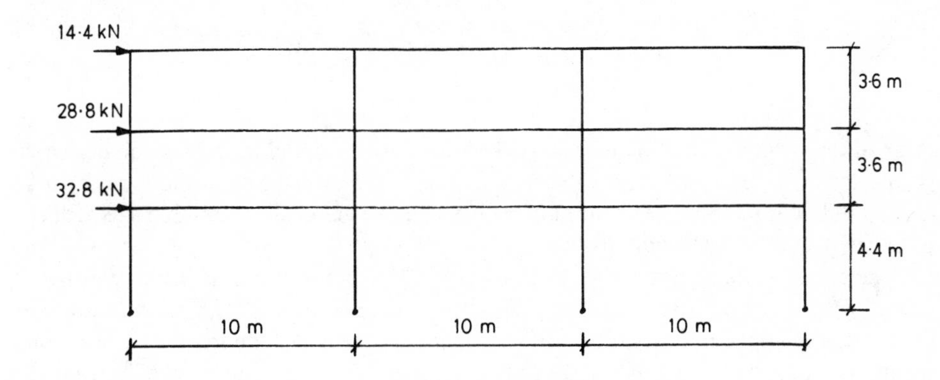


At first floor this will produce a moment diagram thus:

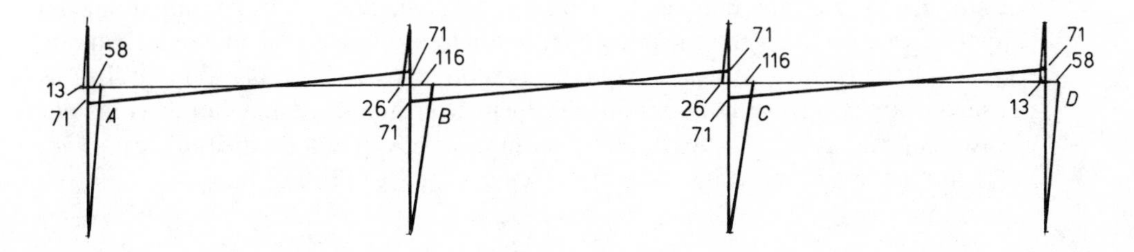

The effective diagram is:

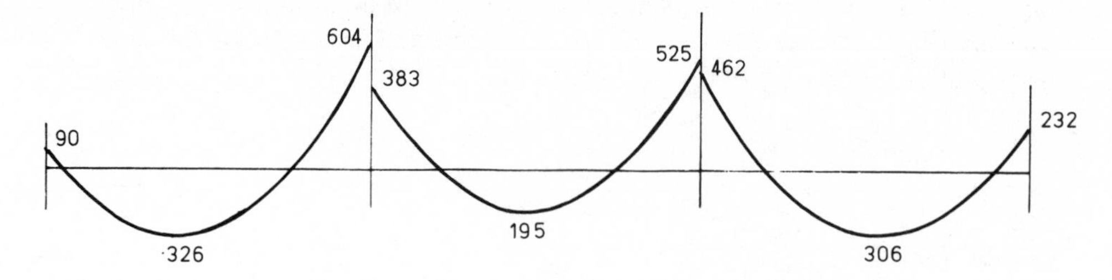


As the wind can also act in the opposite direction the envelope is given in Fig. 3.10.

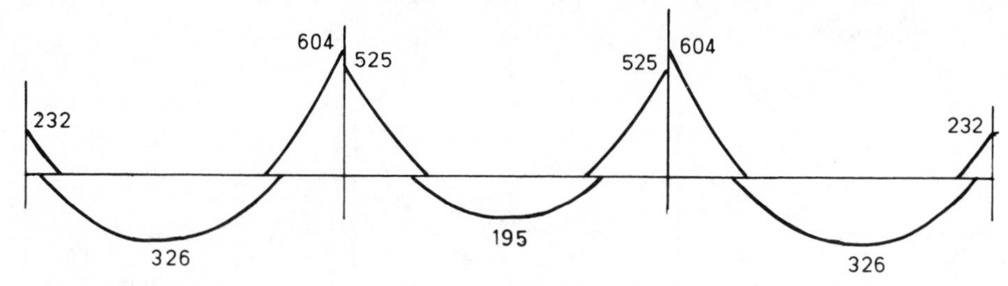


FIG. 3.10 Complete subframe – lateral and vertical loads.

Comparing this with the elastic moment envelope from section 2 of Example 3.1, it is apparent that although the support moments are similar, the span moments from load combination (1) are greater and would be used to determine the reinforcement required. Although the column moments from load combination (3) are greater, the direct load will be less and the columns will have to be designed on the worst combination. This will be done in Chapter 11.

---

## 3.4 Summary

From other examples that have been worked out the only conclusion that can be drawn is that for an unbraced frame subject to lateral loads all load combinations must be considered. With computers and plotters being more generally available this will not make very much additional work.

Concerning redistribution the main question is whether or not any advantage is gained. Using a computer program and plotter will give all the details. From the brief examples shown it can be seen that the main advantage is at supports where reduction in number of bars is always a help to construction. It will mean, however, that the designer must consider what he is trying to achieve and in the construction the positioning of bars is critical.

Without redistribution there is the additional safety factor that maximum moments cannot be reached at span and support simultaneously. A decrease in strength at one section will therefore be assisted by the other section. With 30% redistribution there is not such a large factor and this would give the most economic design, but this may not always be the most practical. Many design offices have decided to use 15% redistribution as normal practice, which seems to be a good solution.

# COVER 4

One of the criticisms of CP110 was that cover to reinforcement was spread throughout the Code, and it was decided to collect all the requirements together and put them in the Code where the designer would require them – between analysis of structure and analysis of sections.

As mentioned in Chapter 1, durability is vitally important for reinforced and prestressed concrete structures. Together with fire resistance, durability depends on the amount of concrete cover to the reinforcement as well as the quality of materials and workmanship.

## 4.1 Nominal cover

Concrete cover to reinforcement is always referred to in terms of nominal cover and is to all reinforcement, including links. This is the dimension used in design and indicated in the drawings, but it is realized that this dimension can never be maintained 100% in an actual structure. There must be tolerances. The actual cover to all reinforcement should never be less than the nominal cover minus 5 mm.

On the other hand, as given in Section 7 of Part 1, where reinforcement is located in relation to only one face of a member the actual cover should not be more than the nominal cover plus:

5 mm on bars up to and including 12 mm size;
10 mm on bars over 12 mm up to and including 25 mm;
15 mm on bars over 25 mm size.

A steel fixer has therefore to be careful to avoid too much cover as well as too little cover.

The nominal cover should: (a) comply with the recommendations for bar size, aggregate size and concrete cast against uneven surfaces; (b) protect the steel against corrosion; (c) protect the steel against fire; and (d) allow for surface treatments which may reduce the cover.

## 4.2 Bar size

There are two reasons for relating cover to bar size. The first is to ensure proper placing and compaction of the concrete around the bars. The second is to ensure a good bond to the bar. Although this variable does not appear in the bond provisions in the Code, bond strength is a function of the ratio of cover to bar size. The nominal cover to a main bar should not be less than the size of the main bar if it is a single bar, or the equivalent size where bars are in pairs or bundles. The equivalent size of a pair or a bundle is the size of a single bar of the cross-sectional area equal to the sum of the cross-sectional areas of the pair or bundle. A bundle of bars is defined as a group of three or four bars in contact. Two bars in contact are defined as a pair. Whether as a pair or bundle, a single bar of equivalent area will be used for design purposes. It must be noted, however, that even at

**Table 4.1** Equivalent bar sizes

| | *Two bars* | | *Three bars* | | *Four bars* | |
|---|---|---|---|---|---|---|
| *Size* (mm) | *Total area* ($mm^2$) | *Equivalent size* (mm) | *Total area* ($mm^2$) | *Equivalent size* (mm) | *Total area* ($mm^2$) | *Equivalent size* (mm) |
| 6 | 56.6 | 9 | 84.9 | 10 | 113 | 12 |
| 8 | 101 | 11 | 151 | 14 | 210 | 16 |
| 10 | 157 | 14 | 236 | 17 | 314 | 20 |
| 12 | 226 | 17 | 339 | 21 | 452 | 24 |
| 16 | 402 | 23 | 603 | 28 | 804 | 32 |
| 20 | 628 | 28 | 943 | 35 | 1260 | 40 |
| 25 | 982 | 35 | 1470 | 43 | 1960 | 50 |
| 32 | 1610 | 45 | 2410 | 55 | 3220 | 64 |
| 40 | 2510 | 57 | 3770 | 69 | 5030 | 80 |

Equivalent sizes have been rounded off to the nearest whole number.

laps there should never be more than four bars in contact. This will be discussed more fully in Chapter 8. The equivalent sizes for pairs and bundles are shown in Table 4.1. There are occasions when the nominal cover to links will require a greater cover to main bars, and this should always control.

In the past it has been the normal procedure to fix the spacers to the main bars and in many cases the cover to the links has been much less than intended. This has been in spite of the requirement that nominal cover has been specified to all reinforcement, including links. It is hoped that this procedure will change and spacers will be fixed to the links and not the main bars in beams and columns.

## 4.3 Aggregate size

The nominal cover should not be less than the nominal maximum size of aggregate. This will help the proper flow of concrete around the reinforcement and hence the compaction of the concrete.

## 4.4 Uneven surfaces

Concrete cast against uneven surfaces generally applies to concrete in foundations and is more related to durability. To ensure an adequate minimum cover it is recommended that where concrete is cast directly against earth the specified nominal cover should generally be not less than 75 mm. Where adequate blinding is provided (this will normally be the underside of a base) the nominal cover should be not less than 40 mm.

## 4.5 Cover against corrosion

This is the key to durability and, as mentioned briefly in Chapter 1, is where the designer really starts in selecting a grade of concrete for his structure. Although Table 3.4 in the Code gives the necessary related requirements regarding cover, grade of concrete etc., the first decision is made using Table 3.2 of the Code, exposure conditions. Five environmental conditions are given with a description of the exposure conditions envisaged. Examples are not given in the Code itself as it was felt that designers should assess their own particular circumstances. The Handbook, however, does give some examples which may assist the designer in making a decision.

Having classified the various parts of the structure, and hence the grade of concrete related to to cover, the next decision is whether or not to vary the grade of concrete throughout the structure. For example, with a mild condition of exposure (internal concrete in shops and offices) a Grade 30 concrete could be used with 25 mm nominal cover to all reinforcement. This grade cannot be used for other conditions of exposure, and for a severe exposure the concrete must be a minimum of Grade C40. It is possible that some small proportion of a structure is classed as severe whilst the remainder is moderate or mild. In this case a change of grade might be used, but generally a consistent grade will be used throughout.

A relaxation on cover to the ends of straight bars is given in clause 3.3.2. It should be emphasized that this only applies to mild exposure. The relaxation can be useful when calculating bearing lengths for precast floor units. As the Handbook points out, there is a possibility of rust staining during construction which will not affect the efficiency of the unit but may affect the appearance. Tables 3.2 and 3.4 of the Code have not been reproduced here as the author considers the reader should refer to the Code itself and become familiar with the requirements given therein.

## 4.6 Cover as fire protection

Cover for protection against corrosion may not be sufficient for protection against fire. In Part 1 of the Code, Table 3.5 and Figure 3.2 give values which will ensure that the fire requirements are satisfied. These values are based on recommendations given in Section 4 of Part 2 and the reader is strongly advised to study this section of the Code to try to understand how the more simplified approach in Part 1 has been derived.

In using Table 3.5 there are some important points to note.

First, in Section 4 of Part 2, cover to reinforcement in beams and columns is related to the main reinforcement whereas Table 3.5 uses nominal cover to all reinforcement as for durability. A notional allowance for stirrups of 10 mm has been made.

Secondly, the nominal covers in Table 3.5 relate specifically to the minimum member dimensions given in Figure 3.2. If smaller members are used then increased covers may be necessary; these are given in Section 4 of Part 2. For example, a two-hour fire resistance beam has a minimum width of 200 mm from Figure 3.2 and a nominal cover of 40 mm for simply supported and 30 mm for continuous beams. Table 4.3 in Part 2 gives the same net requirements for a simply supported beam. With continuous beams the minimum width can be 150 mm with a nominal cover of 40 (i.e. 50 minus 10), but if the width is increased to 200 mm then the nominal cover becomes 30 (i.e. 50 minus 10, minus 10, as given in Table 4.1 of Part 2).

This does not always happen, as using a one-hour fire resistance will demonstrate. Table 3.5 appears to have adopted a 20 mm minimum nominal cover to line up with Table 3.4, durability.

Thirdly, Table 3.5 refers to beams, floors and slabs as being simply supported or continuous. Part 1 does not explain the difference but Part 2 says that continuous will be 'where the designer has made provision for fixity in resistance to normal loads by the provision of reinforcement properly detailed and adequately tied to adjacent members'. It is assumed that the normal design and detailing rules would be satisfactory, but the 1978 report *Design and Detailing of Concrete Structures for Fire Resistance* – Interim Guidance by a Joint Committee of the Institution of Structural Engineers and the

Concrete Society (Institution of Structural Engineers, April 1978) does give an understanding of the requirements for good detailing.

Fourthly, note 2 below Table 3.5 draws attention to where the nominal cover exceeds 40 mm. CP110 caused a great many problems here by saying that 'supplementary reinforcement consisting of expanded metal lath or wire fabric or a continuous arrangement of links at not more than 200 mm centres should be incorporated in the concrete cover at a distance not exceeding 20 mm from the face'. The use of fabric had practical difficulties in keeping the fabric in place and compacting the concrete. Although its use is still included in the Code, four other alternatives are now given in Part 2 and it is hoped that these will be used.

Finally, a distinction is made between a rib in a ribbed floor and a beam. This is given in 4.2.7 of Part 2 and was inserted to ensure that ribs did not move too far apart (beams need minimum links for shear, ribs in floors do not – see later chapters).

## 4.7 Selection of appropriate cover

In choosing the appropriate cover for a particular structural member the designer must take the greatest nominal cover derived from:

1. bar size;
2. environmental conditions;
3. fire resistance.

# 5 STRENGTH OF SECTIONS – ULTIMATE LIMIT STATE

Having found the maximum moments at ultimate limit state we now have to determine the areas of reinforcement which will provide resistance moments at least as large as the design moments. The analysis of a cross-section to determine its moment of resistance at ultimate limit state is based on the following assumptions.

(a) CONCRETE

1. The strain distribution in the concrete in compression is derived from the assumption that plane sections remain plane.
2. The stress distribution in the concrete in compression may be derived from the stress–strain relation in Fig. 2.1 of the Code with $\gamma_m = 1.5$. Alternatively, the simple stress block shown in Fig. 3.3 of the Code may be used.

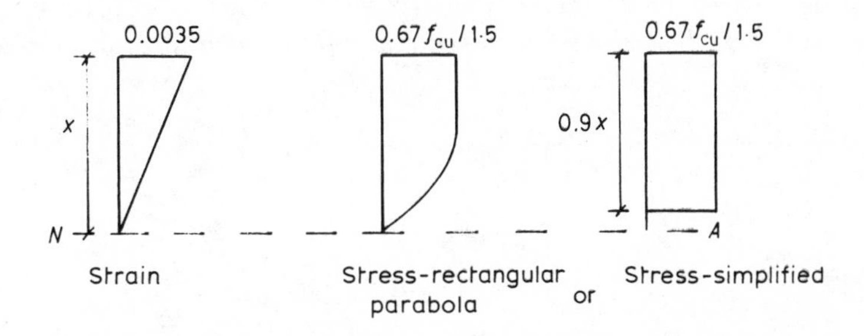


It should be noted that in developing some of the equations in clause 3.4.4.4 the maximum stress value is retained as $0.67 f_{cu}/1.5$ rather than the approximation of $0.45 f_{cu}$ which has been used elsewhere in the Code. This will be made clear when these are discussed.

3. The tensile strength of the concrete is ignored.

(b) REINFORCEMENT

4. The strains in the reinforcement, whether in tension or compression, are derived from the assumption that plane sections remain plane.
5. The stresses in the reinforcement are derived from the stress–strain curve in Fig. 2.2 of the Code with $\gamma_m = 1.15$. As was pointed out earlier the maximum design stress is the same for compression as for tension, which is 400 N/mm$^2$ for grade 460 reinforcement.

6. Where a section is designed to resist only flexure, the lever arm should not be assumed to be greater than 0.95 times the effective depth. This is intended principally for beams and slabs where the tension reinforcement is in the bottom. The concrete in the top of a beam or slab is not as well compacted as that in the bottom. To ensure a designer does not rely on a few millimetres of this concrete a minimum depth is given of approximately $0.1d$, but stated in terms of the lever arm.

These are the basic assumptions and in the actual design to find the amount of reinforcement required we can use: (a) design charts; (b) design formulae; or (c) strain compatibility.

## 5.1 Design charts

These are in Part 3 of the Code and have been prepared using the rectangular-parabolic stress block for the concrete and the bilinear stress–strain curve for the reinforcement. The charts are for rectangular sections reinforced in tension only (singly reinforced), or in tension and compression (doubly reinforced). Appendix A of Part 3 gives notes on the derivation of the charts and in the equation for the concrete compressive force at failure it can be seen that $0.45f_{cu}$ has been used rather than $0.67f_{cu}/1.5$. As in the previous charts, the full area of concrete in compression has been assumed even where compression reinforcement is present. The singly reinforced charts are for reinforcement Grade 250 and Grade 460 with different concrete grades.

The doubly reinforced charts are for a particular grade of concrete, reinforcement Grade 460, and a particular value of $d'/d$.

A typical doubly reinforced chart is shown in Fig. 5.1 and the use of the chart is quite straightforward, but these are some important points which must be borne in mind.

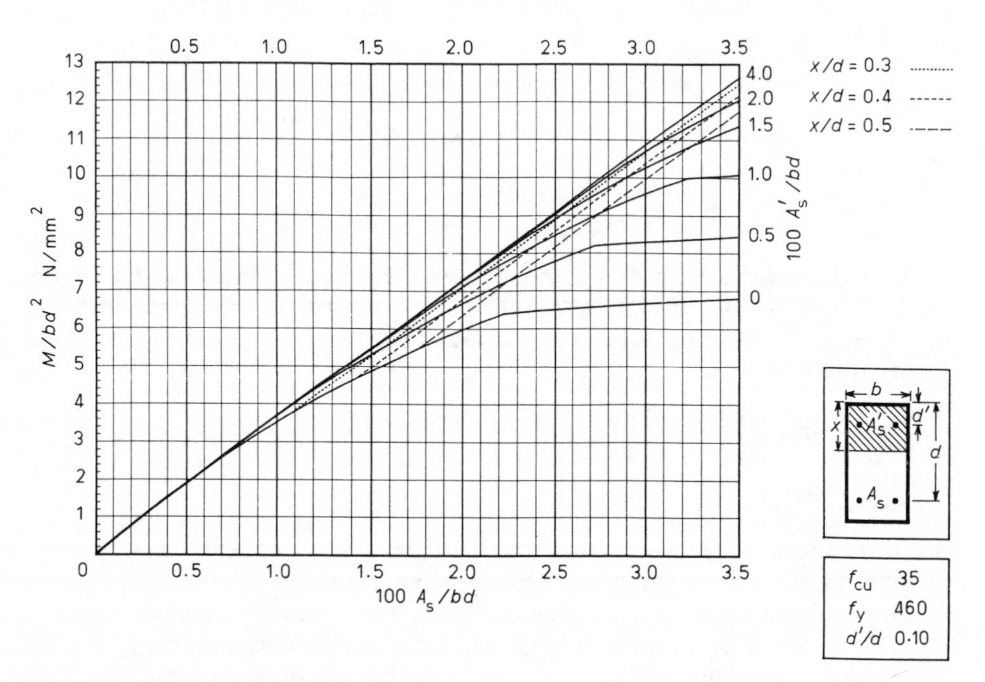


FIG. 5.1 Doubly reinforced beams.

1. Although the design charts appear to apply only to solid rectangular sections such as rectangular beams and solid slabs, they can be used for Tee or Ell beams where the neutral axis is within the flange.
2. On the chart there are three different types of dotted line and these are noted as when the neutral-axis depth factors are 0.3, 0.4, 0.5. These lines are very important when redistribution has been carried out. As explained in Chapter 3, the percentage redistribution is expressed in terms of the maximum moment at the section considered. So, if the amount of redistribution is 30%, i.e. $\beta_b=0.7$, then the maximum depth allowed for the neutral axis is $0.3d$. This means that we cannot use any part of the chart to the right of the line for $x/d=0.3$. Similarly for 20% redistribution we cannot go to the right of the line for $x/d=0.4$ and for 10% redistribution we cannot go to the right of the line for $x/\mathrm{d}=0.5$.
3. For small amounts of redistribution below 10% a designer would need a further line for $x/d=0.6$ (i.e. $\beta_b$ equals 1) and then interpolate. This is being very accurate and a more convenient method is to use a line joining the kinks on the graphs. This line is for $x/d=0.636$, and beyond this the curves flatten out. So interpolation can be carried out if required. It is unlikely that a designer would be using the flat part of the curve as the amount of tension reinforcement increases very rapidly with very little increase in moment of resistance.
4. A line is drawn for $100A_s'/bd=0$, i.e. singly reinforced, and this can be used for interpolation for $100A_s'/bd=0.2$, the minimum amount of compression reinforcement.
5. At the lower end of the chart where the lines merge, take the lowest value of $100A_s'/bd$ of the merging lines.

### 5.1.1 Example of use of chart

Using $f_{cu}=35, f_y=460, d'/d=0.10$, read from the chart the percentage of reinforcement for $M/bd^2=6$ having 0, 10%, 20%, 30% redistribution.

(a) NO REDISTRIBUTION

The results can be tabulated thus:

| | Compression (%) | Tension (%) |
|---|---|---|
| (i) | 0.0 | 2.0 |
| (ii) | 0.5 | 1.74 |
| (iii) | 1.0 | 1.66 |
| (iv) | 1.5 | 1.64 |

Other factors may influence the choice, but the arrangement (i) gives the minimum total percentage.

(b) 10% REDISTRIBUTION

We cannot go beyond the line for $x/d=0.5$ so the only choices we have are arrangements (ii), (iii), (iv) in the table above.

(c) 20% REDISTRIBUTION

Restriction is now the line for $x/d = 0.4$ and the arrangements are:

| | Compression (%) | Tension (%) |
|---|---|---|
| (i) | 0.4 | 1.78 |
| (ii) | 0.5 | 1.74 |
| (iii) | 1.0 | 1.66 |
| (iv) | 1.5 | 1.64 |

(d) 30% REDISTRIBUTION

Restriction is now the line for $x/d = 0.3$ and the arrangements are:

| | Compression (%) | Tension (%) |
|---|---|---|
| (i) | 1.0 | 1.66 |
| (ii) | 1.5 | 1.64 |

The more redistribution that has been carried out the more likely is the need for compression steel. This is to be expected, as for the same moment of resistance and the depth of the compression block reducing, we must add in compression reinforcement to keep the compression force up, even though the lever arm is in fact increasing.

## 5.2 Design formulae

These are given in clauses 3.4.4.4 and 3.4.4.5 and are based on the simplified rectangular stress block for concrete. Clause 3.4.4.4 deals with rectangular sections, but can also be used for flanged sections where the neutral axis comes within the flange. The equations are now presented in a format for calculating areas of reinforcement required, both for tension and compression. Unfortunately they do not give an equation whereby, knowing the area of reinforcement and section size, the moment of resistance of the section can be calculated. This will be given later in this section.

For the reinforcement, it is assumed that the maximum design stress of $0.87f_y$ has been reached in tension and compression. To achieve this for the tension reinforcement the maximum depth of the neutral axis is limited to $d/2$. From the strain profile it can be seen that $\varepsilon_s$, the strain in the tension reinforcement, can never be less than 0.0035. This is greater than 0.002, the value at which the reinforcement reaches its maximum design stress.

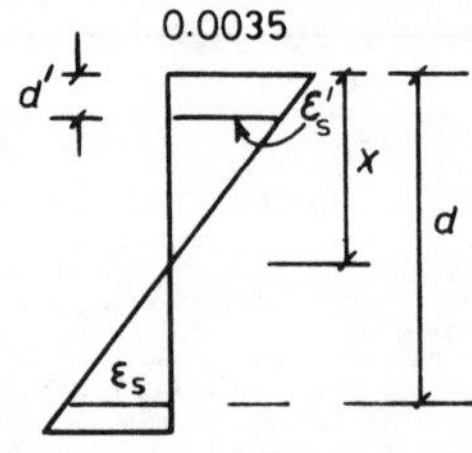


For the compression steel, however, for $\varepsilon_s'$ to be not less than 0.002 the value of $d'/x$ must not be greater than 0.43. If $d'/x$ does exceed this value, which can happen if the highest amount of redistribution has been carried out, then the actual stress should be calculated.

The Code uses the non-dimensional expression $M/bd^2f_{cu}$ and refers to this as $K$, with an upper limit for singly reinforced sections referred to as $K'$.

### 5.2.1 Singly reinforced rectangular sections

The equations can be derived as follows:

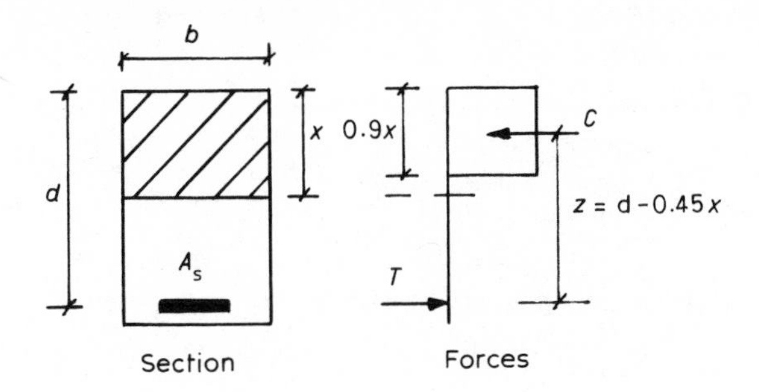


Concrete force $C=(0.67/1.5)f_{cu}b(0.9x)=0.402f_{cu}bx$ (5.1)

Steel force $T=0.87f_yA_s$. (5.2)

By equating forces we can obtain $x$, and from $z=d-0.45x$ we can obtain an expression for $z$ in terms of $A_s$ such that

$$z=d(1-0.97f_yA_s/f_{cu}bd). \quad (5.3)$$

This equation is the one referred to earlier which is not in the Code.

Taking moments for the tensile force about the centre of compression, we obtain

$$M=0.87f_yA_sz. \quad (5.4)$$

Hence a moment of resistance can be found.

As stated previously, the Code equations are primarily concerned with calculating the area of reinforcement required for a particular design moment. From equations (5.3) and (5.4), by eliminating $A_s$ we can arrive at the expression given in the Code for calculating $z$ as

$$z/d=[0.5+\sqrt{(0.25-K/0.9)}] \quad (5.5)$$

where

$$K=M/bd^2f_{cu}.$$

Having calculated $z$ we can now find the reinforcement required from equation (5.4) rearranged as

$$A_s=M/0.87f_yz.$$

The upper limit for $K$ (i.e. $K'$) for singly reinforced sections is derived as follows.

Taking moments for the compressive force about the centre of tension, we obtain

$$M=(0.67/1.5)f_{cu}b(0.9x)z. \quad (5.6)$$

For redistribution not exceeding 10% the maximum moment of resistance is obtained when $x=d/2$.

$$M'=(0.67/1.5)f_{cu}b(0.45d)(d-0.45d/2)=0.156f_{cu}bd^2. \quad (5.7)$$

Where redistribution exceeds 10%, 3.2.2.1 requires that

$$x\leqslant(\beta_b-0.4)d.$$

Substituting this value in equation (5.6) we have

$$M' = (0.67/1.5) f_{cu} b(0.9)(\beta_b - 0.4)d[d - 0.45(\beta_b - 0.4)d]$$

$$= [0.402(\beta_b - 0.4) - 0.18(\beta_b - 0.4)^2] f_{cu} bd^2. \quad (5.8)$$

If $M' = K' f_{cu} bd^2$, then

$$K' = 0.156$$

where redistribution $\leqslant 10\%$

and

$$K' = 0.402(\beta_b - 0.4) - 0.18(\beta_b - 0.4)^2$$

where redistribution $> 10\%$.

Note that we have used $M$ for the design moment and $M'$ for the moment of resistance of a singly reinforced section.

The value for $z/d$ obtained from equation (5.5) must never be taken as greater than 0.95 when used to calculate the area of reinforcement.

For designers with calculators or small computers these calculations can all be done very quickly, but the author has found that tables relating $M/bd^2$ (not $M/bd^2 f_{cu}$) to the percentage of reinforcement required can also be extremely useful. These, of course, have to be related to a specific grade of concrete. They can also be used to find the moment of resistance of a section knowing the percentage of reinforcement. The tables have been published separately in *Design Data for Rectangular Beams and Slabs to BS8110: Part 1* (available from E. & F. N. Spon) and also cover doubly reinforced rectangular sections, which will now be explained.

### 5.2.2 Doubly reinforced rectangular sections

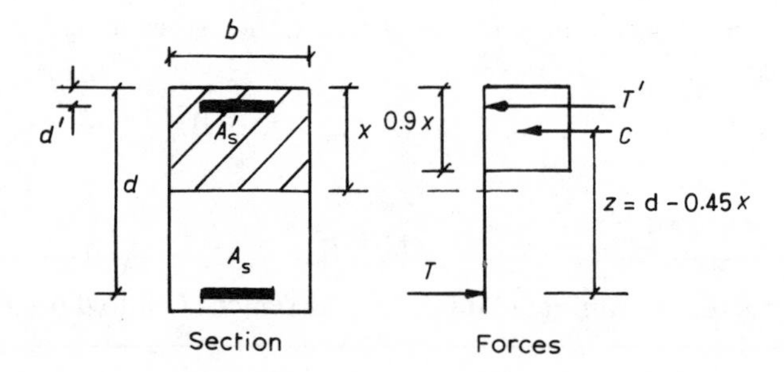


When the design moment is greater than $K' f_{cu} bd^2$, compression reinforcement is required and so we have an extra force, $T' = 0.87 f_y A'_s$, providing the limiting value of $d'/x$ is not exceeded.

The neutral axis depth is now fixed and depends on the amount of redistribution carried out.

The area of tension reinforcement required is in two parts. The first part is to balance the compression force in the concrete ($C$) and the second part to balance the force in the compression steel ($T'$).

At the limiting value of $K'$ the area of reinforcement required to balance the concrete force is $K' f_{cu} bd^2/z$, so, equating forces,

$$0.87 f_y A_s = K' f_{cu} bd^2/z + 0.87 f_y A'_s \quad (5.9)$$

or, as in the Code,

$$A_s = K'f_{cu}bd^2/0.87f_y z + A_s'. \qquad (5.10)$$

The value of $z$ is obtained from equation (5.5) where $K'$ is used instead of $K$.

Obviously $A_s'$ must be obtained before $A_s$ can be calculated and this is found by taking moments for the compression forces about the centre of tension.

$$M = K'f_{cu}bd^2 + 0.87f_y A_s'(d-d'),$$

i.e.

$$A_s' = (M - K'f_{cu}bd^2)/0.87f_y(d-d')$$

$$= (K-K')f_{cu}bd^2/0.87f_y(d-d'). \qquad (5.11)$$

It is important to remember that if the maximum amount of redistribution is carried out the maximum value for $x/d = 0.3$. If $d'/d = 0.2$ then $d'/x = 0.67$, which is well in excess of 0.43. The stress in the compression reinforcement will then be reduced to 233 N/mm$^2$ for grade 460 steel and the value of $A_s'$ in equation (5.11) is increased. This will also affect equation (5.10).

In fact, it might be better to rewrite equations (5.10) and (5.11) as

$$A_s = K'f_{cu}bd^2/0.87f_y z + f_{sc}A_s'/0.87f_y \qquad (5.10a)$$

and

$$A_s' = (K-K')f_{cu}bd^2/f_{sc}(d-d') \qquad (5.11a)$$

where $f_{sc} = 0.87f_y$ for $d'/d < 0.43$.

In the *Manual for the Design of Reinforced Concrete Building Structures*, prepared by a Joint Committee of the Institution of Structural Engineers and the Institution of Civil Engineers (Institution of Structural Engineers, October 1985) it is suggested that where $d'/x$ is greater than 0.43 then $f_{sc} = 700(1 - d'/x)$.

### 5.2.3 Singly reinforced flange sections where neutral axis falls below the flange

Although the Code says the neutral axis is below the flange, what it really means is that $0.9x$ is greater than the depth of the flange, $h_f$. Also, in deriving the equations it assumes that $x = d/2$. The equations have been derived as follows:

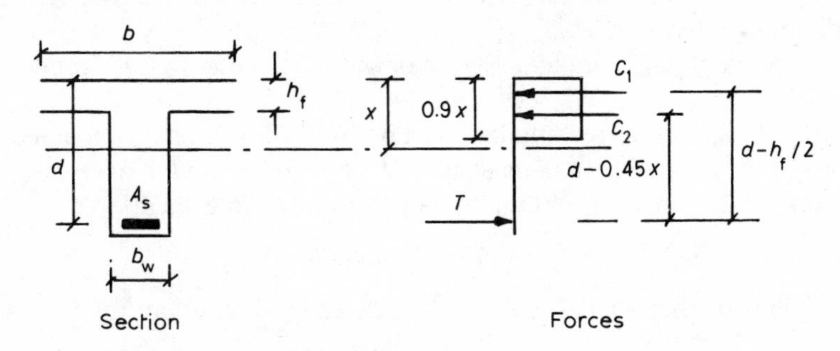


$C_1$ = force in flange outside web $= 0.45f_{cu}(b-b_w)h_f$
$C_2$ = force in web $= 0.45f_{cu}b_w(0.9x)$
$T = 0.87f_y A_s$.

Assuming $x = d/2$ and taking moments for the compressive force about the centre of the tension force, we find

$$M' = 0.45f_{cu}h_f(b - b_w)(d - h_f/2) + 0.45f_{cu}b_w(0.45d)(d - 0.45d/2)$$

$$= f_{cu}bd^2[(0.45h_f/d)(1 - b_w/b)(1 - h_f/2d) + 0.157b_w/b]. \quad (5.12)$$

Note that the term in square brackets is the factor $\beta_f$ (equation (2) of BS8110) except that the coefficient 0.157 is taken as 0.15. Values of $\beta_f$ are tabulated in Table 3.7 of the Code.

Taking moments for the tensile force about the centre of the flange

$$M = 0.87f_yA_s(d - h_f/2) - 0.45f_{cu}b_w(0.45d)(0.45d/2 - h_f/2)$$

$$= 0.87f_yA_s(d - h_f/2) - 0.1f_{cu}b_wd(0.45d - h_f).$$

So

$$A_s = \frac{M + 0.1f_{cu}b_wd(0.45d - h_f)}{0.87f_y(d - 0.5h_f)}. \quad (5.13)$$

This equation (5.13) corresponds to equation (1) of BS8110.

Equation (1) of BS8110 may be used provided the design ultimate moment is less than $\beta_f f_{cu}bd^2$ and that not more than 10% redistribution has been carried out.

If the design ultimate moment exceeds $\beta_f f_{cu}bd^2$ or if more than 10% redistribution has been carried out, the section should be designed by direct application of the assumptions given in 3.4.4.1.

When the neutral axis depth is $h_f/0.9$, the rectangular stress block takes in the whole of the flange, but none of the web. At this value the moment of resistance of the concrete in compression is given by

$$M_f = 0.45f_{cu}bh_f(d - 0.5h_f). \quad (5.14)$$

So, for any value of $M$ less than $M_f$ the section can be designed as a rectangular section of width $b$ to obtain the area of tension reinforcement, $A_s$.

If $M$ is greater than $M_f$ it can be seen that the moment resisted by the web is

$$M_w = M - C_1(d - 0.5h_f)$$
$$= M - 0.45f_{cu}(b - b_w)h_f(d - 0.5h_f)$$
$$= M - M_f(b - b_w)/b$$
$$= M - (1 - b_w/b)M_f.$$

So the approach would be to calculate $M_f$ and proceed as follows, where $M_f$ is obtained from equation (5.14).

If $M \leqslant M_f$, design as a rectangular section of width $b$ to obtain area of reinforcement $A_s$.

If $M > M_f$, design as a rectangular section of width $b_w$ for a moment $M_1 = M - (1 - b_w/b)M_f$ to obtain area of tension reinforcement $A_{s1}$ and, if necessary, area of compression reinforcement $A'_{s1}$. Total area of tension reinforcement,

$$A_s = A_{s1} + 0.45f_{cu}(b - b_w)h_f/0.87f_y. \quad (5.15)$$

This approach is neither approximate nor restricted in application (unlike 3.4.4.5).

### 5.2.4 Minimum percentages of reinforcement

Table 3.27 of the Code gives minimum percentages of compression and tension reinforcement for various types of member.

There are occasions when we have to start from first principles and this will be done using strain compatibility.

However, before discussing this we will design some of the sections for one of the analyses previously carried out. We shall do this using the complete subframe in section 2 of Example 3.1. So that we can compare the answers with the chart in Fig. 5.1 we shall take $f_y = 460$ and $f_{cu} = 35$. To illustrate redistribution we shall use the redistributed envelope of Fig. 3.6.

## EXAMPLE 5.1

Assume a nominal cover of 35 mm to the top bars, and 40 mm to the bottom and side bars.

The moments to be considered are as follows:

| | Moment (kN m) | $\beta_b$ |
|---|---|---|
| 1. Exterior support | 201 | $\frac{201}{221} = 0.91$ |
| 2. Near middle of end span | 479 | $\frac{479}{420} \not> 1.0$ |
| 3. First interior support | 465 | $\frac{465}{664} = 0.70$ |
| 4. Middle of interior span | 341 | $\frac{341}{314} \not> 1.0$ |

Note that for positions (2) and (4) the redistributed moment is greater than the elastic moment so $\beta_b$ is taken as 1.0.

In the following calculations, position (1) will be treated using the Code equations and then compared with values from the design tables referred to earlier. The reader can verify these by using the design chart. Other positions will use values from design tables only.

### Exterior support $M = 201$ kNm

With 35 mm cover to top bars, assume $d = 600 - 50 = 550$ mm. $\beta_b = 0.91$ so $x$ is not greater than $(\beta_b - 0.4)d = 0.5d$. The section here will be rectangular, 600 mm deep by 300 mm wide.

$$K = \frac{M}{bd^2 f_{cu}} = \frac{201 \times 10^6}{300 \times 550^2 \times 35} = 0.063.$$

This is less than $K'$ $(=0.156)$ so compression reinforcement is not required.

$z = \{0.5 + \sqrt{(0.25 - 0.063/0.9)}\}d = 0.924d$ (i.e. $<0.95d$)
$x = (d - z)/0.45 = 0.169d$ (i.e. $<0.5d$)

$$A_s = \frac{M}{0.87 F_y z} = \frac{201 \times 10^6}{0.87 \times 460 \times 0.924 \times 550}$$

$= 988$ mm$^2$. Suggest $2/25\phi$ (980 mm$^2$)

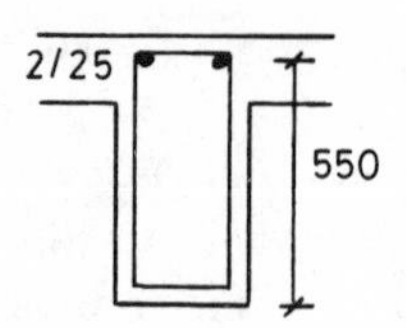


Using tables, $M/bd^2 = 2.21$ (i.e. $< 0.156 \times 35$).
$100A_s/bd = 0.60$, so $A_s = 990\ \text{mm}^2$.

It can also be seen from the tables that $z/d = 0.923$ and $x/d = 0.170$.
From the design chart the answer would probably be read as 0.6%.

## Middle of end span $M = 479$ kN m

The section here will be a Tee section, where $b = b_w + 0.2l_z = 300 + 1400 = 1700$ mm.
$d = 600 - 55 = 545$ mm.
$M_f = 0.45 \times 35 \times 1700 \times 150(545 - 75) \times 10^{-6}$
$= 1888$ kN m, i.e. $> M$, so treat as rectangle of width 1700 mm.

$M/bd^2 = 479 \times 10^6/(1700 \times 545^2) = 0.95.$

From tables $100A_s/bd = 0.25$ and the lever arm is restricted to $0.95d$.
$A_s = 2316\ \text{mm}^2$. Suggest $3/32\phi$ ($= 2410\ \text{mm}^2$).

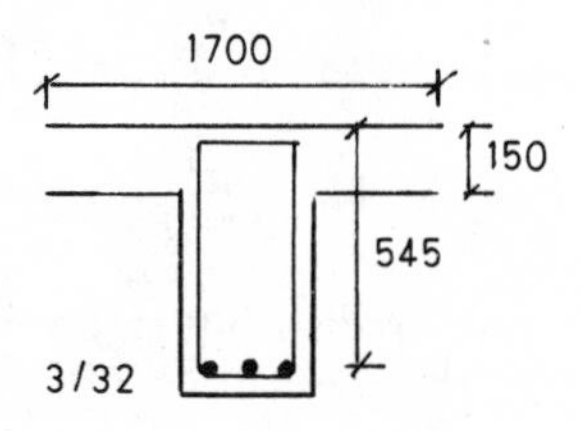


## Interior support $M = 465$ kN m

$\beta_b = 0.7$ or 30% redistribution.
The maximum value for $M/bd^2 = 3.64$ (see tables).
$M/bd^2 = 465 \times 10^6/(300 \times 550^2) = 5.12$, so compression reinforcement required.
$d' = 55$ mm, so $d'/d = 0.10$.

From tables $100A'_s/bd = 0.41$ and $100A_s/bd = 1.46$.
So $A'_s = 677\ \text{mm}^2$ (suggest $2/25 = 980\ \text{mm}^2$)
$A_s = 2409\ \text{mm}^2$ (suggest $3/32 = 2410\ \text{mm}^2$).

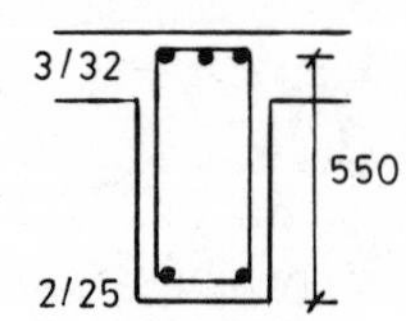


There is no need to check $x/d$ or the value of $d'/x$ as these have been taken into account when preparing the tables.

As the area of tension reinforcement provided is almost identical to that required, the tables would not give any increase in the moment of resistance over that required. From the chart, however, by using the actual percentages of reinforcement provided, it will be found that there is a slight increase in the moment.

### Interior span $M = 341$ kN m

From the calculations relating to the middle of the end span it can be seen that $M$ is less than $M_f$, so we can use a rectangular section as in that case.

$M/bd^2 = 341 \times 10^6/(1700 \times 545^2) = 0.68$.

From tables, $100A_s/bd = 0.18$.

So $A_s = 1668$ mm$^2$. Suggest 4/25 (1960 mm$^2$) or 2/32 + 1/20 (1920 mm$^2$) which will suit the detail better.

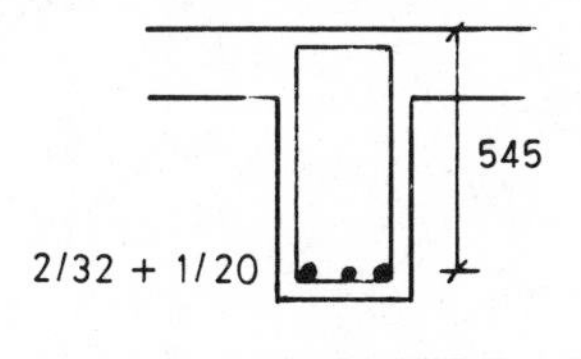


## 5.3 Strain compatibility

If we have a non-rectangular section we cannot use design charts or formulae and so we cannot find the area of reinforcement required for a given ultimate moment or vice versa. We therefore have to go to first principles and use strain compatibility. But even with this method it is only possible to find the ultimate resistance moment based on a given steel area. The basic principle of strain compatibility is that for a given section (including the reinforcement) one can find a neutral-axis depth so that the total compression force equals the total tension force and hence the ultimate moment.

The forces in the reinforcement are found by using a linear strain profile. Determine the strains in the reinforcement and from the stress–strain profile for the particular reinforcement the stress in the reinforcement can be found. An example illustrating the method is as follows.

### EXAMPLE 5.2

Find the ultimate moment capacity of the beam section given below:

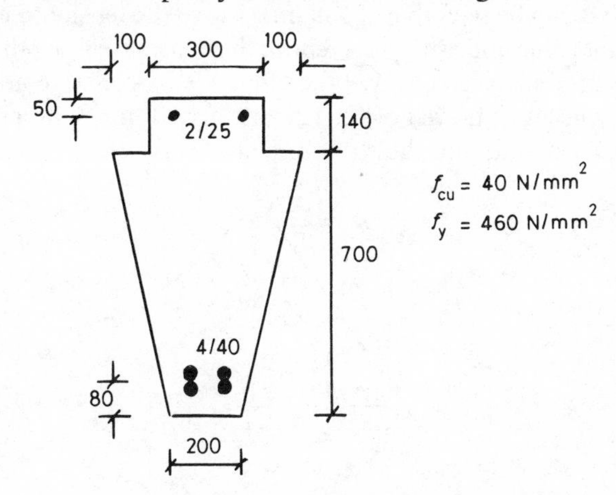

We could use a rectangular parabolic stress block and divide the section into vertical segments, but this would be a very long and tedious process, so we use the alternative rectangular stress block with a uniform compression stress of $0.45 \times 40 = 18$ N/mm$^2$. The stress–strain curve for reinforcement is shown in Fig. 2.2 of the Code where the limiting design stress in tension and compression is 400 N/mm$^2$, as stated earlier in this chapter.

We now have to assume a neutral-axis depth and draw the strain profile with the maximum compression strain in the concrete of 0.0035 at the top edge.

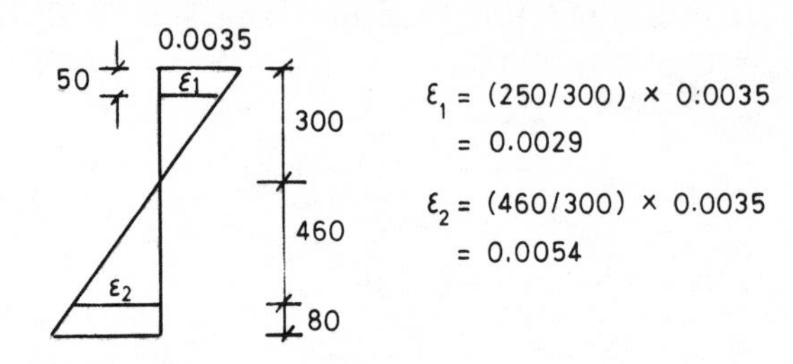


For $\varepsilon_1$ (compression) and $\varepsilon_2$ (tension) it can be seen from the stress–strain curve that these give the limiting design stresses. The area of the section above the neutral axis can be divided into a trapezium and a rectangle which have effective areas of $61.4 \times 10^3$ mm$^2$ and $42.0 \times 10^3$ mm$^2$ respectively for compressive stresses in the concrete.

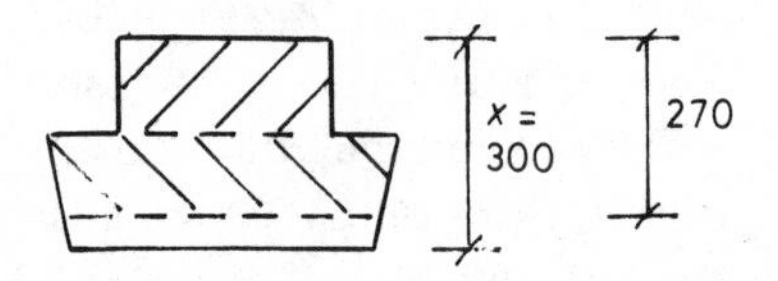


The total compressive force is therefore

| | | |
|---|---|---|
| Concrete | $(61.4+42) \times 10^3 \times 18 \times 10^{-3}$ | $= 1861$ kN |
| Reinforcement | $982 \times 400 \times 10^{-3}$ | $= 393$ kN |
| | | 2254 kN |
| Total tension | $5030 \times 400 \times 10^{-3}$ | $= 2012$ kN. |

As the total compression is larger than the total tension we have assumed too large a neutral-axis depth. There are several ways of adjusting the value of $x$ and it is easier if all the reinforcement is at its limiting stress even after adjustment. By reducing the neutral-axis depth it can be seen that $\varepsilon_2$ will increase so the tension reinforcement will still be at its limiting value and, unless we reduce the neutral-axis depth by one third, so will the compression reinforcement. We therefore need a concrete area which when multiplied by 18 will give an area of $(2012 - 393) \times 10^3$ mm$^2$. It can be found that $x = 267$ mm is near enough and the strain profile is:

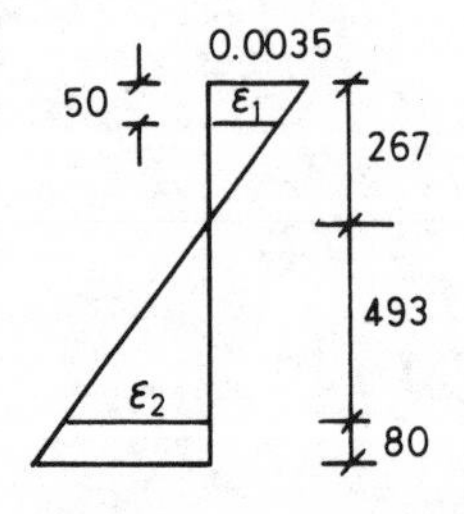

With $x = 267$, $\varepsilon_1 = (217/267) \times 0.0035 = 0.0028$, so compression steel is confirmed as being at its limiting value.

As the compression and tension forces are approximately equal we find the ultimate resistance moment by taking moments about the centre of compression or centre of tension, the latter usually being easier. The force in the upper rectangle is 756 kN and in the trapezium 862 kN. The force in the compression steel is 393 kN, so

$$M_u = 756(760 - 70) \times 10^{-3} + 862(760 - 189.25) \times 10^{-3} + 393(760 - 50) \times 10^{-3}$$
$$= 1293 \text{ kN m}.$$

---

# 6 SERVICEABILITY LIMIT STATE OF DEFLECTION

Having analysed the sections at ultimate limit state and calculated the necessary reinforcement, the designer could carry on and calculate the necessary curtailing points, check bond and anchorage, calculate for shear and any other requirements at ultimate limit state. Before doing this, however, the author considers a more satisfactory procedure is to check that the serviceability limit states are satisfied. Many designers check that the section will meet the shear requirements before proceeding any further, and although this calculation may not appear on the calculation sheets at this stage, it is a procedure to be recommended.

As stated earlier in describing the criteria, Part 1 of the Code does not give any numerical values for acceptable deflection, but in clause 3.4.6.3 for deflection of beams it does quote values. These values are taken from Part 2 in clause 3.2 and are as follows.

### *Vertical loads*

For appearance, a deflection of span/250 will usually become noticeable and so the final long-term deflection should be limited to this value. Although it says deflection what it means is the distance below a horizontal line joining the level of the supports.

For efficiency the Code has three items.

1. Damage to non-structural elements. Here we have the damage to finishes and partitions after construction. For brittle materials the movement should not exceed span/500 or 20 mm, whichever is the lesser. For non-brittle materials the movement should not exceed span/350 or 20 mm, whichever is the lesser.
2. Construction lack of fit. All elements should fit together properly.
3. Loss of performance. Even with the limits given above, ponding on a slab may not be acceptable.

### *Wind loads*

Here the Code says reference should be made to specialist literature, but for damage to non-structural elements the relative lateral deflection in any one storey should not exceed storey height/500.

Although the procedures are very similar for slabs, we will deal with beams in this chapter, leaving the slabs until later.

As the Code says in clause 3.4.6.1 deflections may be calculated but in all normal cases the deflection will be satisfactory if members conform to the limiting span/effective depth ratios as obtained from the various factors in the Code. These factors can be fairly easily determined at the design stage. It is quite obvious, therefore, that span/effective depth ratios will be used in the majority of cases and calculations will only be carried out for special cases where (a) the designer wishes to exceed the span/effective depth ratio; (b) particularly stringent deflection control is required; and (c) the structure is abnormal in any way.

From this it follows that calculations are only worth doing if they are carried out with some care and in special cases. This is emphasized by the fact that the procedure for the calculations is included in Part 2 of the Code. To keep in line with the Code the procedure will be dealt with in Appendix 1 of this book.

If we now revert to the deemed-to-satisfy ratios, deflections have been controlled fairly well in the past by limiting the span/depth ratios depending on the type of reinforcement and the permissible stress in the concrete. The method was simple to apply and was based on practical experience rather than pure calculation. Excessive deflections were relatively uncommon, but with the reduction in safety factors and the increasing use of stronger materials, engineers have been forced to consider the influence of other variables on the permissible span/depth ratios.

The influence on these ratios caused by variations in the design parameters was assessed analytically. By studying the analysis of some hundreds of beams, the number of significant variables was reduced to eight. These are:

1. support conditions
2. span
3. characteristic strength of steel
4. percentage of tension reinforcement
5. ratio of (permanent load)/(total load)
6. ratio of (compression steel)/(tension steel)
7. whether the deflection is critical
8. environment (e.g. effects of temperature, creep and shrinkage).

In the original proposals put forward there were four tables, each considering two of the variables in the above lists, and a permissible span/effective depth was arrived at as the product of four numbers, one from each table. The recommendations were for ratios of span to effective depth rather than span to overall depth. This recognizes the fact that, at the levels of steel stress now permitted, sections carrying their service loads will normally be cracked. The stiffness of the member will therefore be a function of the transformed section, rather than the concrete or gross section, and this is a function of the position of the tension steel, i.e. by the effective depth.

It was realized, however, that any simplification or reduction in the number of variables to be considered would be welcomed. Also more work had shown where considerable improvements could be made. A reappraisal of the variables was undertaken both by analytical methods and by considering some work on the long-term behaviour of beams. The final ratio will still be for span to effective depth and the variables are:

1. span and support conditions for rectangular sections (cf 1 and 2 in original)
2. percentage of tension reinforcement and working stress level of this reinforcement (cf 3 and 4 in original list)
3. percentage of compression reinforcement (cf 6 in original list)
4. other factors
5. flanged beams.

The allowable ratio is found by taking the basic ratio from (1) and multiplying this by factors from (2), (3), (4) and (5) where these are applicable.

We shall examine these factors individually.

## 6.1 Span and support conditions for rectangular sections

The basic ratios given in Table 3.10 of the Code apply to all spans where the only requirement is that the deflection should not exceed span/250. For spans up to 10 m these ratios should normally avoid damage to finishes and partitions, as they will limit deflection occurring after construction to span/500 or 20 mm whichever is the lesser.

At 10 m span the total deflection is 40 mm, of which 20 mm can be after erection of partitions and 20 mm before.

For spans greater than 10 m, where it is necessary further to restrict the deflection to avoid damage to finishes or partitions the values in Table 3.10 should be multiplied by 10/span, except for cantilevers where the deflection should be calculated. For a span of 20 m the basic ratio is halved, if the deflection is further restricted.

## 6.2 Percentage of tension reinforcement and service stress

Deflection is influenced by the amount of tension reinforcement and its stress and the basic ratios are modified accordingly. The area of reinforcement and its stress are taken at the centre of the span (or at the support for a cantilever). Table 3.11 of the Code gives the modification for tension reinforcement, but it should be noted that the amount of tension reinforcement is not involved directly. We use the value $M/bd^2$ where $M$ is the design moment at ultimate limit state, not serviceability limit state.

There is obviously a direct link between $M/bd^2$ at ultimate limit state and the amount of reinforcement required, so this is now used instead of the percentage of reinforcement as in CP110. The stress in the steel, however, is the service stress and this depends on (1) whether more reinforcement has been provided than required and (b) whether redistribution has been carried out. If neither has been done then it is assumed that the service stress is $5f_y/8$, which is very close to $0.87f_y/1.4$. If (a) has been done then the service stress is reduced; if (b) has been done the service stress is increased. By this stage the designer will know all these facts and the BS8110 system has no advantage over the CP110 system. Where it does help is at the preliminary stage. The designer has assessed an effective depth, has calculated $M/bd^2$, and by assuming the service stress is $5f_y/8$ (as given in the table), can get a modification factor very quickly. This enables the designer to obtain an allowable span/effective depth ratio, and hence an acceptable effective depth, at an earlier stage.

For example, with a simply supported rectangular beam, if the effective depth is span/22 and Grade 460 steel is used, $M/bd^2$ cannot exceed 2.0 without providing more steel than is required.

The values in Table 3.11 of the Code have been obtained from equation (7) which is given below the table. For values between those tabulated one may either interpolate linearly, or use the formula. It should be noted that the factor has an upper limit of 2.0. This is so that ratios of span to effective depth would not go beyond those of which there was practical experience.

## 6.3 Percentage of compression reinforcement

For the compression reinforcement the modification factor is presented in terms of the percentage of compression reinforcement. In obtaining this percentage, $b$ will be taken as the effective flange width for a Tee or Ell section. As concrete dries out it tends to shrink, but this shrinkage is partly restrained in the vicinity of the reinforcement. Thus in a singly reinforced beam shrinkage is restrained on the tension face, but not on the compression face. The result is a curvature which is additive to that due to loading. The inclusion of reinforcement in the compression face will reduce this and so account is taken of this effect.

It is important to note that although the term 'compression reinforcement' is used, it means any reinforcement in the compression zone, even those not effectively tied with links.

Table 3.12 of the Code gives the modification factors for amounts of compression reinforcement up to 3%, together with the formula from which the values have been obtained.

By taking account of reinforcement in the compression zone the final ratio of span to effective depth can be increased.

## 6.4 Other factors

The last two items in the original list of variables have now been omitted. It was felt that designers would not know the creep and shrinkage properties of the mix at the design stage and so the variable has been removed. (In the criteria for this limit state it says that account should be taken of the effects due to creep and shrinkage and allowance made for this.)

It does not define 'normal', but it goes on to say that if the free shrinkage strain is expected to be greater than 0.00075 or the creep coefficient greater than 4, the permissible span/effective depth ratio should be reduced, but a reduction of more than 15% is unlikely. So, by implication, action is only required when there is a good reason to expect a concrete with abnormally high shrinkage or creep, and even then the modification factor is only 0.85.

## 6.5 Flanged beams

For Tee or Ell sections, such as in a monolithic beam and slab construction, there is a further modification factor as referred to earlier when discussing the factor for tension reinforcement. If one compares the stiffness of a flanged beam with that of a solid rectangular section of the same width as the flange it is realized that as concrete in the tension zone will contribute some stiffness, the effect is much less in a flanged beam. The modification factor has been introduced to allow for this and varies between 0.8 and 1.0, depending on the ratio of the width of the web to the width of the flange.

Table 3.10 of the Code includes values for the basic span/effective depth ratio for flanged beams where $b_w/b$ is less than or equal to 0.3, by multiplying the ratios for rectangular sections by 0.8. For values of $b_w/b$ between 0.3 and 1.0 one must interpolate. The factor can be obtained from the formula

$$\text{Modification factor} = 2(b_w/b - 0.1)/7 + 0.8.$$

---

## EXAMPLE 6.1

---

Check that the exterior span in section 2 of Example 3.1 complies with the serviceability limit state of deflection by using the tables for the span/effective depth ratio.

From this section and the design of the section in Chapter 5 we have the following data:
Span: 10.0 m continuous.

Section: Tee beam, flange width 1700 mm, width of web 300 mm, effective depth 545 mm.
Reinforcement: $A_{s,req} = 2316$ mm$^2$; $A_{s,prov} = 2410$ mm$^2$.
No redistribution.

## Check

1. $b_w/b = 300/1700 = 0.176$, i.e. $< 0.3$.
   Basic ratio from Table 3.10 = 20.8.

2. Modification factor for tension reinforcement.
   $M/bd^2 = 0.95$.

$$f_s = \frac{5}{8} \times 460 \times \frac{2316}{2410} = 276 \text{ N/mm}^2.$$

   From Table 3.11 (by interpolation) or equation (7), modification factor = 1.46.

3. Modification for compression reinforcement. Although there will be link carriers in the top of the beam it is not normally necessary to take these into account.

4. Creep and shrinkage. Assuming normal aggregates, no modification factor.

Allowable span/effective depth ratio = $20.8 \times 1.46 = 30.4$.
Actual ratio = 10 000/544 = 18.4.

Actual is less than allowable, so the section is satisfactory. From this calculation it appears that we could have chosen a section with smaller depth. Before doing so, however, it is advisable to examine the calculations for the strength of sections at the supports. If we reduce the effective depth $d$, we increase $M/bd^2$ very rapidly. This will rapidly increase the amount of tension reinforcement, and compression reinforcement may also be introduced, adding to the congestion at the column–beam intersection. This does not mean the depth of the beam should not or could not be reduced. It is just that the difference between actual and allowable ratios is not quite what it seems.

# SERVICEABILITY LIMIT STATE OF CRACKING

# 7

As with deflection, Part 1 of the Code does not give any numerical values for crack widths. It states in clause 2.2.3.4.1 that cracking will normally be controlled by compliance with the detailing rules but, where specific attention is required, one must go to clause 3.2.4 in Part 2.

In Part 2 it says that for appearance and corrosion the calculated maximum crack width should not exceed 0.3 mm. For loss of performance such as watertightness it says that other limits may be appropriate. For this particular case one would need to refer to BS8007 where the limits are 0.1 and 0.2 mm.

Also in Part 2, clause 3.8, is the procedure for calculating crack widths. Although it is not intended to calculate crack widths in this chapter (see Appendix 2) a basic understanding of what is happening should be of interest to designers.

The width of flexural cracks at a particular point on the surface of a member depend primarily on three factors:

1. the proximity to the point under consideration of reinforcing bars perpendicular to the cracks
2. the proximity of the neutral axis to the point under consideration
3. the average surface strain at the point under consideration.

The formula given in Part 2 of the Code is as follows. As can be seen, it gives a relationship between crack width and these three principal variables.

Surface crack width

$$w_{cr} = \frac{3a_{cr}\varepsilon_m}{1 + 2[(a_{cr} - c_{min})/(h - x)]}$$

where $a_{cr}$ is the distance from the point under consideration to the surface of the nearest longitudinal bar, $c_{min}$ is the minimum cover to the longitudinal bar, and $\varepsilon_m$ is the average strain at the level considered.

The formula gives acceptably accurate results in most normal design circumstances, but it should be emphasized that cracking is a semi-random phenomenon and that an absolute maximum crack width cannot be predicted. The formula is designed to give a width of crack which has an acceptably small chance of being exceeded. Thus an occasional crack slightly larger than the predicted width should not be considered as cause for concern. But if a significant number of the cracks in a structure exceed the calculated width, reasons other than the structural nature of the phenomenon should be sought to explain their presence.

The crack width formula has been arrived at through a study of previous crack formulae and the results of a large number of tests on beams and slabs. Taking into account the fact that the probability of a member being subjected to its design load for any significant length of time is low and also that members are not generally subjected to uniform conditions of bending over any great length so that the only cracks which have a serious chance of being critical are those close to the critical sections, it is felt that the chances of the specified width being exceeded by a single crack will be about 1 in 1000. This is considered acceptable.

Generally, however, it will not be necessary to do calculations as the Code gives rules which, if followed, will satisfy the criteria for beams in normal internal or external conditions of exposure. These rules are based on the crack width formula as given above.

The rules are generally referred to as the bar spacing rules, as they are mainly concerned with the spacing of the main bars in the tension areas, and can be illustrated diagrammatically as shown in Fig. 7.1.

For $a_b$ it should be noted that this is the clear distance between bars. This varies depending on the amount of redistribution carried out at ultimate limit state and on the characteristic strength of the reinforcement. Quite obviously, if redistribution has been carried out reducing the design moment (indicated by the minus sign in Table 3.30 of the Code) the service stress in the reinforcement is higher and the bars are therefore to be closer together. On the other hand if moment has been added at the section (indicated by the plus sign in Table 3.30) then the bars can be moved further apart. This

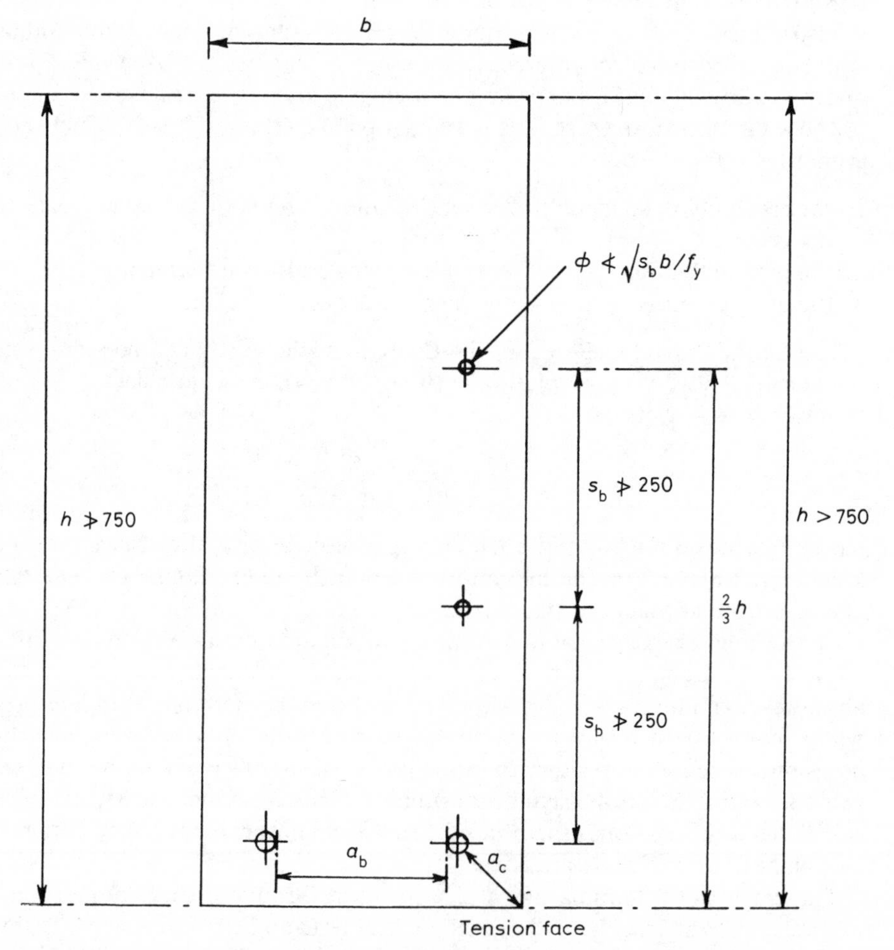


FIG. 7.1 Bar spacing rules. (1) $a_b \ngtr$ value specified in Table 3.30 of the Code. (2) $a_c \ngtr \frac{1}{2} \times$ value specified for $a_b$. (3) If $h > 750$, add longitudinal bars as indicated.

happens for all types of reinforcement except Grade 250, where the maximum clear distance is 300 mm, at zero or positive redistribution. Although it is the service stress which is being used this has been taken into account in deriving the values in Table 3.30 by assuming the service stress is $5f_y/8\beta_b$ as for deflection. Where reinforcement has been provided which is more than is required, the service stress $f_s$ can be calculated as for deflection and the formula in clause 3.12.11.2.4 applied. When determining the percentage of redistribution this should be taken as $(\beta_b - 1) \times 100$, $\beta_b$ being defined as for deflection. Thus if there has been a reduction in moment the value will be negative.

In applying the above rules any bar with a diameter of less than 0.45 times the maximum bar size in the section should be ignored except for those in the side faces. By section is meant the tension area where one is considering the cracks. This ratio can be illustrated as follows:

| Maximum bar size | Minimum bar size to be considered |
|---|---|
| 40 | 20 |
| 32 | 16 |
| 25 | 12 |
| 20 | 10 |

So if one has two $32\phi$ bars and the spacing is greater than allowed, one cannot put in an intermediate $10\phi$ bar to reduce the distance between bars; it must be a $16\phi$ bar.

For bars up the side face in sections over 750 mm deep, the diameter should not be less than $\sqrt{(s_b b/f_y)}$ where $s_b$ is the spacing of the bars, not the clear distance, and $b$ is the lesser of the breadth of the section and 500 mm. If we take $s_b$ at the maximum value of 250 mm and $f_y = 460$ N/mm$^2$, then the diameter must not be less than $0.74\sqrt{b}$. For $b = 300$ mm this means $\phi = 12.8$ so use a $16\phi$ bar. To reduce the bar size to 12 we must have $s_b$ not greater than 220 mm.

An important point to note about the rules is that they do not apply to members where the cover exceeds 50 mm.

As mentioned at the beginning of the chapter, calculations will not be done very often. Instead, the above rules will be applied. In selecting the number and diameter of bars for the reinforcement required at ultimate limit state, the designer must therefore have due regard to the bar spacings and the distance to the corner of the beam. For the spacings themselves, this will be relatively easy to apply. With a monolithic beam and slab construction, however, where is the corner of the beam when considering the tension bars over a support? Is it the corner of the theoretical rectangular section that has been assumed in the strength of sections? If so, then the author considers this distance as irrelevant. In determining the crack width formula due regard was given to the tension stiffening effect of the concrete at the level of the tension reinforcement. With the top flange much wider than the rib this effect must be much greater. This would also affect the spacing of the main bars, but as information is lacking on this aspect it is suggested that the bar spacing rules are still applied to the main bars. These rules could be applied to the distance to the adjacent slab bars in the top of the slab rather than to a hypothetical corner.

## EXAMPLE 7.1

Check that the reinforcement provided for the continuous beam in Example 5.1 complies with the bar spacing rules. To avoid looking back through the previous chapters a summary of the details is given below, together with the clear distances required by Table 3.30 of the Code.

| | Reinforcement | Percentage redistribution | Clear distance (mm) | Corner distance (mm) |
|---|---|---|---|---|
| 1. Exterior support | 2/25 (top) | −9 | 148 | 74 |
| 2. End span | 3/25 (bottom) | +14 | 186 | 93 |
| 3. Interior support | 3/32 (top) | −30 | 115 | 57.5 |
| 4. Interior span | 2/32 + 1/20 (bottom) | +9 | 178 | 89 |

## Exterior support

Beam is the same width as column (300 mm) and beam bars will be positioned inside column bars. Assuming 45 mm cover to column bars and that these bars are $32\phi$, the side cover to the beam bars is 77 mm.

The clear distance between the beam bars will be $300-2\times 77-2\times 25=96$ mm, i.e. $<148$ mm.

There are two points to note: (a) If the theoretical corner distance is calculated it will be found that this is greater than 74 mm, but, as discussed earlier, this is not relevant. (b) The side cover appears to exceed 50 mm, but as we are in the flange of the beam this again is not relevant.

## End span

The side cover here is 40 mm, so the clear distance between the bars $=(300-2\times 40-3\times 32)/2=62$ mm $<186$ mm.
The corner distance $=\sqrt{(56^2+56^2)}-16=63$ mm $<93$ mm.

## Interior support

Here the columns are 400 mm wide. Allowing 40 mm cover to $32\phi$ column bars, these bars will project 22 mm into the beam space. As we are allowing 40 mm side cover to the beam bars it will be this latter dimension which controls. The spacing between the bars will be as for the end span above.

## Interior span

The spacing here between the bars will be 6 mm more than in the example of the end span above and will be satisfactory.

Although the calculations have been done here to prove that the original selection of bars is satisfactory, it must be appreciated that in making the original selection reference was made to Table 3.30 of the Code to ensure an arrangement of bars which complied with the rules.

# 8 BOND AND ANCHORAGE

## 8.1 Minimum distance between bars

We have discussed previously the cover to bars, or groups of bars, and have noted that this is not only for durability but for satisfactory bond. We also said that sufficient cover should be allowed for proper compaction of the concrete. Again, to allow for proper compaction and to enable the development of proper bond we have to consider the minimum distance between bars.

The recommended distances between bars are given in clause 3.12.11.1 of Part 1, and are in terms of $h_{agg}$, the maximum size of the coarse aggregate. The horizontal distance should be not less than $h_{agg} + 5$ mm, the vertical distance not less than $2h_{agg}/3$. However, it goes on to say that if the bar size (which means the equivalent size for pairs or bundles) exceeds $h_{agg} + 5$ mm, a spacing of less than the bar size should be avoided.

The vertical distance applies where there are two or more rows and an obvious requirement in this case is that the gaps between corresponding bars in each row should be vertically in line. Assuming that the maximum size of coarse aggregate is 20 mm, the minimum horizontal distance is 25 mm. This would apply to individual bars up to $25\phi$ or groups of bars up to an equivalent diameter of $25\phi$, e.g. $2/16\phi$ bars in contact. Over this size we would need the bar diameter (or equivalent diameter). For the vertical distance the minimum should be $2 \times 20/3 = 13.3$, but 16 mm would be a practical dimension. This would appear to be suitable for bar sizes up to $25\phi$ (i.e. $h_{agg} + 5$ mm), but if we follow the Code recommendations we would then need a 32 mm gap for $32\phi$ bars. This does not seem a very satisfactory progression and it is suggested that if the bar size is greater than $2h_{agg}/3$, the bar size should be used.

The distance can then be summarized very simply as horizontal = greater of bar size and $h_{agg} + 5$, vertical = greater of bar size and $2h_{agg}/3$, where 'bar size' means the equivalent bar size for groups of bars.

Some engineers may think these larger distances rather excessive and minimum spacing is best determined by experience or proper works tests. The main point to remember is that the concrete must be able to move around the reinforcement in order to be fully compacted.

## 8.2 Minimum percentages of reinforcement in beams and slabs

These have now been collected together for various types of member in Table 3.27 of Part 1. For tension reinforcement it will be seen that they are related to the overall depth of the members, rather than the effective depth, as in CP110.

In solid slabs this now means that we have the same minimum reinforcement in both directions, but as calculations for areas of reinforcement are generally in terms of effective depth, a further check is required to ensure that the minimum reinforcement is provided. Note also that for flanged beams the percentage changes depending on the ratio of $b_w/b$ for the web in tension. Although the notation for this clause says that $b$ is the breadth of the section it means the effective width of the flange. The breadth or effective breadth of the rib is $b_w$. For compression reinforcement in a flanged beam where the web is in compression (e.g. in a monolithic beam and slab construction over a support) and compression reinforcement is required, the minimum percentage is $0.2\%\ b_w h$.

The transverse reinforcement in the flanges of flanged beams has now been reduced to 0.15% of the longitudinal cross-sectional area of the flange. This means that the area required in the top surface of the flange is $1.5\ h_f\ \mathrm{mm}^2/\mathrm{m}$ and is to be provided across the full effective width of the flange.

## 8.3 Maximum percentages of reinforcement in beams and slabs

The maximum percentage for tension and compression reinforcement is given as 4% of the gross cross-sectional area of the concrete.

## 8.4 Anchorage bond

To prevent bond failure we need a sufficient length of bar beyond any section to develop the necessary force at that section. It is assumed that the bond stress is uniform over the effective anchorage length. We know this is not completely accurate, but providing the cover and bar spacing requirements are complied with, the values obtained from equation (49) in clause 3.12.8.4 as uniform bond stresses will be satisfactory. The anchorage bond stress is taken as the force in the bar divided by its effective surface anchorage area, and so

$$f_b = F_s / \pi \phi_e l,$$

where $f_b$ is the bond stress, $F_s$ is the force in the bar or group of bars, $\phi_e$ is the 'effective' bar size (note previous comments on groups), and $l$ is the anchorage length. As we classify bars by equivalent diameter, we can say the force in the bar is $\pi \phi^2 f/4$ where $f$ is the stress in the bar. Rewriting the equation we find that the anchorage length

$$l \nless f\phi / 4 f_{bu},$$

where $f_{bu}$ is the design ultimate anchorage bond stress derived from equation (49), which says

$$f_{bu} = \beta \sqrt{f_{cu}}$$

and $\beta$ is obtained from Table 3.28.

Values for $\beta$, and hence the design ultimate anchorage bond stresses, are for plain bars, deformed bars and fabric. Deformed bars are divided into Type 1 and Type 2, where these are defined in BS4449 and BS4461. It should be noted, however, that the higher allowable stresses for deformed bars do not apply in beams without minimum links. As the majority of beams must be provided with minimum links (see Chapter 9) this is unlikely to occur. If it does, then the values for plain round bars should be used, irrespective of the type of bar used. This requirement does not apply to slabs.

From the formula and allowable bond stresses, lengths of bars required for different stresses and different concrete grades can be calculated and tabulated. The usual procedure is to give the required length at the design stresses for ultimate limit state as a factor times the bar size. Table 3.29 of the Code does this for anchorage bond lengths and lap lengths and is very comprehensive.

For quick reference, Table 8.1 gives values for anchorage bond lengths for ultimate design stresses for deformed bars with $f_y = 460\ \mathrm{N/mm}^2$. These have been obtained by

**Table 8.1** Values of $K$ for bars at ultimate design stresses.

| *Bar type* | *Design stress* (N/mm²) | *Grade of concrete* | | | |
|---|---|---|---|---|---|
| | | 25 | 30 | 35 | 40 *or more* |
| Type 1 | 400 | | | | |
| Tension $K_1$ | | 51 | 46 | 43 | 40 |
| Compression $K_2$ | | 41 | 37 | 34 | 32 |
| Type 2 | 400 | | | | |
| Tension $K_1$ | | 40 | 37 | 34 | 32 |
| Compression $K_2$ | | 32 | 29 | 27 | 26 |

substituting $0.87\,f_y$ for $f$ in the equation above. For tension, $l = K_1\phi$, for compression, $l = K_2\phi$, where $K_1 = 100/f_{bu}$ (tension) and $K_2 = 100/f_{bu}$ (compression).
It should be noted that in Table 8.1 the grade of concrete does not exceed 40. The Code does not give such a limit, but the author feels that as with shear it would appear advisable to do this.

For stresses lower than ultimate design stresses a linear interpolation can be used.

If we have a group of bars and all the bars are being anchored at the same time, the anchorage length will be a factor times the equivalent diameter. For example, a pair of $25\phi$ bars, deformed Type 2, in Grade 35 concrete, would require a full anchorage bond length of $34 \times 35 = 1190$ mm in tension and $27 \times 35 = 945$ mm in compression.

Local bond stress calculations are no longer required. As explained in the Handbook this has seldom been a crucial factor and seemed an unnecessary refinement. Provided the bars are given an appropriate embedment length or other end anchorage, local bond stresses may be ignored.

## 8.5 Curtailment of bars

When curtailing bars other than by the simplified rules we first have to decide which bars we would like to curtail. Then on the bending moment diagram we draw a line, to scale, at a distance from the datum equal to the moment of resistance of the continuing bars assuming they are fully stressed. The point where this line cuts the bending moment diagram is referred to as the theoretical cut-off point (TCP).

In clause 3.12.9.1 the Code states that in every flexural member every bar should extend, except at supports, beyond the TCP for a distance equal to the greater of:

1. the effective depth of the member; or
2. twelve times the bar size.

This applies to compression and tension reinforcement, but for reinforcement in the tension zone we have to satisfy, in addition, one of the following conditions:

3. The bars extend a full anchorage bond length beyond the theoretical cut-off point.
4. At the point of physically cutting-off the bars (PCP) the shear capacity is at least twice the actual shear force, i.e. the actual shear at the PCP is not more than half the shear capacity at this point.
5. At the physical cut-off point the continuing bars provide double the area required for flexure, i.e. the actual bending moment at the PCP is not more than half the moment at the TCP.

The conditions stated above can be illustrated diagrammatically as shown below. *A* is the theoretical cut-off point (TCP), *B* is the physical cut-off point (PCP).

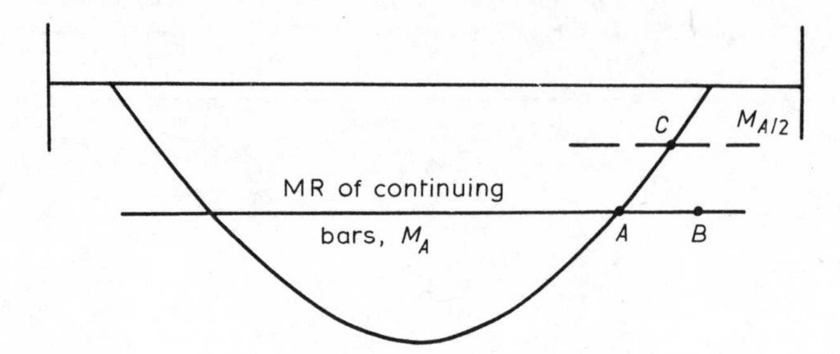


The requirements, as we are in a tension zone, are therefore:

1./2. *AB* is the greater of *d* and $12\phi$.

In addition the least distance of:

3. *AB* equals the full anchorage bond length;
4. at *B*, actual shear < half shear capacity;
5. at *B*, moment < half moment at *A*.

Condition (1)/(2) is the absolute minimum distance and the additional conditions will only apply if the least value of (3), (4) and (5) is greater than the minimum distance.

Condition (3) is easy to find from Table 3.29 of the Code. Condition (4) is not easy to apply until we have done the shear calculations. As will be seen in Chapter 9 the shear capacity depends on the amount of tension reinforcement and this depends on the curtailment. So it can be used as a final check if required, but at this stage it is proposed to ignore this condition.

Condition (5) can be relatively simple to apply by drawing an additional line at a resistance moment equal to half that at *A* and seeing how this intersection with the bending moment diagram (point *C*) compares with *B*. The procedure would be:

1. Locate point *B* by making *AB* the greater of *d* and $12\phi$.
2. If point C comes within *AB*, then point *B* controls.
3. If point *C* comes outside *AB*, extend *AB* to point *C* or a full anchorage bond length, whichever comes first.

This is the full procedure, but it can be shortened if required by making *AB* a full anchorage bond length. Provided this is greater than *d* or $12\phi$ the conditions will be satisfied.

It must also be remembered that when we are curtailing we have to make sure that the distance from the point of maximum stress in the bar (i.e. the point of maximum bending) to point *B* is not less than the anchorage bond length for that stress. Using condition (3) will always satisfy this requirement, but using condition (5) with a very steep bending moment diagram may not.

When referring to the bending moment diagram it must be remembered that in most cases this will be a bending moment envelope as all arrangements of ultimate load must be considered.

Curtailing substantial areas of reinforcement at the same position should be avoided as the section properties will change considerably and could lead to problems with cracking.

At a simply supported end of a member clause 3.12.9.4 gives the requirements for anchoring the tension reinforcement:

1. An effective anchorage length equivalent to twelve times the bar size beyond the centre line of the support. No bend or hook should begin before the centre line of the support.
2. An effective anchorage length equivalent to twelve times the bar size plus $d/2$ from the face of the support. No bend or hook should begin before $d/2$ from the face of the support.
3. For slabs, if the shear stress at the face of the support is less than half the allowable value, $v_c$, a straight length of bar beyond the centre line of the support equal to the greater of one third of the support width and 30 mm.

For condition (1) we can use a 90° bend with a $3\phi$ internal radius in mild steel or high-yield steel to provide the required anchorage.

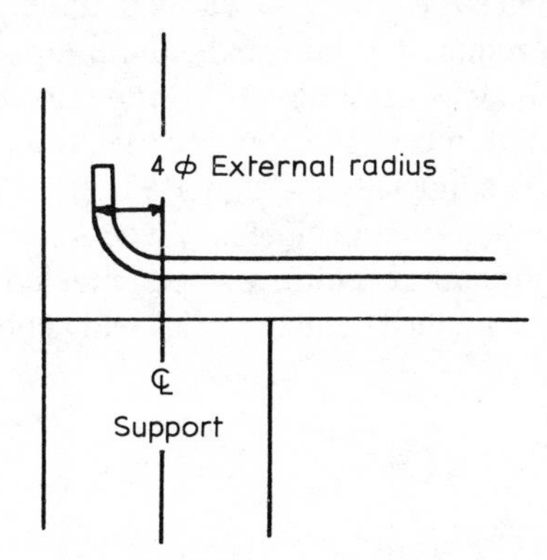


With mild steel, or high-yield reinforcement not greater than size 20, this means a minimum support width of $2(4\phi + \text{cover})$ is required. With a moderate exposure requiring 30 mm cover and a $20\phi$ bar this means a minimum width of 220 mm. For high-yield bars greater than size 20 the internal radius becomes $4\phi$ and the minimum support width is $2(5\phi + \text{cover})$.

Condition (2) is useful if the supports are wide compared with the effective depth of the supported member.

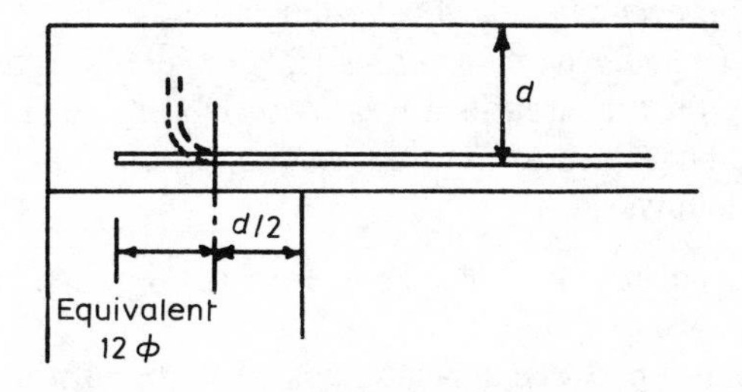


Where the width of the support is greater than the effective depth the bend can start earlier than in (1) and with small-diameter bars it may be possible to use a straight bar without a bend.

As a generalization we can say that where the width of the support is less than the effective depth of the spanning member try (1); if greater try (2).

Owing to the size of bars condition (1) may not be possible, in which case we should try to use smaller bars. For slabs, not beams, condition (3) may be adopted and this can be illustrated as follows:

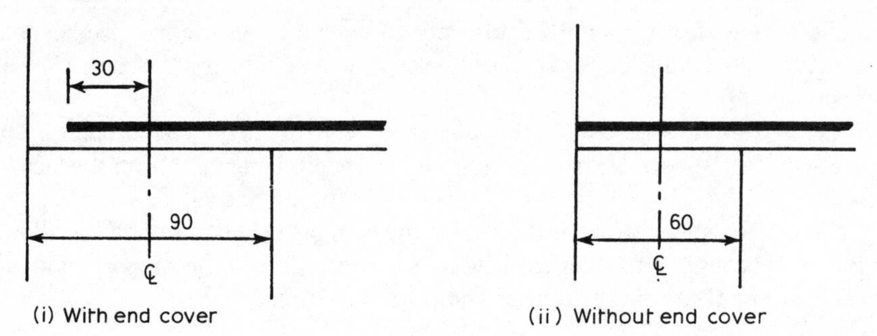


If end cover is required, as in (i), it can be found quickly that the minimum value for $a$, the support width, is six times the cover. So with 15 mm end cover the dimension is 90 mm. If end cover is not required, as in (ii), the minimum support width is 60 mm.

In determining the value for $v_c$ in condition (3) the tension reinforcement must be effective, but no additional enhancement should be used even though the section is less than $2d$ from the face of the support (see Chapter 9).

The general recommendations for curtailment given so far would apply where a full analysis had been carried out and a bending moment envelope prepared. Where such a procedure is not necessary simplified detailing rules will apply and these are described later in this chapter.

## 8.6 Lapping of reinforcement

### 8.6.1 Tension

When deformed bars are lapped we have to consider situations where research has indicated that greater lengths should be provided. In the first edition of CP110 there was a global increase of 25% on the anchorage length in all cases when the bars are in tension. Further investigations have shown that top bars, lateral distances between lapped bars and cover or corner bars need further consideration. Before describing these it should be borne in mind that lapping bars in regions of high tensile stress in the bars is generally avoided wherever possible. In certain areas, such as stability ties and in columns, this cannot be avoided and clause 3.12.8.13 increases the anchorage length by a factor depending on the circumstances. Where bars are of different size the lap length and cover will be related to the smaller bar.

The factors are as follows:

1. 1.4 if the minimum cover to the lapped bars from the top of the section as cast is less than twice the bar size.
2. 1.4 if the clear distance between adjacent laps is less than the greater of 75 mm and six times the size of the lapped reinforcement or if a corner bar is being lapped where the minimum cover to either face is less than twice the bar size.
3. 2.0 if conditions (1) and (2) apply.

It should be noted that for condition (2), even if both requirements apply there is only a single factor of 1.4. The reasoning is that the member may tear along the plane of the lapping bars or across the corner. One may happen but not both.

Condition (1) takes into account the fact that with small amounts of concrete above the top bars the bond is not as good as for bottom bars.

The factors can be illustrated as shown in Fig. 8.1.

Where the lapping bars are size 25 and larger and the nominal cover is less than one and a half times the smaller bar size (or the bar size for equal bars) then transverse links of at least one quarter of the smaller bar size should be provided at a maximum spacing of 200 mm. So in columns, at the laps, if 25$\phi$ bars are being used and the cover is less than 37.5 mm, the links would need to be spaced at 200 mm as compared with 300 mm (i.e. twelve times bar size) in the remainder of the columns.

The minimum lap lengths for both tension and compression bars where the stress is nominal is the greater of 15 times the bar size and 300 mm. This also applies to cases where main bars are stopped and bars are lapped to act as link carriers.

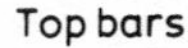


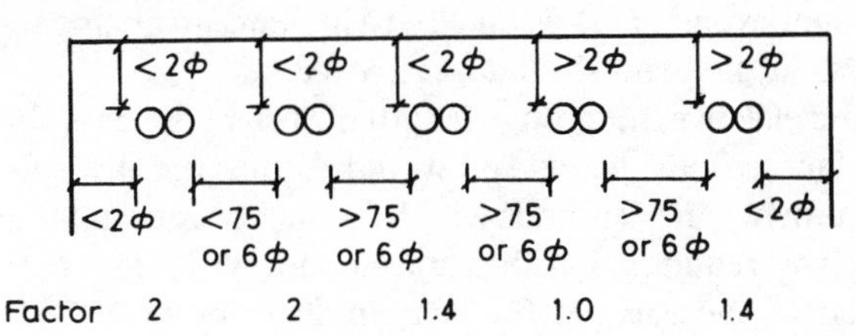


Bottom bars

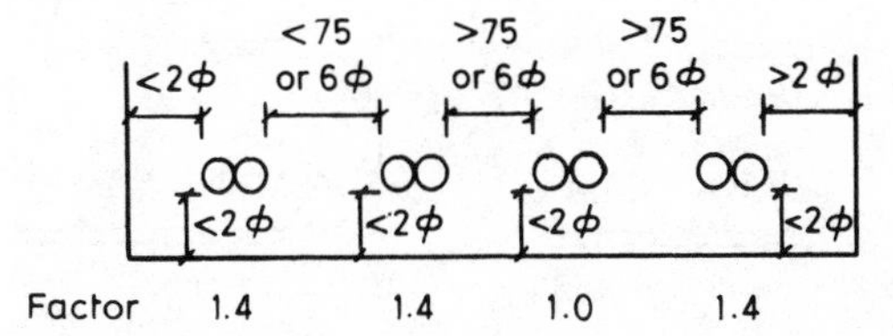


FIG. 8.1 Factors for lapping bars.

### 8.6.2 Compression

For compression laps the lengths are much simpler and are to be at least 25% greater than the compression anchorage length necessary to develop the stress in the reinforcement. When unequal size bars are being lapped, the length can be based on the smaller bars.

Values for full design stress are given in Table 3.29 of the Code, and a summary was given earlier in this chapter for the anchorage bond factors.

## 8.7 Anchorage lengths of hooks and bends

The effective anchorage length of a hook or bend starts at the beginning of the bend (point *A*) and finishes at a point four times the bar size beyond the end of the bend (point *B*); see clause 3.12.8.23.

### 8.7.1 180° hook

For a 180° hook, the effective length is eight times the internal radius with a maximum of 24 times the bar size or the actual length of bar, whichever is greater. So with an

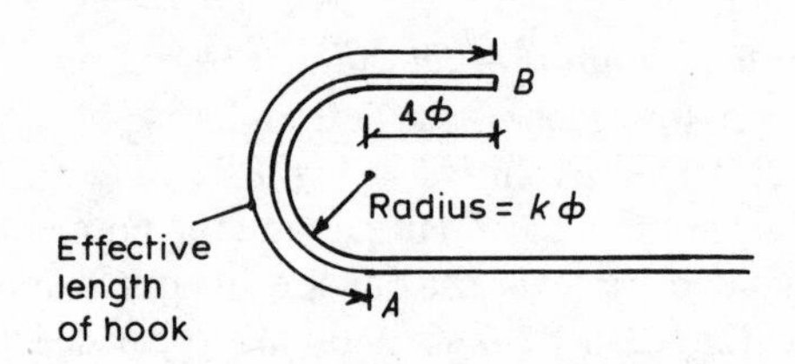


internal radius of more than three times the bar size, as occurs with high-yield reinforcement for bars of size 25 and above, the maximum value is 24 times the bar size or the actual length of bar, whichever is greater. The latter requirement does not control until the internal radius is six times the bar size.

If a required anchorage length does not extend beyond the end of the hook (i.e. beyond point *B*) then it is not necessary to check the stress in the inside of the bends of the hook.

Extending the bar beyond point *B* can also be included in an anchorage length, but in this case the stress in the bend should be checked.

For example, the effective anchorage length of both of the bars shown below is $24\phi$. Diagram (i) is in fact two 90° bends and would require the stress to be checked in the first bend (see comments on 90° bends which follow), but not in the second bend if only a $24\phi$ anchorage was required. For the same anchorage length, however, it would not be necessary to check the stress in the bend in diagram (ii).

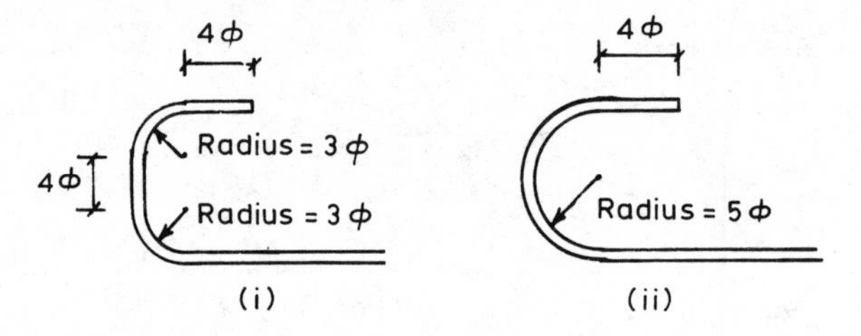


## 8.7.2 90° bend

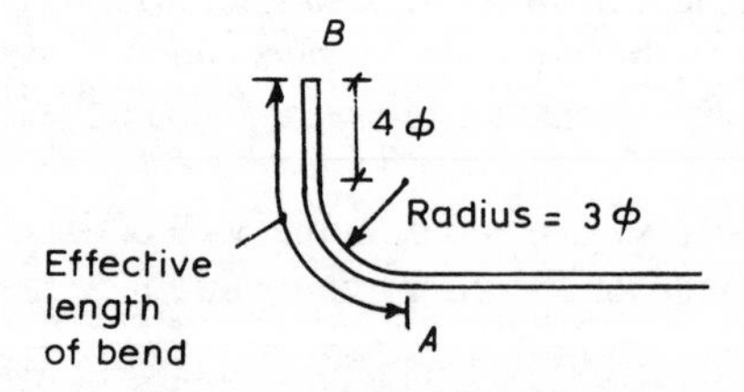


For a 90° bend, the effective length is four times the internal radius with a maximum of twelve times the bar size or the actual length of bar, whichever is greater. So with an internal radius of more than three times the bar size the maximum value is twelve times the bar size until the internal radius is five times the bar size, when the actual length of bar becomes greater.

As with 180° hooks, the stress in the bend need not be checked if the required anchorage length does not extend beyond *B*.

Extending the bar beyond *B* to provide a required anchorage length, however, would mean that the stress in the bend would have to be checked.

## 8.8 Bearing stresses inside bend, clause 22.8.25

As previously discussed under anchorage values of hooks and bends, it is not necessary to check bearing stresses inside a bend on a bar which does not extend or is assumed not to be stressed beyond a point four times the bar size past the end of the bend.

In other cases where a longer length is required the bearing stress inside the bend has to be checked. If $f_b$ is the bearing stress inside the bend then from the Code equation

$$f_b = F_{bt}/r\phi,$$

where $F_{bt}$ is the tensile force due to ultimate loads in a bar or group of bars in contact at the start of the bend; $r$ is the internal radius of the bend; and $\phi$ is the size of bar or the size of the equivalent bar for a group of bars. The stress should not exceed $2f_{cu}/(1+2\phi/a_b)$, where $a_b$ is defined as the distance centre to centre between bars (or groups of bars) perpendicular to the plane of bend, or $\phi$ plus the cover when dealing with bars adjacent to the face of a member. This can be illustrated as follows:

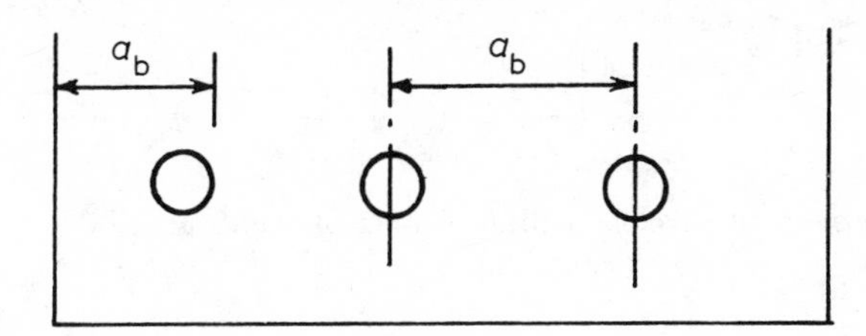


If we take $f_t$ as the tensile stress in the bar, then

$$F_{bt} = (\pi\phi^2/4)f_t,$$

and so

$$f_b = (\pi\phi^2/4)f_t/r\phi.$$

Hence

$$r = \pi f_t\phi/4f_b.$$

As

$$f_b \leqslant 2f_{cu}/(1+2\phi/a_b)$$

then

$$r \geqslant \pi f_t\phi(1+2\phi/a_b)/8f_{cu},$$

i.e.

$$r \geqslant K\phi,$$

where

$$K = \frac{\pi f_t(1+2\phi/a_b)}{8f_{cu}}.$$

For a particular grade of concrete, $K$ can be found for different values of $\phi$ depending on $a_b$ and $f_t$. Values of $K$ for concrete Grades 25, 30, 35 and 40 and bar sizes 16, 20, 25 and 32 are given in Appendix 3.

## EXAMPLE 8.1

In the previous chapters we have been using a continuing design example and this will now be taken a stage further. To avoid referring back the bending moment envelope is reproduced here as Fig. 8.2.

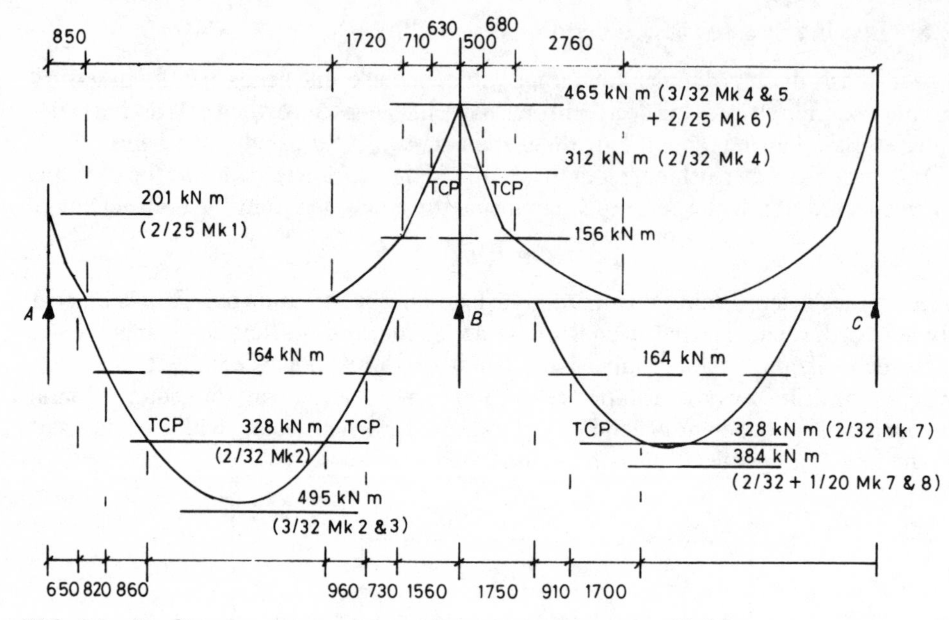


FIG. 8.2 Bending moment envelope with resistance moments added.

The moments of resistance of the bars provided at the critical sections have been added, as have the moments of resistance of continuing bars where curtailment is proposed, i.e. at TCPs. It should be pointed out here that if the designer is providing more than two bars at a critical section and proposes to do curtailment by calculation, not by eye, it is useful to find the moments of resistance referred to above when actually designing the section. This will now be done.

End span
Reinforcement provided $=3/32\phi$ (bar marks 2 and 3)
Moment of resistance $=495$ kN m
Moment of resistance of $2/32\phi$ (mark 2) $=328$ kN m

Interior support
Reinforcement provided $=3/32\phi$ (tension) (bar marks 4 and 5) $+2/25\phi$ (compression) (bar mark 6)
Moment of resistance $=465$ kN m
Moment of resistance of $2/32\phi$ (mark 4) tension only $=312$ kN m.

Interior span
Reinforcement provided $=2/32\phi+1/20\phi$ (bar marks 7 and 8)
Moment of resistance $=384$ kN m
Moment of resistance of $2/32\phi=328$ kN m (mark 7)

The first criterion for extending a bar beyond a proposed position is that it must be the greater of the effective depth and $12\phi$. We have two effective depths in this run of beams and it is proposed to use the greater dimension of 550 m. As can be seen this will always be greater than $12\phi$. A full anchorage bond (FAB) length for $32\phi$ bar is $34\times 32=1088$ mm, say 1090 mm.

BEARING STRESSES INSIDE BEND, CLAUSE 22.8.25

### Bar mark 1 (exterior support)

Extend bars 550 mm beyond point of contraflexure, i.e. to a distance of 1400 mm from centre line of support.

### Bar mark 2 (end span)

These are the continuing bars and are not curtailed. Stop at face of column each end.

### Bar mark 3 (end span)

Extend left-hand end by 860 mm and right-hand end by 960 mm beyond TCP. In both cases this is where the moment is half that at TCP and is less than a full anchorage bond length.

### Bar mark 4 (interior support)

These are the continuing bars and should extend a distance of 550 mm beyond the point of contraflexure. Into span *AB* this means a distance of $3060 + 550 = 3610$ mm from centre line of support.

Into span *BC* this would be $3940 + 550 = 4490$ mm from centre line of support. As similar bars would be coming in the same distance from support *C* there would only be a small gap between the ends. We suggest the bars are continued and are lapped in the centre of the span for the minimum lap of $15 \times 32 = 480$ mm.

### Bar mark 5 (interior support)

Extend left-hand end by 710 mm and right-hand end by 680 mm beyond TCP, i.e. to where moment is half that at TCP.

### Bar mark 6 (interior support)

These are 25$\phi$ compression bars in the bottom and these should lap with the bottom bars in the adjoining beams. For compression this will be $1.25 \times 27$ times bar size. Table 3.29 in the Code gives a numerical value of 34 times the bar size, i.e. $34 \times 25 = 850$ mm.

If these bars are used for internal ties they will be in tension and the lap length may be greater.

### Bar mark 7 (interior support)

These are the continuing bars and will be from the inside of the column face at *B* to the inside of the column face at *C*.

**Bar mark 8 (interior span)**

This is a single 20$\phi$ bar and will extend a full anchorage bond length of 1090 mm beyond the TCP at both ends.

We now have to consider the anchorage of bars mark 1 into the exterior column. From the calculations in determining these bars it can be seen that they will be at the full design stress. The tension anchorage length required is 34$\phi$ and this cannot be provided by a standard bend or even a 180° standard hook. So it will be necessary to check the bearing stress in the bend. The distance centre-to-centre of bars, $a_b$, is $96+25=121$ mm.

From the tables in Appendix 3 the internal radius required is approximately 6.4$\phi$ (by interpolation), so we suggest 7$\phi$. The actual length along the centre line of the bar, round the bend, is approximately 12$\phi$, so we would need to extend the bar down the column a further 22$\phi$ from the end of the bend.

It is suggested that a better solution would be to use a hairpin-shaped bar as shown below. The bottom leg of the hairpin would then lap with the two 32$\phi$ (mark 2) which effectively take these bars into the support, and can also be used for the internal tie bars in the beams, if required.

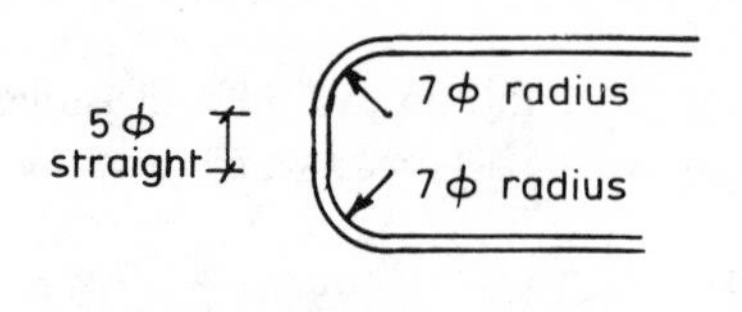


## 8.9 Simplified rules for detailing

It was mentioned briefly in Chapter 3 that continuous beams in frames could be analysed by ignoring the columns and treating the supports as simple. Providing the criteria as set out in clause 3.4.3 are met there is no need to carry out a complete analysis and curtailment is done by following rules given in clause 3.12.10.2.

The criteria are:

1. characteristic imposed load $Q_k$ does not exceed the characteristic dead load $G_k$;
2. loads should be substantially uniformly distributed over three or more spans;
3. variations in span length should not exceed 15% of the longest.

Table 3.6 of the Code gives coefficients for the bending moments and shear forces at the critical sections. In section 1 of Example 3.1 we compared the results for moments with those obtained from a complete analysis. It was also pointed out in this example, and is repeated here, that in using Table 3.6 no redistribution of moments can be carried out.

In the criteria listed above it does not state that the beams have a uniform cross-section, but the Code assumes that in limiting the variations in span, this would be the case.

Calculations for areas of reinforcement at critical sections will then be carried out in the same way as in Chapter 5. If we turn now to clause 3.12.10.2 it does not specifically refer to clause 3.4.3 nor does it restrict the ratio of the dead and imposed loads. Obviously they are related for continuous spans, but the detailing rules can also be applied to single-span simple supported beams and cantilever beams. The criterion for ratio of loads does not apply in these two cases.

Figure 3.24 of the Code shows the rules in diagrammatic form and the first definition we need is the effective span, $l$. For detailing, this can sometimes be different from the effective span used in analysis. For detailing, $l$ is equal to the clear distance between supports plus the effective depth of the supported member, or the distance between centres of supports, whichever is the lesser. For beams this difference will occur only if the support (e.g. a column) is wider than the effective depth of the beam.

From Fig. 3.24 of the Code it can be seen that although the dimensions for curtailing bottom bars are to the centre line of the 'effective support', the top bars are related to the face of the support.

Another important point to note is that the above rules assume the external support does not have restraint – the moment obtained from Table 3.6 of the Code is zero. We have already discussed the fact that many designers use these rules when external columns are involved. Diagram (a) in Fig. 3.24 says the bottom steel can be curtailed at $0.1l$ from the centre of the support but does not suggest what to do about the top reinforcement. Although the author recommended a calculation to determine the moment and hence the reinforcement required (see Chapter 3), the proposal used in his previous book to accompany CP110 is still valid. Curtail and anchor the bottom reinforcement as for a simple support (diagram (b) in Fig. 3.24). The top reinforcement should be approximately half the bottom reinforcement in the span, properly anchored into the column and projecting $45\phi$ from the face of the column into the span. 'Properly anchored into the column' may cause problems, but quite a neat arrangement would be provided by having ⊏ bars (hairpins) projecting from the column, having an area of half the span reinforcement and lapping with the bottom reinforcement which stops at the face of the column. The bends could be standard radius.

# 9 SHEAR

Shear is to be considered at ultimate limit state only. No requirements are made for serviceability limit states. Numerical values are given in the Code for shear resistance depending on the percentage of tensile reinforcement at the section considered, the grade of concrete and the effective depth of the section. The partial safety factor $\gamma_m$ has been absorbed into the numerical values.

Reliable and agreed methods exist for the design of reinforced concrete members in pure bending. The method of strain compatibility uses equilibrium equations, a compatibility condition (plane sections remain plane), and a failure criterion (a maximum compressive strain which must not be exceeded) to produce an accurate assessment of a member's ultimate strength.

Strictly speaking these methods only apply where there is constant bending moment, such as the central portion of a beam loaded with two point loads.

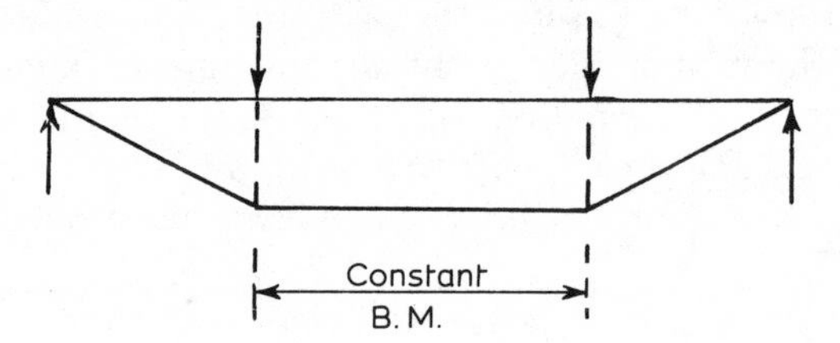


Where the bending moment is changing and a shear force is introduced, the equilibrium equations are complicated by the presence of the shear force and a new vertical equilibrium equation is required. The compatibility conditions must be altered to include shear displacements and the failure criterion must allow for concrete in states of biaxial, and in some cases triaxial, stress.

A satisfactory design method which fulfils all these requirements has not so far been achieved, and Codes of Practice have therefore concentrated on producing reliable empirical methods of adding shear reinforcement to a structure to ensure that it has an adequate factor of safety at all points. Many people have carried out tests and put forward theories, more than in any other field.

With all this in mind the Institution of Structural Engineers set up a Shear Study Group in 1965 'to consider the available information on shear in concrete, to decide what further tests are required, and to put forward a research programme which will eventually enable a relationship to be established between design formulae and various modes of failure that can occur'. Subsequently the Group was asked to make proposals for clauses to be incorporated in the Unified Code. The Group's report was published in 1969 and design clauses were prepared.

From the review of existing work it was clear that no shear theory was correct in all cases, and that earlier British Codes such as CP114 were in need of revision for two main reasons:

1. The permissible shear stresses for concrete were too high. This means that the requirement for nominal shear reinforcement in beams may not have been adequate in all cases.
2. The truss analogy method of designing shear reinforcement by assuming that it resists the whole of the shear ignores the resistance of the concrete to inclined thrust and of the longitudinal steel to the extension of shear cracks. Test results have consistently shown that actual shear strengths are much higher than those calculated by this approach.

The effect of applying a shear force to a reinforced concrete beam in bending is quite well known. The usual arrangement for investigating shear failure is thus:

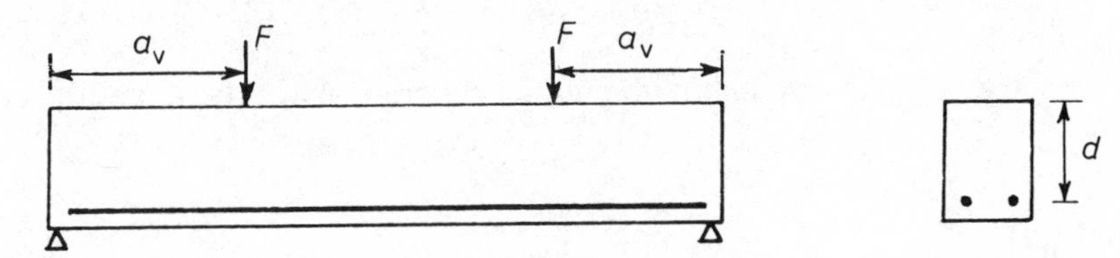


It has the advantage of combining two test conditions – pure bending between the two loads and a constant shear force in the end sections. The section is rectangular and the depth to the reinforcement is $d$. No shear reinforcement is included. In certain cases, the ultimate strength of such beams can be considerably less than the strength in pure flexure. This can best be illustrated by using an interaction diagram between moment capacity and the $a_v/d$ ratio. In other words, for a constant cross-section and reinforcement by varying the span and the points of loading we can find the ratio of the failure moment to the full flexural moment; we call this the moment capacity. Plotting this against $a_v/d$ we get the diagram shown in Fig. 9.1.

The exact values at which the line changes direction vary depending on the amount of reinforcement and on the concrete strength, but for about 2% reinforcement and

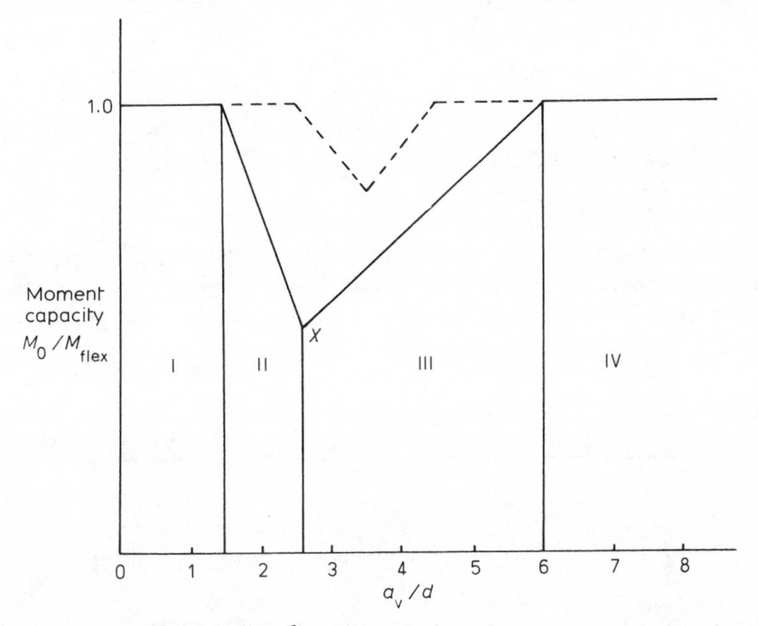


FIG. 9.1 Moment–$a_v/d$ interaction diagram.

concrete Grade 30 we get values for $a_v/d$ at the change points of about 1.5, 2.6, 6 with $X$ about 0.5. With low percentages, say 0.25%, the values are about 2.5, 3.5, 4.5 and with $X$ about 0.8. We consider each section individually.

(a) GROUP IV

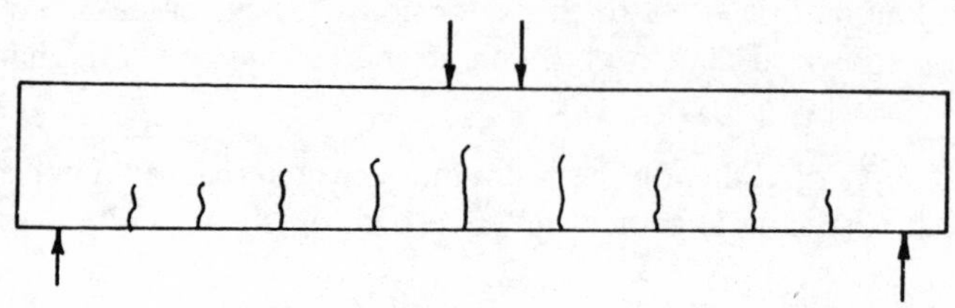

Here we have a pure flexural failure due to the concrete above the top of the cracks crushing.

(b) GROUP III

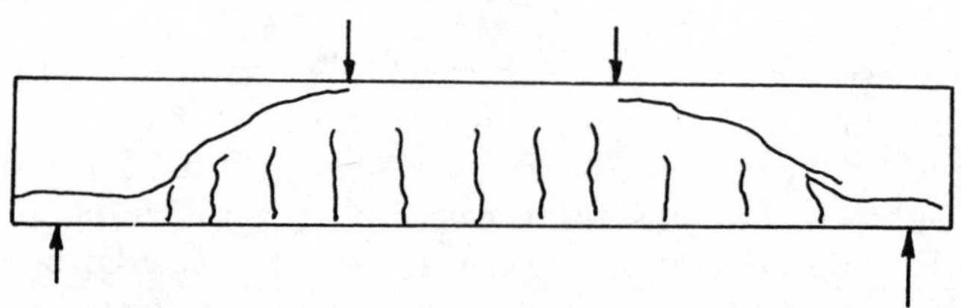

Here we have what is generally called diagonal tension but sometimes shear compression. The cracks in the end sections turn over at the top and then one large one starts from the reinforcement and goes to the load point. As the bond between the reinforcement and concrete fails this is sometimes called a shear bond failure.

(c) GROUP II

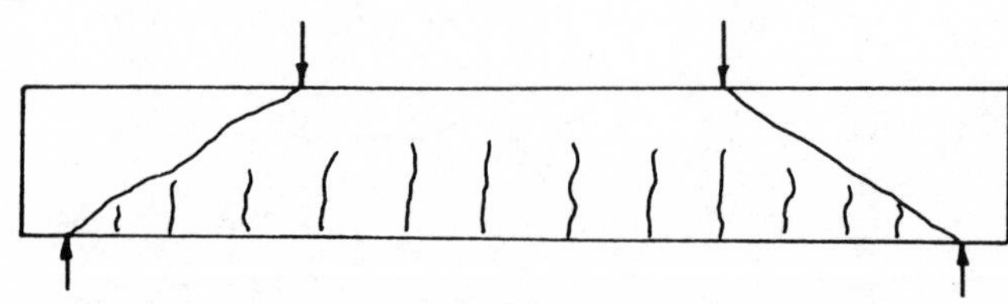

Here we have shear compression but this is sometimes called diagonal tension failure or sometimes called shear proper. The failure line goes from the support to the load. This time there is no bond failure and the result is analogous to the cylinder splitting test.

(d) GROUP I

Here we have a small $a_v/d$ ratio, more like a corbel, and the failure is along a line joining the load to the face of the support.

The $a_v$ dimension was called the *shear span* and defined by the Shear Study Group as the length of the beam over which the shear has the same sign, e.g. the length from a simple support to a single point load in a span. It should be noted that $a_v$ is used for different dimensions in the Code, and this will be redefined as we come to it.

So for a given concrete section of a certain grade concrete we have two main variables – the percentage of tensile reinforcement and the shear span ratio.

The Shear Study Group concluded from the results of tests that for beams without shear reinforcement the difference between the commencement of shear cracking and failure is small, except for short shear spans. The main problem was defining when shear cracking took place, but it was decided that for the majority of practical cases where shear span/effective depth is greater than 2 the ultimate load would be taken as the load causing the first shear crack.

Various equations have been proposed for determining the shear cracking load and most of them define it as a function of the cube strength and the ratio of longitudinal tensile reinforcement. The importance of these two factors is borne out by plotting test results as shown in Fig. 9.2.

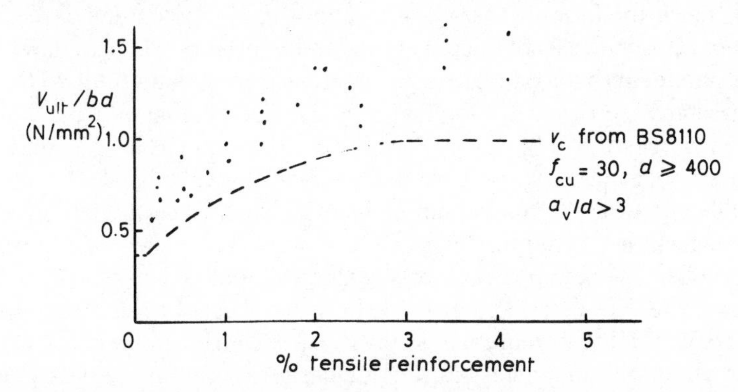


FIG. 9.2 Test results for ultimate shear stress in beams. (Taken from a report by the Shear Study Group of the Institution of Structural Engineers.)

In plotting these results, $V_{ult}$ is the load causing the first shear crack, and the effective depth $d$ is used instead of the lever arm which had previously been used. This is because the effective depth is a much more definitive dimension and test results can be directly compared. The concrete strength varies, the upper points being due to the higher strengths, but for the same strength concrete the value of $V_{ult}/bd$ increases with the increase in tensile reinforcement.

The shear stress $v$ at any cross-section in a beam should be calculated from

$$v = V/b_v d,$$

where $V$ is the design shear force due to ultimate loads, $b_v$ is the breadth of the section which, for a flanged beam, should be taken as the average width of the rib below the flange, and $d$ is the effective depth.

This equation for shear stress strictly applies to beams of uniform depth where the tension bars are parallel to the compression face. In members of varying depth the equation should be modified to

$$v = \frac{V \pm (M \tan \theta_s)/d}{b_v d}$$

where the negative sign is used when the moment is increasing numerically in the same direction as the effective depth of the section (see Fig. 9.3).

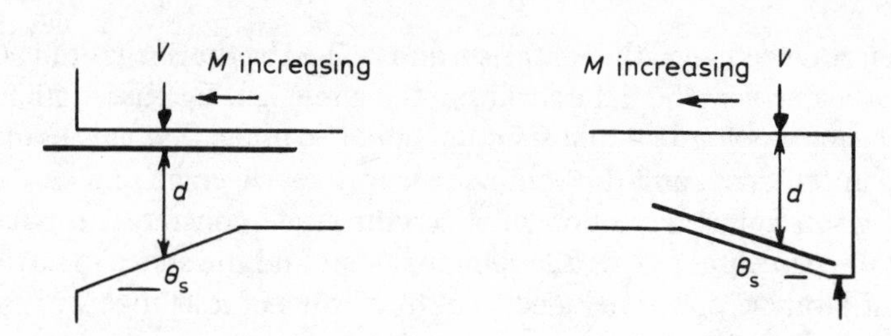


FIG. 9.3 Shear stresses at haunches. (a) $v=[V-(M/d)\tan\theta_s]/b_v d$. (b) $v=[V+(M/d)\tan\theta_s]/b_v d$.

Table 3.9 of the Code gives values of $v_c$, the design concrete shear stress. These values depend on the percentage of effective longitudinal tension reinforcement, the effective depth of the reinforcement, and use of a Grade 25 concrete. The notes below the table are extremely important.

Note 2 gives the formula from which the tabulated values have been obtained and points out that if the formula is used one cannot use a percentage of reinforcement greater than 3.0, nor an effective depth greater than 400 mm. There is, however, a very significant statement that if one is using a concrete grade greater than 25 (but even for concrete greater than Grade 40 you can only use the value of 40), the values in the table can be increased by $(f_{cu}/25)^{1/3}$. For Grade 40 this means a 17% increase.

So $v_c$ is a critical value for when 'design' links should be calculated. The word 'design' is used to differentiate these from minimum links which must be provided in all beams of structural importance. Minimum links can be omitted in members of minor structural importance when the actual shear stress is less than half $v_c$.

Table 3.8 of the Code describes when and how to provide shear reinforcement and in Note 2 it is reported that minimum links provide a shear resistance of 0.4 N/mm$^2$. The lower limit when one must start to provide design links is when $v$ is greater than $(v_c+0.4)$ N/mm$^2$. The upper limit for $v$ is the lesser of $0.8\sqrt{f_{cu}}$ and 5.0 N/mm$^2$. If the design shear stress is larger than this one must then change the section.

Clause 3.4.5.8 recognizes the fact that if the normal shear plane of failure is forced to be inclined more steeply than 30° the shear force required to produce failure will be increased. The concrete shear stress $v_c$, in the region close to supports can therefore be increased to $2v_c d/a_v$, where $a_v$ is the distance from the face of the support at the section where shear is being considered. This enhancement cannot be carried out indiscriminately. There is still an upper limit of $0.8\sqrt{f_{cu}}$ or 5 N/mm$^2$.

This procedure and calculation of the reinforcement as given in clause 3.4.5.9 are rather time consuming and will probably only be used for such items as corbels or pile caps or where concentrated loads are applied close to supports.

The normal procedure will be to use the simplified approach given in clause 3.4.5.10. This applies to beams carrying generally uniform load or where the principal load is located further than $2d$ from the face of the support. The actual shear stress, $v$, will be calculated at a distance $d$ from the face of the support. The concrete shear stress, $v_c$, will also be calculated at this point but no enhancement factor will be applied. Any shear reinforcement required at this section will be provided up to the face of the support.

We have said already that $v_c$ is a critical value, and one of the most important factors in calculating it is the amount of effective tension reinforcement, $A_s$. This is the area of longitudinal tension reinforcement which continues for a distance at least equal to $d$ beyond the section being considered, in each direction. At end supports it may not always be possible to do this, but provided the tension reinforcement is properly anchored in accordance with clause 3.12.9 the full area may be used.

At a monolithic beam–column junction where the beam has been designed on the assumption that the column provides a simple support but nominal steel has been provided to control cracking there are two alternative methods of calculating $v_c$. If the main tension steel in the bottom has been anchored in accordance with 3.12.9.4 one should use this area. If this anchorage has not been provided then one should use the area of top steel, but remembering that this steel must extend a distance at least $3d$ into the span from the face of the support – see clause 3.4.5.4.

As $v_c$ depends on the amont of effective longitudinal reinforcement, the value will change where the amount of reinforcement changes. It will also change from top to bottom or vice versa at points of contraflexure. So to calculate the amount of shear reinforcement required, the concrete shear stress at all these points will need to be known. Let us consider one beam in a run of beams, carrying a uniformly distributed load, as shown in Fig. 9.4.

At point $X$, the point of contraflexure for the lower part of the diagram, there are different areas of reinforcement in the top and in the bottom. The shear resistance will change from top to bottom, based on tension reinforcement. If all the bars are the same diameter and $d$ remains the same, the resistance to the left of $X$ is based on three bars and to the right of $X$ on two bars. At point $Y$, where the other two bottom bars become effective, the resistance is based on four bars.

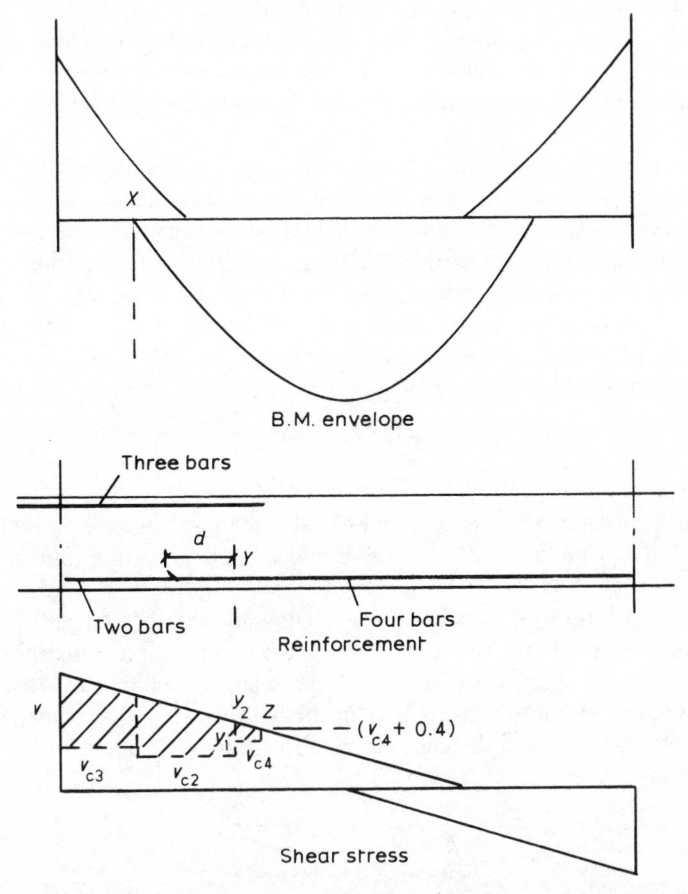


FIG. 9.4 Continuous beam with uniformly distributed load.

If we now draw the shear stress diagram and superimpose the appropriate resistances $v_{c2}$, $v_{c3}$, $v_{c4}$, the shaded area is to be covered by shear reinforcement. Point $Z$ is where the actual shear stress is $(c_{c4}+0.4)$.

To be able to do these calculations we need a bending moment envelope and full details of curtailment. If the simplified rules for calculating bending moments and shear forces are used we do not have an envelope. The procedure is also tedious and a simpler but conservative approach will normally be used. In this approach, where curtailment details are not necessary, we assess what will be the *minimum effective tension* reinforcement in a span. For the above example it is two bars from the bottom reinforcement. We calculate $v_c$ appropriate to two bars and add the value provided by minimum links (not less than 0.4 N/mm$^2$). This now gives the 'maximum' shear stress the beam will take. By multiplying this total value by $bd\times10^{-3}$ we obtain the 'maximum' shear force the beam will carry. If this is greater than the actual maximum design shear force then minimum links are all that is needed. If the maximum design shear force is greater than the value we have calculated then we use the formula

$$A_{sv}\geqslant b_v s_v(v-v_c)/0.87f_{yv}$$

to determine the links we need. The value we calculated using minimum links is still useful as it will enable us to find where minimum links finish and where design links start.

The formula given above for determining the links can be adapted in various ways, bearing in mind that the two values wanted are $A_{sv}$ and $s_v$.

One way is to tabulate values of $A_{sv}/s_v$ for various bar sizes and spacing; one then calculates $b_v(v-v_c)/0.87f_{yv}$ and selects from the table a bar diameter and spacing that give a value not less than this. Only one table is needed as the relevant value of $f_{yv}$ is inserted.

Another method is to tabulate values for $0.87f_{yv}A_{sv}/s_v$ and calculate $b_v(v-v_c)$; this will require two tables, one for 460 steel and one for 250 steel.

Table 1 of Appendix 4 gives values for the first method, and Table 2 gives values for the second method. In both methods the minimum links are determined by putting $(v-v_c)$ equal to 0.4 N/mm$^2$. Having selected the diameter and spacing, the actual shear resistance of these links can be calculated and used instead of 0.4 N/mm$^2$. An equation not given in BS8110 is the value of the shear force or shear stress for a particular area of links at a particular spacing. These are

shear force, $$V=0.87f_{yv}A_{sv}d/s_v$$

shear stress, $$v=0.87f_{yv}A_{sv}/b_v s_v.$$

So for the shear resistance stress of given links at a given spacing we multiply the value from Table 1 of Appendix 4 by $0.87f_{yv}/b_v$ or divide the appropriate value from Table 2 by $b_v$.

The spacing of links in the direction of the span should not exceed $0.75d$. At right angles to the span, the horizontal spacing should be such that no longitudinal tension bar is 150 mm from a vertical leg. The horizontal spacing should in any case not exceed $d$. So with a beam with four tension bars in the bottom, at 150 mm centres, one link would be sufficient provided the effective depth is at least 450 mm.

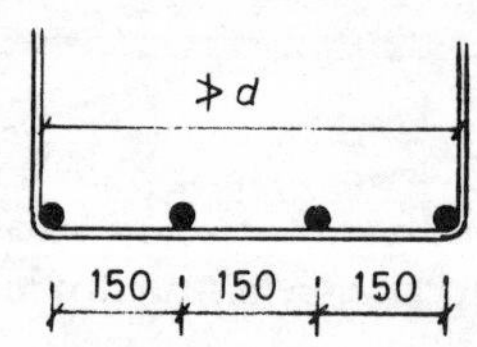

Before calculating the links required in the continuing example from the previous chapters, the shear force envelope to be used should be discussed. If redistribution has not been carried out then the shear force envelope associated with the bending moment envelope will be used. However, if redistribution has been carried out, it should be appreciated that this will affect the shear force envelope. A reduction in moment at only one end of a span will reduce the shear force at that end but increase it at the other end.

## EXAMPLE 9.1

Using the results of the analysis in section 2 of Example 3.1 we can produce a shear force envelope as shown below. Note that in span $AB$ the left-hand side of the envelope is obtained from the redistributed moments and the right-hand side from the elastic moments.

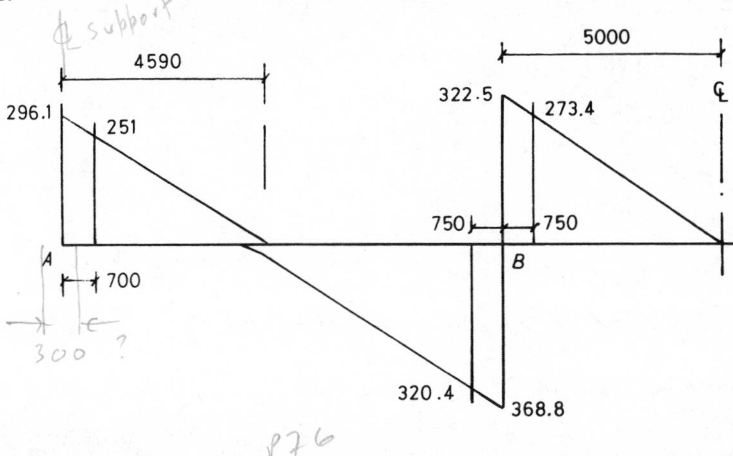


The shear forces will be taken at an effective depth of 550 mm from the face of the support, i.e. a distance of 700 mm from $A$ and 750 mm from $B$. These are indicated on the diagram. From the curtailment drawing in Chapter 8 it can be seen that at these positions the minimum effective tension reinforcement is $2/32\phi$.

$100A_s/b_vd = (100 \times 1610)/(300 \times 550) = 0.975.$

From the equation for $v_c$ in Table 3.9 of the Code we find that $v_c = 0.70$ N/mm$^2$.

For minimum links $b_v(v - v_c) = 300 \times 0.4 = 120$. Table 2 of Appendix 4 suggests $8\phi$ at 300 centres which will give a shear resistance stress of $134/300 = 0.45$ N/mm$^2$.

Shear force with minimum tension reinforcement and minimum links $= (0.70 + 0.45) \times 300 \times 550 \times 10^{-3} = 190$ kN.

### End span

Maximum load $= 64.5$ kN/m.

Shear force of 190 kN occurs at

$(296.1 - 190)/64.5 = 164$ m from $A$

and

$(368.8 - 190)/64.5 = 2.77$ m from $B$.

Near support *A*,

$v = (251 \times 10^3)/(300 \times 550) = 1.52\ \text{N/mm}^2$.

$b_v(v - v_c) = 300(1.52 - 0.70) = 246$.

From Table 2 of Appendix 4, provide 8$\phi$ at 150 centres.

Near support *B*,

$v = (320.4 \times 10^3)/(300 \times 550) = 1.94\ \text{N/mm}^2$.

$b_v(v - v_c) = 300(1.94 - 0.70) = 372$.

From Table 2 of Appendix 4, provide 8$\phi$ at 100 centres.

## Centre span

Shear force of 190 kN occurs at

$(322.5 - 190)/64.5 = 2.05$ m from *B*.

Near support *B*,

$v = (273.4 \times 10^3)/(300 \times 550) = 1.66\ \text{N/mm}^2$.

$b_v(v - v_c) = 300(1.66 - 0.70) = 288$.

From Table 2 of Appendix 4, provide 8$\phi$ at 125 centres.

A suitable arangement of links is indicated below.

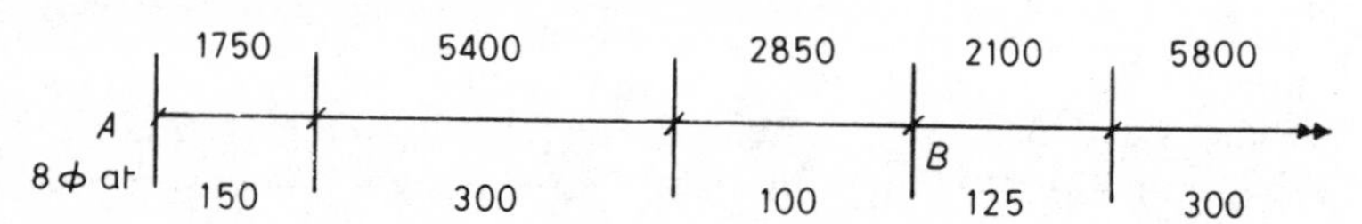


It should be pointed out that a slightly smaller number of links may be required if fuller calculations are carried out as described in clauses 3.4.5.8 and 3.4.5.9 together with the actual amount of tension reinforcement at the critical sections. These sections will be at points of contraflexure and at effective depths from the ends of curtailed bars, in addition to those carried out above. The full treatment can, of course, only be carried out if a bending moment envelope has been prepared. If the simplified rules are used the shear calculations will have to be done as illustrated above.

## 9.1 Bent-up bars

Bent-up bars are not often used for shear resistance, partly because the shear resistance can be provided by links and partly because only half the shear resistance can be provided by the bars.

Clause 3.4.5.6 gives an equation to determine the resistance and refers to Fig. 3.4 of the Code. In the first edition of the Code this figure attracted adverse comments and has been amended. To avoid confusion the revised figure is shown here as Fig. 9.5, but it should be pointed out that the failure plane, *A–A*, does not have to go through the bend

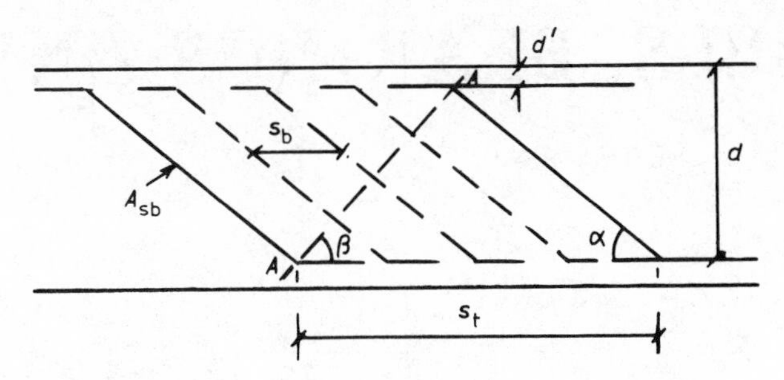


FIG. 9.5 System of bent-up bars.

in the bar. It can be anywhere at the angle $\beta$, and it is the number of inclined bars cut by this plane which determines whether the system is classed as single, double, etc. In Fig. 9.5 the system is quadruple, which can be seen more easily if the line $AA$ is moved slightly to one side.

The Code gives the maximum value of $s_t$ as $1.5d$, and the values of $\alpha$ and $\beta$ as not less than 45°. The traditional way of detailing bent-up bars has been to make $\alpha=45°$, $\beta=67\frac{1}{2}°$, with $s_t=1.41(d-d')$.

Although the designer can work out the shear resistance of the system from equation (4) of the Code, Table 9.1 below gives some typical values using the traditional values for $\alpha$ and $\beta$ given above.

Bars should be checked for stresses inside bends and also for anchorage.

**Table 9.1** Ultimate shear resistance (kN) for bars inclined at 45°

| | $f_y=250$ N/mm$^2$ | | $f_y=460$ N/mm$^2$ | |
|---|---|---|---|---|
| *Bar* | *Single system* | *Double system* | *Single system* | *Double system* |
| 16 | 30.9 | 61.8 | 56.9 | 113.8 |
| 20 | 48.3 | 96.6 | 88.9 | 177.8 |
| 25 | 75.5 | 151.0 | 138.9 | 277.8 |
| 32 | 123.7 | 247.4 | 227.6 | 455.2 |
| 40 | 193.3 | 386.5 | 355.6 | 711.2 |

# 10 CORBELS, BEARINGS AND NIBS

Although many corbels and nibs are constructed *in situ* the Code deals with them in Section 5 of Part 1 under precast and composite construction. This is probably because the majority of members being supported will be precast concrete and the positioning of these members, i.e. the contact area, is very important.

Shear is also an important consideration, and in this context the distance $a_v$ which can be used for enhancing the allowable shear stress $v_c$ must also be the appropriate dimension.

For both corbels and nibs the Code uses the words 'short cantilever' and limits a nib to 300 mm depth. The design approach, however, is quite different. A corbel transfers the load from the supported member to another member which is below the corbel, i.e. a column or a wall. A nib, generally continuous and not an isolated projection, transfers the load to a member above itself. The load on a corbel goes down, whilst the load on a nib goes up.

## 10.1 Corbels

These are defined in clause 5.2.7.1 as short cantilever projections in which (a) $a_v/d$ is less than 1.0, and (b) the depth at the outer edge of the contact area, $d_b$, is not less than $h/2$, where the symbols are illustrated in Fig. 10.1.

The Code uses the wording 'contact area of the supported load' rather than bearing area but, as will be seen for the calculations, it is essential to be able to define exactly the contact area. This can only be done by providing a definite bearing pad.

The basis of the design method is that the concrete and reinforcement may be assumed to act as elements of a strut-and-tie system as indicated in Fig. 10.2. From this it can be seen why the line of action of the load must be located. For two concrete surfaces in contact the line of action would be the leading edge of the corbel and one would be unable to obtain a triangle of forces.

The Code also says that compatibility of strains between the strut and tie at the corbel

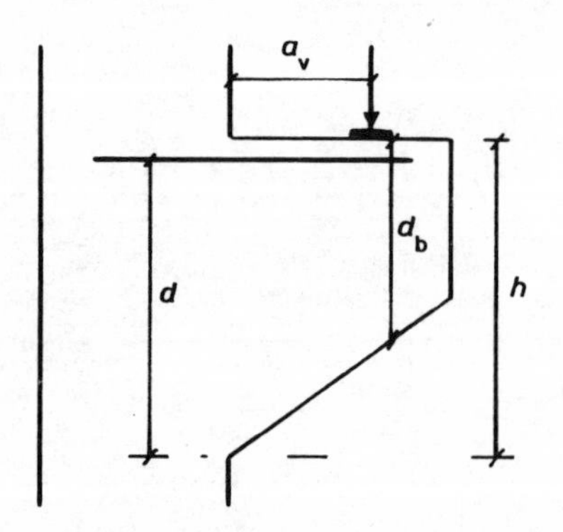


FIG. 10.1 Dimensions of a corbel.

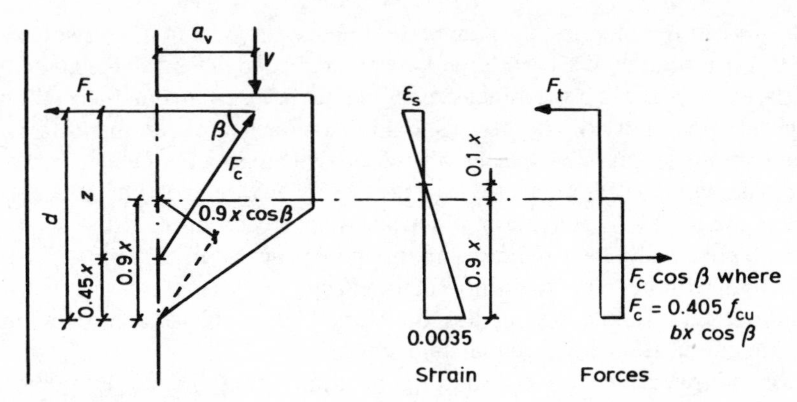


FIG. 10.2 System of forces.

root should be ensured. What this means is not very clear, but the Handbook interprets this in the same way as it did for CP110 – the failure strain of 0.0035 is taken to act at right angles to the face of the supporting member. This would appear to be more for expediency and simplicity in avoiding another term containing the angle $\beta$.

Again, to act as a strut and tie, the depth of the corbel at the face of the support must be sufficient to eliminate the possibility of a shear failure.

So the design shear stress $v$ at the face of the support must be less than the smaller of $0.8\sqrt{f_{cu}}$ and $5\ \mathrm{N/mm^2}$. The value of $v_c$, based on the effective area of tension reinforcement, can also be enhanced by the factor $2d/a_v$. If $v$ is greater than $v_c$, shear links can be provided in accordance with clause 5.2.7.2.3.

It should be noted at this stage that not all designers agree with this procedure and increase $d$ so that $v$ does not exceed $v_c$. They still provide shear links to half the area of tension reinforcement as required by the Code, but do not take them into account for shear resistance. This appears to be rather conservative.

From these requirements the minimum effective depth, and hence the overall depth, at the face of the support can be assessed.

To be able to determine the forces $F_t$ and $F_c$ the value of $z$ or $\beta$ needs to be found. This can be done by assuming a value and carrying out an iterative process or directly from a chart produced by evolving mathematical equations based on the various criteria. This is shown in Fig. 10.3 and will be used in the example that follows.

The value of $F_t$ can now be calculated, together with the strain profile, the value of $\varepsilon_s$

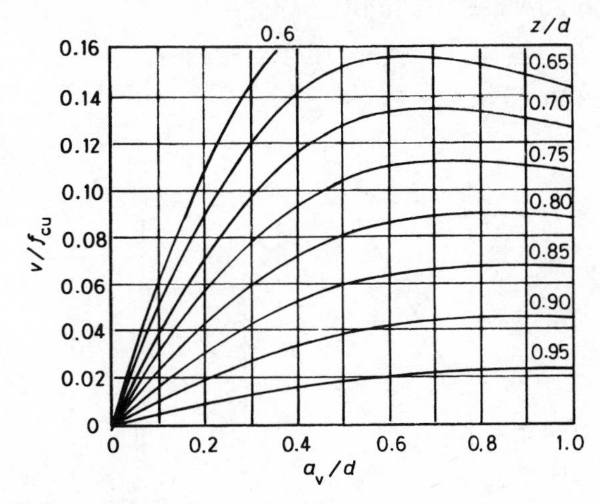


FIG. 10.3 Chart for determining $z/d$.

and the stress in the main reinforcement, and hence the area of main reinforcement required. Concerning the area of this reinforcement the Code says that the magnitude of the resistance provided to horizontal force should not be less than one-half of the design vertical load. It does not say what stress should be assumed, but presumably this should be the maximum design stress. The greater of the two areas of reinforcement would be supplied. The Handbook, however, repeats the CP110 requirement that the area should be not less than 0.4% *bd* and does not appear to have been updated.

As stated previously shear reinforcement is to be provided in the form of horizontal links distributed in the upper two-thirds of the effective depth of the corbel. The area of this reinforcement should be not less than one-half of the area of main tension reinforcement and should be adequately anchored.

The size of the bearing plate transmitting the ultimate design load ($V_u$) to the corbel should be calculated using a bearing stress not greater than $0.8f_{cu}$ as suggested in clause 5.2.3.4, provided the horizontal force at the bearing is less than $0.1V_u$.

## EXAMPLE 10.1

Design a corbel for a 300 mm wide column to support a vertical ultimate design load of 500 kN with the line of action of the load 200 mm from the face of the column.

$f_{cu} = 30$ N/mm², $f_y = 460$ N/mm².

Maximum bearing stress at contact surface $= 0.8 \times 30 = 24$ N/mm².

Assuming effective length of bearing plate is 250 mm,

minimum width $= (500 \times 10^3)/(24 \times 250) = 83.3$ mm, say 85 mm.

Note that the terms length and width are in accordance with clause 1.2.5 of the Code and will be dealt with in the next section of this chapter.

Select a section such that the shear stress at the column face,

$v = V/bd \ngtr 0.8\sqrt{f_{cu}}$.

For Grade 30 concrete maximum $= 4.38$ N/mm².
Minimum effective depth at column face for 300 mm wide corbel,

$d \nless (V/v_u)b = (500 \times 10^3)/(4.38 \times 300) = 380$ mm.

Consider $h = 500$ mm, $d = 500 - (20 + 20/2) = 470$ mm.

$v = (500 \times 10^3)/(300 \times 470) = 3.55$ N/mm².

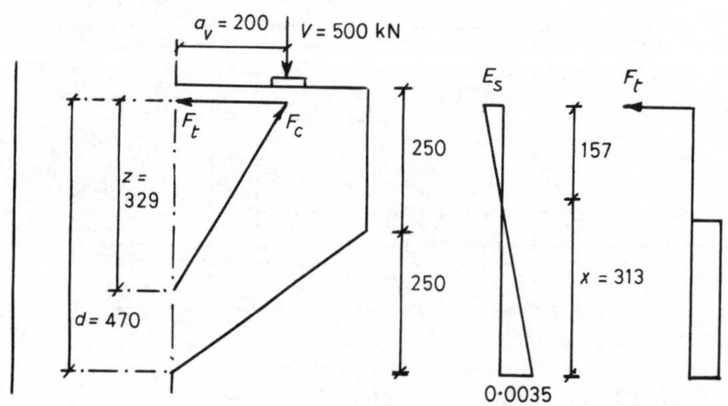

$a_v/d = 200/470 = 0.43$, $v/f_{cu} = 3.55/30 = 0.118$.

From chart $z/d = 0.7$, so $z = 329$.
$x = (d - z)/0.45 = (470 - 329)/0.45 = 313$.
$F_t = V \cot \beta = (500 \times 200)/329 = 304$ kN.

From strain profile, $\varepsilon_s = (157/313) \times 0.0035 = 0.001755$, i.e. $< 0.002$.
$f_s = 351$ N/mm$^2$.

$A_s = (304 \times 10^3)/351 = 866$ mm$^2$, say 3/20$\phi$ (943 mm$^2$).

Minimum area $= (500 \times 10^3/2)/400 = 625$ mm$^2$.

$100A_s/bd = (100 \times 943)/(300 \times 470) = 0.67$.

Allowable shear stress $= 0.59 \times 2/0.43 = 2.73$ N/mm$^2$.

This is less than $v$, so $b(v - v_c) = 300(3.55 - 2.73)$
$= 246$.

From tables, using $f_{yv} = 460$, 10$\phi$ at 250 centres will do this. The depth over which links are to be provided is

$2/3 \times 470 = 313$ mm.

This would mean two links and gives an area of 314 mm$^2$, but minimum area $= 943/2 = 472$ mm$^2$.

Provide 3/10$\phi$ links at 75 mm centres.
The arrangement of reinforcement is shown in Fig. 10.4.

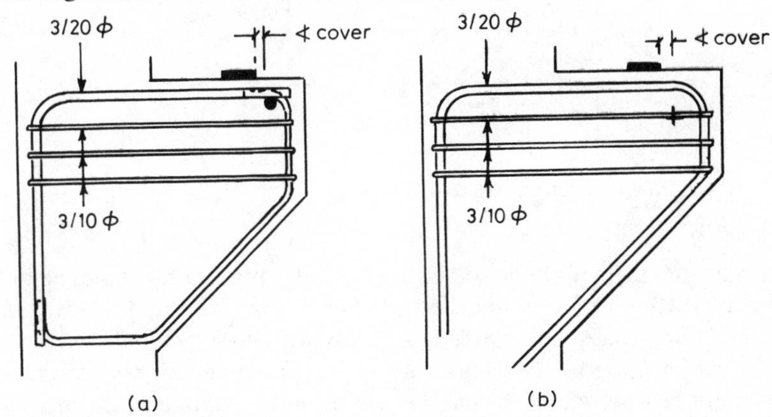


FIG. 10.4 Reinforcement in corbel.

In Fig.10.4(a) the main tension bars are welded to a cross bar of equal size and strength. The distance between the edge of the bearing plate and the inside face of the cross bar should be not less than the cover to the main tension bars. In Fig. 10.4(b) the main tension bars are bent down into the corbel with a standard radius. The start of the bend should be not less than the cover beyond the edge of the bearing plate. The Code suggests that the bend can start at the edge of the bearing plate; however the author does not agree.

It should be noted that the Handbook says the distances to the edge of the bearing plate are related to the bar diameter, but cover would appear to be more relevant.

## 10.2 Bearings

From the previous section on corbels it is fairly obvious that a definitive contact area in a prescribed position is required. The anchorage length of reinforcement in the supported member has also to be taken into account and this may determine the position of the bearing plate.

For other forms of construction, such as continuous concrete nibs, we can have precast members bearing directly on to other concrete members, precast or *in situ*, or through a mortar bedding. In this case, as stated in clause 5.2.3.1, it is important to recognize that the integrity of a bearing is dependent upon two essential safeguards: (a) an overlap of reinforcement in reinforced bearings; (b) a restraint against loss of bearing through movement.

Before describing the allowances for various effects we must first make sure of the terms being used. It is a pity that clause 5.2.3 refers the reader back to clause 1.2.5, as in Fig. 5.4 of the first edition of the Code the term 'width' was used for both dimensions. This might not have happened if the definitions had been given in 5.2.3.

The bearing *length* is the length of the support, supported member or intermediate padding material (whichever is the least) measured along the line of the support. The bearing *width* is the overlap of the support and supported member measured at right angles to the support.

The best way of illustrating this is in diagrammatic form, as shown in Fig. 10.5. The effective bearing length is the least of: (a) bearing length per member; (b) one-half of

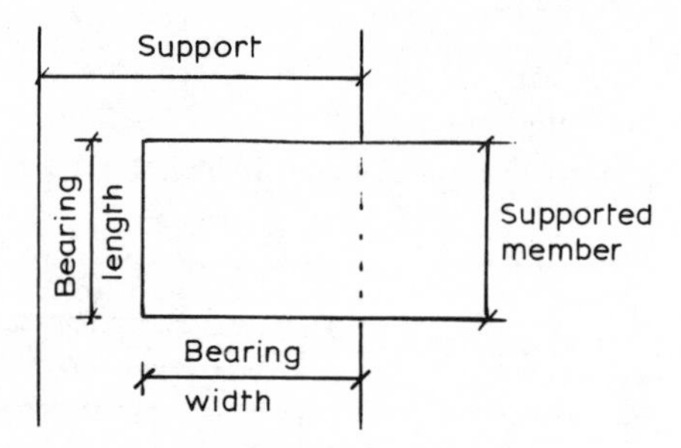


FIG. 10.5 Plan showing bearing length and width.

bearing length per member plus 100 mm; or (c) 600 mm. In the example on corbel design it can now be seen why the effective length of the bearing plate was taken as 250 mm, assuming that the actual length of the plate was 300 mm. For the bearing width we start with the net bearing width to which we add allowances for various effects. The net bearing width, as can be seen from the equation in clause 5.2.3.2, depends on the ultimate bearing stress, which again depends on the surface-to-surface contact. Values are given in clause 5.2.3.4 but again reference to clause 1.2.5 is needed for definitions.

Design ultimate bearing stresses are:

1. Dry bearing. A bearing with no intermediate padding material – $0.4f_{cu}$.
2. Bedded bearing. A bearing with contact surfaces having an intermediate padding of cementitious material – $0.6f_{cu}$.
3. Steel bearing plate cast into the supported member or support and not exceeding 40% of the concrete dimensions – $0.8f_{cu}$.

If the supported member and support have different concrete grades, then the lower grade shall be used for $f_{cu}$. Figure 5.4 of the Code gives a schematic arrangement for various bearing widths. It is reproduced here as Fig. 10.6 in a slightly different form.

The distance $a$, classed as minimum in the Code, is obtained from the sum of the net bearing width (obtained from 5.2.3.2 or 5.2.3.5) plus allowances for spalling (obtained from Tables 5.1 and 5.2).

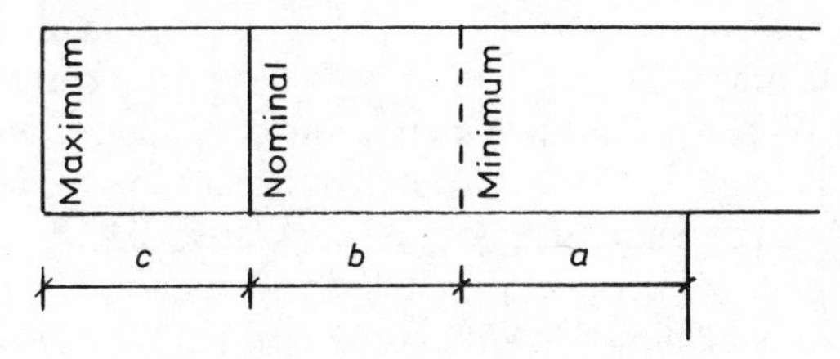


FIG. 10.6 Allowances for bearing.

The distance $b$, obtained from 5.2.4, when added to $a$ is classed as the nominal bearing width and will give the nominal length for the supported member. The minimum anchorage lengths for reinforcement in the supported member will be based on the centre line of this width.

The distance $c$ is again obtained from 5.2.4. When added to $a$ plus $b$ it gives the maximum bearing width and represents the distance the supporting member will project from the face of the main support, e.g. the face of a beam.

Note that all these distances apply at each bearing.

Finally, there is the distinction between an isolated member and a non-isolated member. A non-isolated member is a supported member which, in the event of loss of an assumed support, would be capable of carrying its load by transverse distribution to adjacent members. In this case the calculations carried out to determine the net bearing width using the equation in clause 5.2.3.2 are valid. For an isolated member, the net bearing width should be 20 mm greater than for a non-isolated member. In the previous example for a corbel, it should be borne in mind that if the supported member is classed as isolated then the bearing width given in the calculation should be increased by 20 mm.

## 10.3 Continuous concrete nibs

These can be classified generally as 'narrow bearings' and are projections at the bottoms of beams to carry secondary beams, floor units or brickwork. Depths vary, but when supporting brickwork are usually one or two brick courses deep depending on the amount of brickwork carried. When they are supporting secondary beams or floor units the depth is usually 100 mm or more.

Clause 5.2.8.1 says that a continuous nib which is less than 300 mm deep should normally be designed as a short cantilever slab. It is highly unlikely that a projection from the bottom of a beam will exceed 300 mm in depth. A continuous projection from a reinforced concrete wall may possibly do so, and in this case it would be better to use the corbel approach previously described.

For the short cantilever approach the line of action of the load and the distance $a_v$ are different from the corbel approach. The line of action of the load is assumed to act at the outer edge of the loaded area and this can be at:

1. the front edge of the nib if the nib does not have a chamfer;

2. the upper edge of a chamfer;
3. the outer edge of a bearing pad.

In other words, the line of action should be taken as far as possible from the face of the supporting member.

The distance $a_v$, which is used to calculate the bending moment and also the shear enhancement factor, is taken from the line of action of the load to the nearest vertical leg of the links in the member from which the nib projects. Most designers use the centre line of the leg of the link. The bending moment is the design load multiplied by $a_v$, from which the area of reinforcement can be calculated. The area of reinforcement provided should be not less than the minimum as given in clause 3.12.5. The reinforcement should be positioned in the top of the nib and taken to a point as near to the front face as considerations for cover will allow. As with corbels the bars can be anchored either by welding to a transverse bar of equal strength or by bending through 180° to form loops in the horizontal or vertical plane. With vertical loops the bar size should not exceed 12 mm.

For shear resistance the value of $v_c$ obtained in the normal way may be increased by the enhancement factor of $2d/a_v$.

## EXAMPLE 10.2

An *in situ* reinforced concrete nib is required to carry a series of precast pretensioned hollow floor units, 400 mm wide by 140 mm deep. The floor units are to span a clear distance of 4.5 m and carry a superimposed load of 5.0 kN/m$^2$ (including 1.5 kN/m$^2$ for screeds and finishes). The *in situ* concrete is to be Grade C40 using 20 mm maximum size aggregate and the reinforcement is to be Grade 460. The exposure condition is mild. The precast floor units can be classed as non-isolated members with a self weight of 2.50 kN/m$^2$, and concrete Grade C40.

With the exposure condition and 20 mm aggregate we shall need 20 mm cover to the reinforcement in the nib. If we use vertical loops for the reinforcement in the nib the minimum depth of the nib will be 20 + 20 + 8 times the bar diameter. Assuming 8$\phi$ bars the depth cannot be less than 104 mm, say 105 mm.

It is also suggested that the front edge of the nib should have a 15 × 15 mm chamfer.

Characteristic load carried by floor unit:

| | | |
|---|---|---|
| Dead: | self | 2.50 kN/m$^2$ |
| | screeds | 1.50 kN/m$^2$ |
| | | 4.0 kN/m$^2$ |
| Imposed: | | 3.5 kN/m$^2$. |

Ultimate design load = $4 \times 1.4 + 3.5 \times 1.6 = 11.2$ kN/m$^2$. As the clear distance is 4.5 m we will assume the length of the unit is 4.7 m to calculate the net bearing width, but this will be checked later.

Reaction each end = $4.7/2 \times 11.2 = 26.3$ kN/m run $\times 0.4 = 10.52$ kN for each 400 mm wide unit.

We shall assume a dry bearing, so design bearing stress is $0.4 \times 40 = 16$ N/mm$^2$. Effective bearing length = $400/2 + 100 = 300$ mm.

Net bearing width = $10.52 \times 1000/300 \times 16 = 2.2$ mm, so use minimum width of 40 mm.

Allowance for spalling $= 20$ mm (Table 5.1)
$+0$ (Table 5.2, exposed tendon).

Allowance for inaccuracies $= 25$ mm.

The nominal bearing width is $40 + 20 + 25 = 85$ mm, so the nominal length of the unit is $4.5 + 2 \times 0.085 = 4.67$ mm.

The width of the nib projection is $85 + 25 = 110$ mm.

The line of action of the load $= 110 - 15 = 95$ mm from face of beam, and assuming 20 mm cover to $10\phi$ links in the beam, the distance $a_v = 95 + 20 + 5 = 120$ mm.

Although the effective bearing length of 300 mm is less than the width of the floor unit, it will be assumed that the load will spread for the width of the unit in calculating the moments and shears. Thus:

$M = 26.3 \times 0.12 = 3.16$ kN m/metre run.
Effective depth, $d = 105 - 20 - 4 = 81$ mm.

$M/bd^2 = (3.16 \times 10^6)/(1000 \times 81^2) = 0.48.$

This will give less than the minimum percentage, so

$A_s = (0.13/100) \times 1000 \times 105 = 137$ mm$^2$/m

and spacing not greater than $3 \times 81 +$ bar diameter.

Use $8\phi$ at 250 centres.

$v = (26.3 \times 10^3)/(1000 \times 81) = 0.32$ N/mm$^2$

$100A_s/bd = (201 \times 100)/(1000 \times 81) = 0.25.$

From Table 3.9 of the Code, $v_c = 0.62$ N/mm$^2$.

Allowable shear stress $= 0.62 \times 162/120 = 0.84$ N/mm$^2$.

The detail would be as shown:

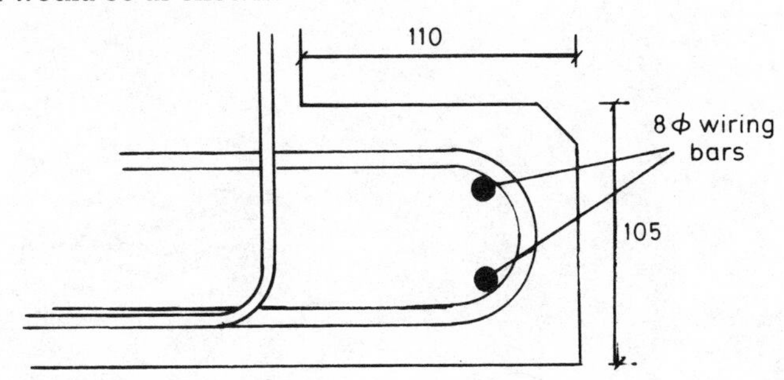


Points to note:

1. The depth of the nib could be reduced by using a bar welded to the ends of horizontal projecting bars, or by using horizontal loops.
2. Additional area of reinforcement is required in main beam to carry the load of 26.3 kN/m to the top of the beam.
3. If reinforced precast floor units are being used, it would be better to use straight bars exposed at the end, provided the shear stress is less than $v_c/2$. If bars with a bend are

required the length of the unit and also the bearing width would have to be recalculated.

---

In general a continuous nib will carry a continuous load such as a floor slab (probably precast) or a height of brickwork, so it can be designed in accordance with the Code as a slab, i.e. per metre length. On occasions, however, it will carry isolated loads such as from a precast beam or a precast double-Tee floor unit. In these cases the length of the nib which can be considered as carrying the load is very important.

From tests carried out it would appear that the failure cracks spread at an angle of 45° from the loaded area. The length which can be considered as carrying the load can, therefore, be taken as the width of the supported member plus twice the distance from the line of action of the load to the face of the supporting beams (see Fig. 10.7).

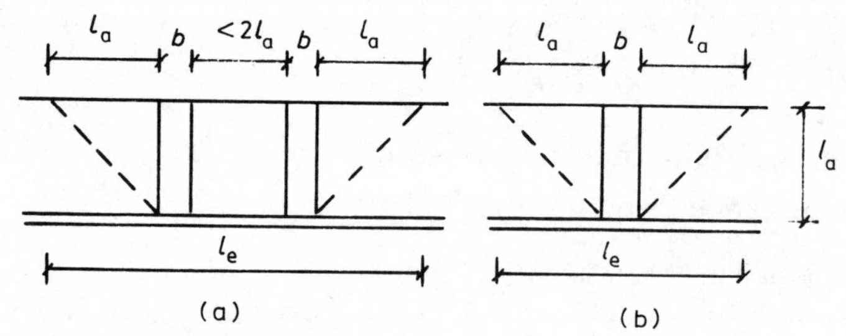


FIG. 10.7 Isolated loads on continuous nibs.

If the clear distance between the loads is less than $2l_a$, as in case (a), then $l_e$ is the distance over all the loads plus $2l_a$. Where the clear distance between the loads is more than $2l_a$, as in case (b), then $l_e$ equals $b+2l_a$. Whichever case is appropriate the reinforcement in the nib will be the same throughout its length.

# COLUMNS 11

For columns, the first need is to define the type of column. In clause 3.8.1.5 the Code gives the following definitions:

*Braced*

A column is considered braced in a given plane if the lateral stability to the structure as a whole is provided by walls or other suitable bracing to resist all lateral forces in that plane.

*Unbraced*

A column is considered unbraced in a given plane where lateral stability in that plane is provided by the column.

A column supporting beams will have two axes for bending in directions generally at right angles, and will therefore have two planes of bending. We therefore have to consider each axis of bending (or each plane of bending) in turn.

As an example, consider the following layout of a framed building.

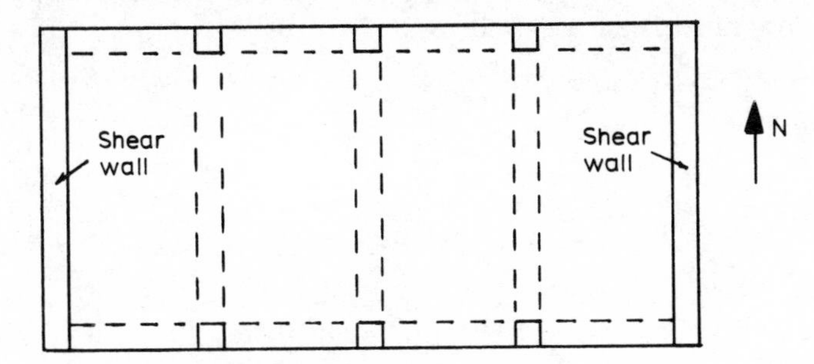


In the north–south direction, stability to the structure is provided by shear walls, so the columns are braced. In the east–west direction stability is provided by the columns, so the columns are unbraced. So, in different directions the column has to be treated differently. In most structures this is unlikely to happen and the column will have one classification only.

Most designers dealing with columns tend to think of bending about an axis rather than bending in a particular plane. It is very useful to establish these axes early on, as calculations to find the effective heights in the two directions are the next step. If we have a rectangular section, the major axis will be the axis about which the section has the larger second moment of area. This will be called the $X$–$X$ axis. Bending about the $X$–$X$ axis will be bending in the $Y$–$Y$ plane.

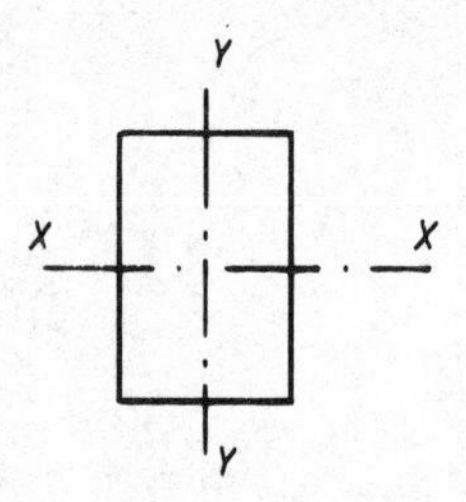


For square or circular columns the second moment of area is the same about both axes and it is up to the designer to nominate the $X–X$ and $Y–Y$ axes.

Failure of a column is due to buckling rather than pure compression and the effective height is a convenient method for dealing with buckling. The effective height can be found from clause 3.8.1.6 or more rigorously from 2.5 in Part 2 of the Code.

In clause 3.8.1.6.1 it states that the effective height in a given plane, $l_e$, is $\beta l_0$ where $\beta$ is obtained from Tables 3.21 and 3.22 of the Code for braced and unbraced columns respectively, and $l_0$ is the clear height between end restraints. The values for $\beta$ are a function of the end conditions of a storey height of a column. The end conditions are defined in terms of a scale of 1 to 4. The smaller the number the greater is the fixity. Although the various conditions are described in the Code, they are shown in diagrammatic form in Fig. 11.1.

The values for $\beta$ have been assessed using typical values for column and beam stiffnesses, but it must not be expected that they will agree every time with calculations carried out in accordance with the formulae in Part 2.

What is important to remember, however, is that the conditions do not necessarily apply where there is a beam on one side of the column only, for example an external column. They may do; it depends on the relative stiffnesses of the beam and column. To avoid possible differences of opinion it may be necessary to go to Part 2, Section 2.5, and use the formulae. These are as follows:

*Braced columns*

Lesser of

$$l_e = l_0\{0.7 + 0.05(\alpha_{c,1} + \alpha_{c,2})\} \leqslant l_0 \quad (11.1)$$

and

$$l_e = l_0\{0.85 + 0.05\alpha_{c,min}\} \leqslant l_0. \quad (11.2)$$

*Unbraced columns*

Lesser of

$$l_e = l_0\{1.0 + 0.15(\alpha_{c.1} + \alpha_{c.2})\} \quad (11.3)$$

and

$$l_e = l_0\{2 + 0.3\alpha_{c,min}\}. \quad (11.4)$$

*Symbols*

$I$ Second moment of area of the concrete section.

$l_e$ Effective height of a column in the plane of bending considered.

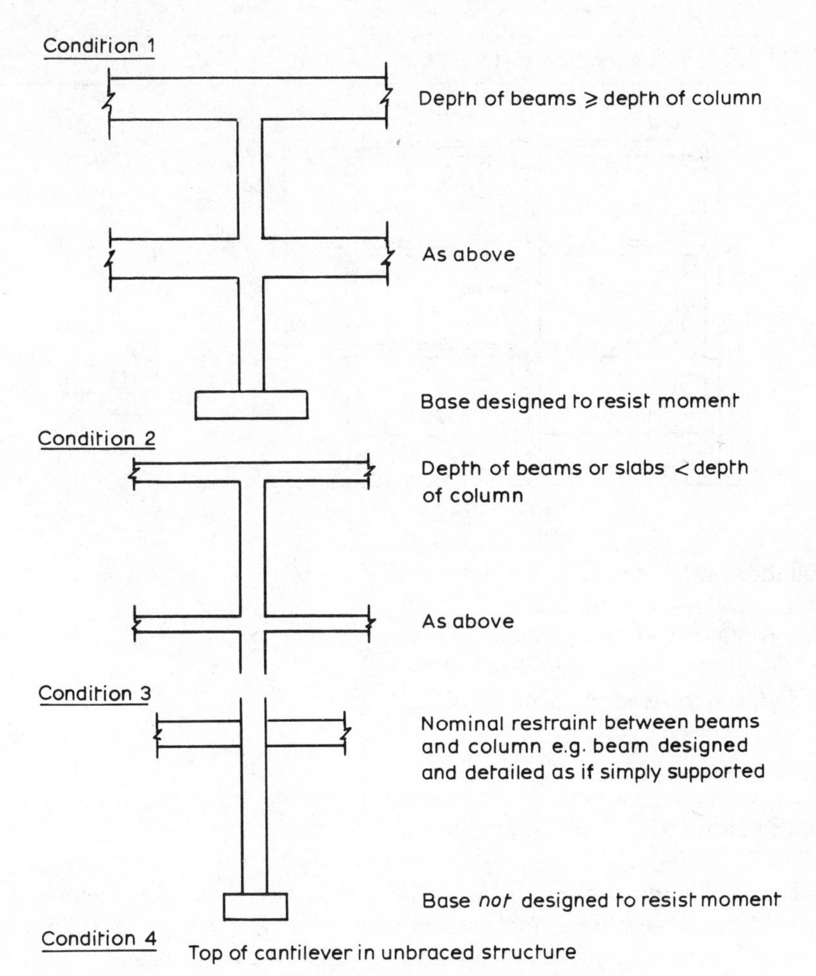


FIG. 11.1 End conditions to determine effective height of column.

$l_0$ Clear height between end restraints.
$\alpha_{c,1}$ Ratio of the sum of the column stiffnesses to the sum of the beam stiffnesses at the lower end of a column.
$\alpha_{c,2}$ Ratio of the sum of the column stiffnesses to the sum of the beam stiffnesses at the upper end of a column.
$\alpha_{c,min}$ Lesser of $\alpha_{c,1}$ and $\alpha_{c,2}$.

In calculating $\alpha_c$, only members properly framed into the end of the column in the appropriate plane of bending should be considered. The stiffness of each member equals $I/l_c$ where $l_c$ is the distance between centres of restraints. At a column base $\alpha_c$ may be taken as 1.0 where the base is designed to resist the restraint moment. If the base is not designed to resist moment, $\alpha_c$ should be taken as 10. For complete restraint, such as at a very large base, $\alpha_c$ can be taken as zero.

Note that in the Code it says the stiffness of each member is $I/l_0$ whereas here it is given as $I/l_c$. The answers should not be very different, but it does not appear to be logical to take centres of supports for beams and clear heights for columns.

The following example illustrates the calculations required, and although they may appear to be a bit tedious they are not difficult to perform.

## EXAMPLE 11.1 Effective heights of columns

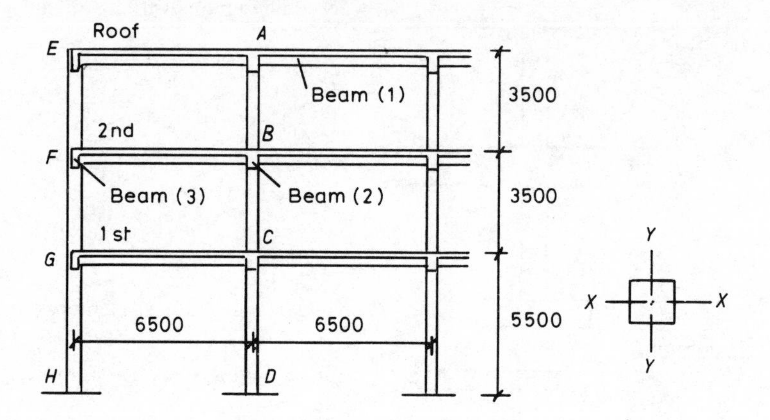


All columns 300 × 300.
Beam 1 is 450 × 250, span 6500.
Beam 2 is 500 × 300, span 5500.
Beam 3 is 500 × 200, span 5500.
Bases designed to resist moment.

### Internal columns

Using end conditions as given in clause 3.8.1.6.2, these would be condition 1 at all junctions. The effective height for a braced column would, therefore, be 0.75 times the clear height, and for an unbraced column, 1.2 times the clear height.

### External columns

The Code does not give any guidance when beams are on one side of a column only, and the designer must either use his judgement or use the equations in Part 2.

The following examples illustrate the procedure.

### Section properties

| Member | $I$ (mm$^4$) | $l_c$ (mm) | $I/l_c$ |
|---|---|---|---|
| Columns (1st–Roof) | $300 \times 300^3/12 = 675 \times 10^6$ | 3500 | $193 \times 10^3$ |
| Column (Found–1st) | $300 \times 300^3/12 = 675 \times 10^6$ | 5500 | $123 \times 10^3$ |
| Beam (1) | $250 \times 450^3/12 = 1900 \times 10^6$ | 6500 | $292 \times 10^3$ |
| Beam (2) | $300 \times 500^3/12 = 3125 \times 10^6$ | 5500 | $568 \times 10^3$ |
| Beam (3) | $200 \times 500^3/12 = 2083 \times 10^6$ | 5500 | $379 \times 10^3$ |

Note. In determining $I$ for the beams, the rectangular section only has been used.

## Ratio of stiffnesses

| | Bending about | |
|---|---|---|
| Joint | X–X axis | Y–Y axis |
| A | 193/(2 × 568) = 0.17 | 193/(2 × 292) = 0.33 |
| B | (2 × 193)/(2 × 568) = 0.34 | (2 × 193)/(2 × 292) = 0.66 |
| C | (193 + 123)/(2 × 568) = 0.28 | (193 + 123)/(2 × 292) = 0.54 |
| D, H | 1.0 | 1.0 |
| E | 193/(2 × 379) = 0.25 | 193/292 = 0.66 |
| F | (2 × 193)/(2 × 379) = 0.25 | (2 × 193)/292 = 1.32 |
| G | (193 + 123)/(2 × 379) = 0.57 | (193 + 123)/292 = 1.08 |

## Effective height factors

(1) Braced columns – equation (11.1) controls

| | Bending about | |
|---|---|---|
| Column | X–X axis | Y–Y axis |
| A–B | 0.7 + 0.05 (0.17 + 0.34) = 0.72 | 0.7 + 0.05 (0.33 + 0.66) = 0.75 |
| B–C | 0.7 + 0.05 (0.34 + 0.28) = 0.73 | 0.7 + 0.05 (0.66 + 0.54) = 0.76 |
| C–D | 0.7 + 0.05 (0.28 + 1.0) = 0.75 | 0.7 + 0.05 (0.54 + 1.0) = 0.78 |
| E–F | 0.7 + 0.05 (0.25 + 0.51) = 0.74 | 0.7 + 0.05 (0.66 + 1.32) = 0.8 |
| F–G | 0.7 + 0.05 (0.51 + 0.57) = 0.75 | 0.7 + 0.05 (1.32 + 1.08) = 0.82 |
| G–H | 0.7 + 0.05 (0.57 + 1.0) = 0.78 | 0.7 + 0.05 (1.08 + 1.0) = 0.8 |

These factors are to be compared with 0.75.

(2) Unbraced columns – equation (11.3) controls

| | Bending about | |
|---|---|---|
| Column | X–X axis | Y–Y axis |
| A–B | 1.0 + 0.15 (0.17 + 0.34) = 1.08 | 1.0 + 0.15 (0.33 + 0.66) = 1.15 |
| B–C | 1.0 + 0.15 (0.34 + 0.28) = 1.09 | 1.0 + 0.15 (0.66 + 0.54) = 1.18 |
| C–D | 1.0 + 0.15 (0.28 + 1.0) = 1.19 | 1.0 + 0.15 (0.54 + 1.0) = 1.23 |
| E–F | 1.0 + 0.15 (0.25 + 0.51) = 1.11 | 1.0 + 0.15 (0.66 + 1.32) = 1.30 |
| F–G | 1.0 + 0.15 (0.51 + 0.57) = 1.16 | 1.0 + 0.15 (1.32 + 1.08) = 1.36 |
| G–H | 1.0 + 0.15 (0.57 + 1.0) = 1.23 | 1.0 + 0.15 (1.08 + 1.0) = 1.31 |

These factors are to be compared with 1.2.

As can be seen from the factors they do not tie up exactly with the values in Tables 3.21 and 3.22 of the Code, but are not greatly different. However, there are significant differences in the external columns bending about the *Y–Y* axis, i.e. where there is a beam on one side only.

It is important to remember that only the flexural stiffnesses of the beams in the plane of bending being considered should be taken into account. The torsional stiffness of any beam at right angles to the plane of bending is ignored.

A small point to note concerns the effective height of a cantilever column. It can only be unbraced by definition, and the effective height is 2.2 times the clear height. As there cannot be any restraints at the free end the clear height can only be taken as the height from the top of foundation to the top of the column.

## 11.1 Slenderness ratios

A column is considered as short when both the ratios $l_{ex}/h$ and $l_{ey}/b$ are less than 15 (braced) and 10 (unbraced). In these definitions $h$ is the dimension of the column section in the plane of bending when bending about the major axis with the effective height of $l_{ex}$. The dimensions $b$ and $l_{ey}$ refer to the minor axis.

The slenderness limits are slightly confusing. For any column, braced or unbraced, the clear distance $l_0$ should not exceed 60 times the minimum thickness of the column (3.8.1.7). For braced columns this will mean that a deflection check is not necessary (3.8.5(a)), but with an effective height equal to the clear height (it cannot be greater) it means that $l_e/b$ can be as high as 60. $b$ is the minimum thickness.

For unbraced columns, we have to consider all the columns at a particular level, and if the average value of $l_e/h$ (note $h$ is in the plane of bending) is not greater than 30, a deflection check is not necessary (3.8.5(b)).

Where one end of an unbraced column is unrestrained (e.g. a cantilever column) clause 3.8.1.8 states that $l_0$ should not exceed $100b'^2/h' \leqslant 60b'$, where $h'$ and $b'$ are the larger and smaller dimensions respectively.

The limits as given are quite high and would induce such large additional moments due to deflection that the practical design would be almost impossible.

## 11.2 Moments and forces in columns

These will usually be calculated at ultimate limit state only, except in the design of foundations, where to determine the size of these we shall need the loads etc. at serviceability limit states. This is because allowable bearing pressures as given in Codes of Practice are for working loads, i.e. serviceability conditions.

To find the moments and forces we have to consider braced and unbraced, short and slender columns. As mentioned in Chapter 3, what we are trying to find are the worst effects, that is a combination of loads and bending moments. If a full frame analysis has been done using all possible load combinations and load patterns the answers are there and have to be sorted out. If a series of subframes has been used for the vertical loads it must be remembered that even a single load case at all levels down the column will not produce a true bending moment diagram. The subframe assumes that the ends of the columns remote from the beam column junctions are fixed. This will automatically give moments at the remote ends of half the moments at the junctions. As an example the diagram below shows the results for an intermediate column storey height where subframe analysis has been carried out.

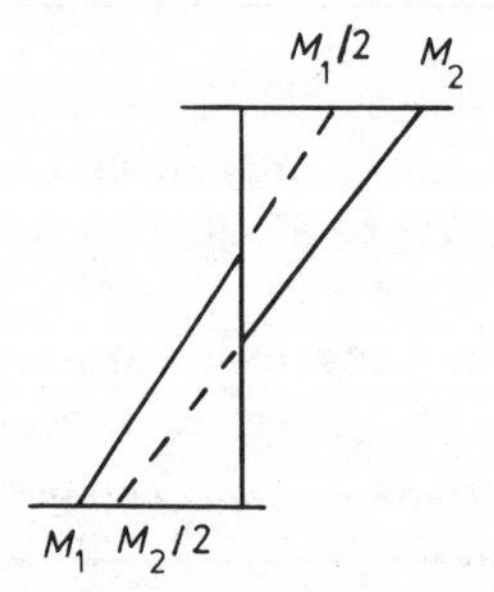


This is obviously not a true diagram, but will generally be satisfactory if only the maximum moment is required. As will be seen later this will apply to all columns except braced slender columns. In this case only one load case bending moment will be used, not a combination. For example, if $M_2$ is the larger moment then the diagram would be taken as $M_2$ at one end and $M_2/2$ at the other end of the column. This agrees with one of the conditions for slender braced columns, as will be seen later.

For unbraced columns the Code suggests that full frame or subframe analyses should be carried out. But with this type of column, if it is slender, additional moments due to deflection will be induced in the column at the beam column junction. This will affect the equilibrium of the joint, but unless the average value of $l_e/h$ for all columns at a particular level is greater than 20, the Code says that although the columns will be designed to resist these additional moments the beams need not be. It is prudent, therefore, to ensure that column sizes are such that the average value of $l_e/h$ does not exceed 20 so that any previous calculations for the design of the beams is not affected.

For braced columns, it is allowable to ignore the columns in a beam analysis. If the columns are supporting symmetrical arrangements of beams, the moments can be ignored, except for a nominal allowance, only the column loads being used. Although not specifically stated, it will be seen later that this only applies to *short* braced columns. If the columns are not symmetrically loaded, and the columns have been ignored in the beam analysis, the moments in the columns may be calculated by simple moment distribution procedures. This procedure assumes that the column and beam ends remote from the junction under consideration are fixed. The columns will have their actual stiffness, but the beams will have only half their actual stiffnesses. This can be illustrated as shown:

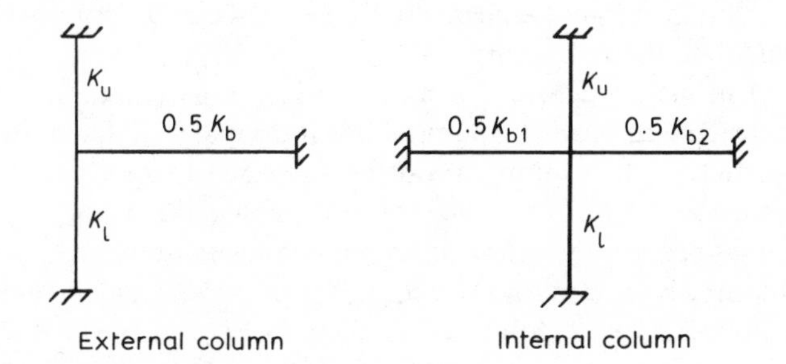


The moments in the columns are as follows:

External column

$$\text{Moment at foot of upper column} = \frac{K_u}{K_l + K_u + 0.5K_b} M_e$$

$$\text{Moment at head of lower column} = \frac{K_l}{K_l + K_u + 0.5K_b} M_e$$

Internal column

$$\text{Moment at foot of upper column} = \frac{K_u}{K_l + K_u + 0.5K_{b1} + 0.5K_{b2}} M_{es}$$

$$\text{Moment at head of lower column} = \frac{K_l}{K_l + K_u + 0.5K_{b1} + 0.5K_{b2}} M_{es},$$

where $M_e$ is the maximum fixed end moment of the beam framing into the column (i.e. fixity at both ends of beam), and $M_{es}$ is the maximum difference in fixed end moments of the beams framing into the column. One beam will have maximum design load and the other minimum design load.

For an external column, the moment in the beam at this junction would be the sum of the column moments.

### 11.2.1 Minimum moment

In clause 3.8.2.3 the Code states that even though in some cases only the axial load is considered, an allowance must be made for this axial load acting at a nominal eccentricity.

In the next clause it then goes on to say that for no section in a column should the design moment be taken as less than that produced by considering the design ultimate load acting at a minimum eccentricity, $e_{min}$. The value for $e_{min}$ should be taken as 0.05 times the overall dimension of the column in the plane of bending considered, but not more than 20 mm. So, for a column dimension of not greater than 400 mm one would use 0.05 times the dimension, but over 400 mm one would use 20 mm.

What this really means is that any and every column must be able to resist a minimum moment of $Ne_{min}$, even if the analysis indicates a smaller moment.

## 11.3 Short columns

For braced and unbraced short columns the general method is to obtain the axial load and moment from analysis. The moment from analysis is compared with the minimum moment of $Ne_{min}$ and the larger value taken. However, if certain criteria are met it is not necessary to calculate the moments.

Equation (38) in clause 3.8.4.3, reproduced below as equation (11.5), does not appear to be restricted to braced columns, but it does say that where, owing to the nature of the structure, a column cannot be subjected to significant moments, the equation can be used. The term significant is not defined, but from the numbers in the equation it would appear to be referring to moments less than $Ne_{min}$ of 0.05 $Nh$. This reasoning is obtained by comparing the equation for pure axial load capacity given in clause 3.8.3.1 (following equation (33)) as $N_{uz}=0.45f_{cu}A_c+0.87f_yA_{sc}$.

By allowing for a minimum moment of 0.05 $Nh$ the value of the axial load capacity is reduced to

$$N=0.4f_{cu}A_c+0.75f_yA_{sc}, \tag{11.5}$$

a reduction of approximately 10%.

Equation (39) in clause 3.8.4.4 goes a stage further, but only applies to braced columns. The criteria now are that the beams have uniformly distributed imposed loads and the beam spans do not differ by more than 15% of the longer. The difference between equations (38) and (39) is very similar to that between equation (38) and $N_{uz}$, so it would appear to cater for moments up to $0.10Nh$.

The problem in using equation (38) is to know how the moments are significant if they are not calculated. If they have been calculated, why not use them? With equation (39) the imposed loads may be uniform, but what is the ratio of imposed load to dead load implied in the criteria? Unless it is very obvious, it is probably better to use the general method for short columns, braced or unbraced.

## 11.4 Slender columns

For slender columns we have to take into consideration the additional moments induced by the lateral deflection of the loaded column. The behaviour, and hence the design moments, of braced and unbraced columns is different and will be dealt with separately, but the calculation for the additional moment is common to both.

The formula for the additional moment given in equation (35) of the Code is

$$M_{add}=Na_{u} \quad (11.6)$$

where $N$ is the ultimate axial load on the column, and $a_u$ is the deflection, taken to be $\beta_a Kh$ (equation (32) of the Code).

The coefficient $\beta_a$ is derived from the expression

$$\beta_a=(1/2000)(l_e/b')$$

where $b'$ is the smaller dimension of the column, except for biaxial bending. Values are given in Table 3.23 of the Code for various values of $l_e/b'$, and although the relationship is not linear, intermediate values can be interpolated on this basis. It should be remembered that $l_e$ and $h$ are always in the plane of bending under consideration, so that when considering biaxial bending one axis will be $l_{ex}$ and $h$ and the other axis will be $l_{ey}$ and $b$.

$K$ is a reduction factor which corrects the deflection to allow for the influence of axial load. It is given in Equation (33) of the Code as

$$K=\frac{N_{uz}-N}{N_{uz}-N_{bal}}\leqslant 1.0. \quad (11.7)$$

It should be pointed out that $N_{uz}$ depends on the amount of reinforcement in the column and so cannot be determined at this stage. Methods of determining $K$ will be described later, but in the initial stage the value will be taken as 1.0.

$N_{bal}$ is defined under *Symbols* in 3.8.1.1, and for a symmetrically reinforced rectangular section may be taken as $0.25f_{cu}bd$, i.e. independent of the reinforcement.

We shall now deal with the methods of determining the total design moments for braced and unbraced slender columns.

### 11.4.1 Braced slender columns

In the case of braced columns, the ends of the columns are fixed in position, but not in direction.

Figure 11.2 reproduces Figure 3.20 of the Code and it should be pointed out that the final diagram is an envelope of possible cases. For example, if $M_{add}/2$ is greater than $M_2$ the moment at the upper end will be reversed. It is also assumed that the column is bent in double curvature.

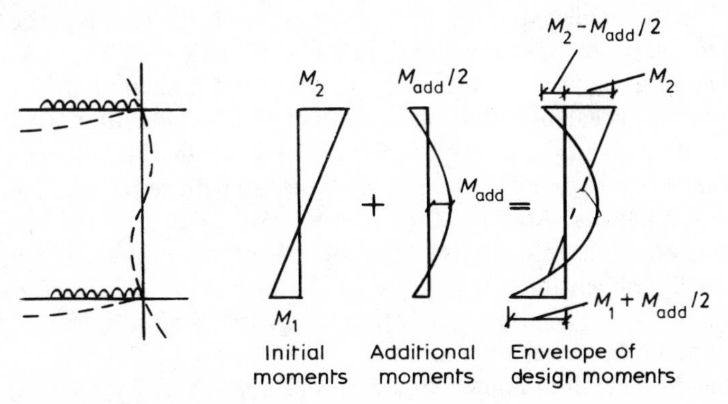


FIG. 11.2 Moments in a braced slender column.

The Code says that the maximum additional moment can be assumed to occur mid-height and so from the diagrams it can be seen that the maximum total moment could occur within the height of the column. For this reason, clause 3.8.3.2 of the Code says that the initial moment, $M_i$, may be taken as $0.4M_1 + 0.6M_2$, where $M_i$ is the smaller initial end moment, assumed negative if the column is bent in double curvature, and $M_2$ is the larger initial end moment, assumed positive.

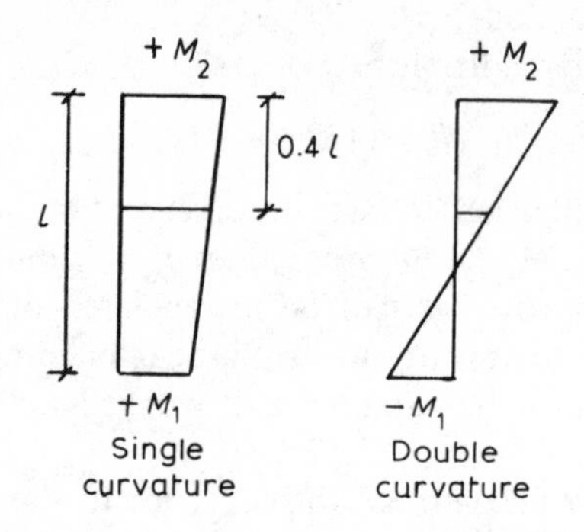


From the expression above it can be seen that the initial moment is taken at 0.4 times the column length from the larger moment (the Code says approximately mid-height).

There is a further restriction that $M_i$ should not be less than $0.4M_2$, which occurs when $M_1$ equals $-M_2/2$.

## 11.4.2 Design moments

From the diagram giving the envelope of moments it can be seen that the design moment will be the greatest of

1. $M_2$
2. $M_i + M_{add}$
3. $M_1 + M_{add}/2$
4. $Ne_{min}$.

This will be given the notation $M_{tx}$ for the major axis and $M_{ty}$ for the minor axis.

Having described what happens to the moments in a braced column, the Code then gives four cases of slender columns which are meant to apply to braced and unbraced structures. As we are dealing with slender braced columns we will deal specifically with these. We have the following cases:

1. Bending about a single axis (major or minor), where $h$, the length of the longer side, is less than $3b$, three times the length of the shorter side.
   (a) For bending about the major axis with $l_e/h$ not greater than 20, the design moment $M_{tx}$ is calculated as above, where $M_{add}$ is a function of $b$.
   (b) For bending about the major axis with $l_e/h$ greater than 20, the column should be treated as for biaxial bending (see (3) below) with zero initial moment about the minor axis, i.e. $M_1$ and $M_2$ (hence $M_i$), $Ne_{min} = 0$.
   (c) Bending about the minor axis is straightforward as the dimension in the plane is the smaller dimension anyway.
2. Bending about a single axis (major or minor) where $h$ is $3b$ or more (but not greater than $4b$ when it becomes a wall).
   (a) For bending about a major axis treat as (1b).
   (b) For bending about a minor axis treat as (1c).

3. Biaxial bending
   Where there are significant moments about both axes we have to consider initial and additional moments about both axes. What is meant by significant is not very clear, but a suggested solution is that if the moments about one axis, either $M_{tx}$ or $M$, are small enough for $Ne_{min}$ to control, then the moment could be regarded as not significant and uniaxial bending as described in (1) and (2) above would be considered. Note that in obtaining the additioal moments these are related to the dimension of the column in the plane of bending being considered.

### 11.4.3 Unbraced slender columns

With unbraced columns the ends are not fixed in either position or direction. In braced columns we can deal with the deflection of a single column, but in unbraced columns we have to bear in mind that all the columns at a particular level will deflect sideways by the same amount. It is necessary, therefore, to calculate the average deflection for all columns in a particular storey height and this can be assessed from

$$a_{uav} = \sum a_u / n,$$

where $n$ is the number of columns involved. Bearing in mind that $a_u$ involves the reduction factor $K$, it is fairly obvious that $K$ will be taken as 1.0 in the initial stages as before, but even when values can be established for individual columns it would not be practical to go right back to the beginning and establish a revised $a_{uav}$. The $K$-factor modification will probably be applied to each individual column as it arises.

Having calculated $a_{uav}$ the Code suggests that any value of $a_u$ which is more than twice $a_{uav}$ should be rejected and a new average calculated based on the modified value for $n$.

As the joints can now move laterally the bending moment diagrams are as shown in Fig. 11.3, and it can be seen that the additional moments will increase the initial moments at the ends of the column rather than within the height of the column.

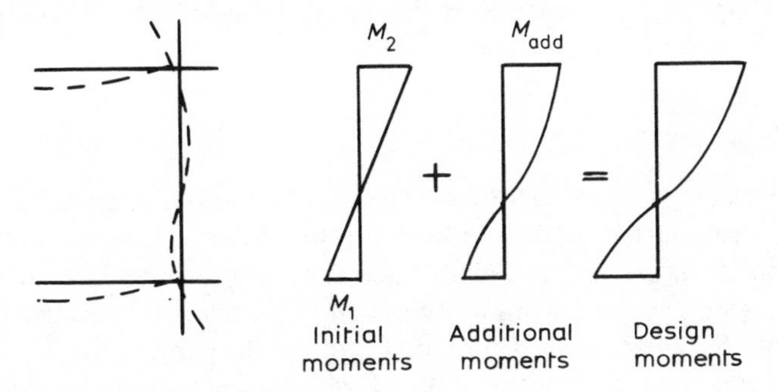


FIG. 11.3 Moments in unbraced slender columns.

The full additional moment will occur at whichever end has the stiffer joint, that is the joint which has the lower $\alpha_c$ value. When using Table 3.22 of the Code it will be the end with the smaller condition number. If the conditions are the same this cannot be done and the full additional moment will be added to the larger moment.

We can now summarize the various cases for slender unbraced columns on similar lines to those for braced slender columns.

1. Bending about a single axis (major or minor), where $h$, the length of the longer side, is less than $3b$, three times the length of the shorter side.

(a) For bending about the major axis with $l_e/h$ not greater than 20, the design moment is either at the top or bottom of the column. If $M_2$ is greater than $M_1$ and occurs at the stiffer joint, then

$$M_t = M_2 + M_{add}.$$

Remember that $M_{add}$ is a function of $b$, and $M_t$ cannot be less than $Ne_{min}$.

(b) For bending about the major axis with $l_e/h$ greater than 20, the column should be treated as for biaxial bending (see (3) below) with zero initial moment about the minor axis. As the additional moments are added at the top and bottom of the column, where the end is connected monolithically to other members, then these members should be designed to withstand the additional moments applied by the ends of the column in addition to those calculated using normal analytical methods.

(c) Bending about the minor axis is straightforward except where $l_e/b$ (where $b$ is in the plane of bending) is greater than 20, then see (1b) above.

2. Bending about a single axis (major or minor) where $h$ is $3b$ or more.
   (a) For bending about a major axis treat as (1b).
   (b) For bending about a minor axis treat as (1c).
3. Biaxial bending
   The comments given in (c) for braced columns still apply, but the total moments about each axis will be at the top and bottom of the column.

Mention has been made of biaxial bending and also the reduction factor $K$, but as both of these items need the use of design charts, they will be described after consideration of design of column sections.

### 11.4.4 Design of column sections

Once the axial load and appropriate moment have been calculated the necessary reinforcement can be calculated by using: (a) design charts, for symmetrically reinforced rectangular columns; and (b) strain compatibility, for non-rectangular columns with unsymmetrical reinforcement.

#### (a) DESIGN CHARTS

These have been prepared using the rectangular parabolic stress block for concrete, and the bilinear stress–strain curve for reinforcement, as for beams. The charts for rectangular columns are in Part 3 of the Code and a typical chart is shown in Fig. 11.4.

Each chart is for a particular grade of concrete, a particular characteristic strength of reinforcement and a particular $d/h$ ratio (i.e. the positioning of the reinforcement). If one knows $N/bh$ and $M/bh^2$ the area of reinforcement can be found from the appropriate chart.

It should be noted that $A_{sc}$ is the total area of reinforcement and this is divided equally between the faces parallel to the axis of bending and remote from the axis of bending. Any reinforcement in the depth of the section is not taken into account.

The 'kink' in the chart corresponds to the one in the stress–strain curve for steel (Fig. 11.5). At point $A$, the reinforcement will have reached its maximum design stress. So, along the design curve above point $A$, the concrete will have reached its maximum design strain, but the stress in the reinforcement will vary from compression to tension. Below point $A$, the stress in the reinforcement will be constant at its maximum design stress.

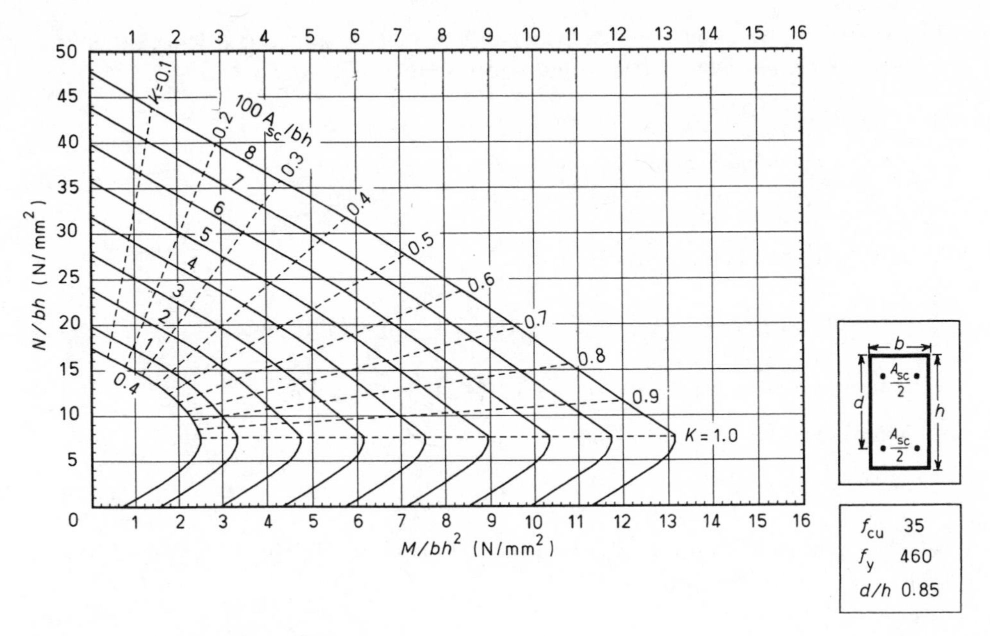


FIG. 11.4 Rectangular columns.

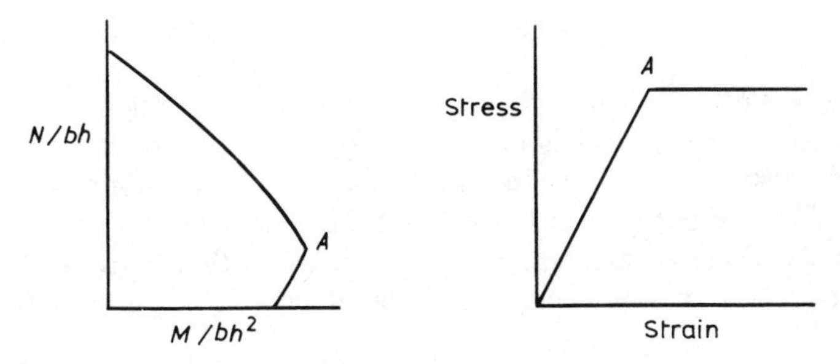


FIG. 11.5 Relationship of 'kink' on design chart to stress–strain curve.

It should also be noted that in preparation of the design charts, the gross area of the concrete has been used at all times. No reduction has been made for the area of reinforcement.

An important factor in reading the charts at small percentages of reinforcement is to appreciate that the smallest percentage line is for $100A_{sc}/bh = 0.4$.

## EXAMPLE 11.2

In a braced frame the spans of the beams are 8.0 m, carrying uniformly distributed load, and at ultimate limit state produce a maximum difference in fixed end moments, $M_{es}$, of 232 kN m. The axial load at the level to be considered is 1200 kN. The columns are 300 mm square and have the same stiffness above and below, but the beams are twice as stiff as the columns. Using Grade 35 concrete and $f_y = 460$ calculate the area of reinforcement required. Assume the columns are short.

(a) As we meet the requirements for a short braced column carrying a symmetrical arrangement of beams, try equation (39) of the Code.

$N=0.35f_{cu}A_c+0.67A_{sc}f_y$

$1200\times10^3=0.35\times35(300\times300-A_{sc})+0.67A_{sc}\times460.$
Hence $A_{sc}=329$ mm$^2$ but not less than 0.4% = 360 mm$^2$.

(b) Alternatively, calculate the moment:

$$\text{Moment in column}=\frac{1}{1+1+\frac{1}{2}(2+2)}\times232=58 \text{ kN m}.$$

Minimum design moment $=0.05\times1200\times0.3=18$ kN m.

$N/bh=(1200\times10^3)/(300\times300)=13.3$

$M/bh^2=(58\times10^6)/(300\times300^2)=2.1.$

Assume cover is 30 mm (i.e. 20 mm to 10$\phi$ links) and 20$\phi$ bars, then $d=260$ and $d/h=0.87$. Use 0.85 chart.

From chart

$100A_{sc}/bh=0.8,$

so $A_{sc}=720$ mm$^2$.

This shows a considerable increase over the previous calculation, and is due to the difference between the fixed end moments for loaded and unloaded spans. So, although we appear to satisfy the requirements for a simple formula, a column such as we have just calculated would not have sufficient reinforcement.

If the calculated design moment had been 36 kN m, that is twice the minimum moment, the amount of reinforcement from the design chart would be 0.4%.

---

### 11.4.5 Adjustments to additional moments in slender columns

In using the design chart for the last example, it will have been noticed that lines have been drawn on the chart representing $K=0.1$ to 1.0 in intervals of 0.1. In the design of slender columns these lines can play a very important part.

This factor $K$ was defined at the beginning of the section on slender columns and although it can be calculated for any section, it is unlikely to be used with other than symmetrically reinforced columns – hence the lines on the charts. It cannot be determined until the column reinforcement has been provided, either actual or assumed.

'Actual' means that one knows the reinforcement one is putting in. 'Assumed' means that one is carrying out an iterative procedure to determine the area of reinforcement needed.

In the latter case the procedure is as follows:

1. Calculate the additional moment assuming $K=1.0$;
2. Calculate total design moment;

3. Calculate $N/bh$, $M/bh^2$ and from appropriate chart determine $K$;
4. Use $K$ to calculate amended additional moment;
5. Repeat stages (2) and (3);
6. If new $K$ is virtually the same as before the process is finished; read off $100A_{sc}/bh$;
7. If new $K$ is substantially different go to stage (4), bearing in mind that you modify the basic additional moment, not an amended one;
8. Repeat the process until the condition in stage (6) is reached. Note: do not use a total design moment of less than $Ne_{min}$.

The following example illustrates the procedure and also what happens when one actually knows the reinforcement in a column. Bearing in mind that the vertical axis of the design chart gives the values of $N_{uz}/bh$ (i.e. $M/bh^2=0$) it should be remembered that the charts do not make a reduction in the concrete area for the reinforcement whereas the equation does. The differences are not significant and should not cause any problems, but for the purist the equation will give the more correct answer.

## EXAMPLE 11.3 Design of slender braced column

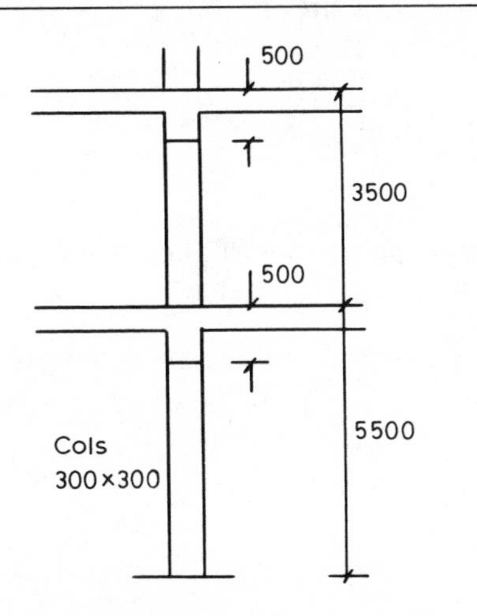


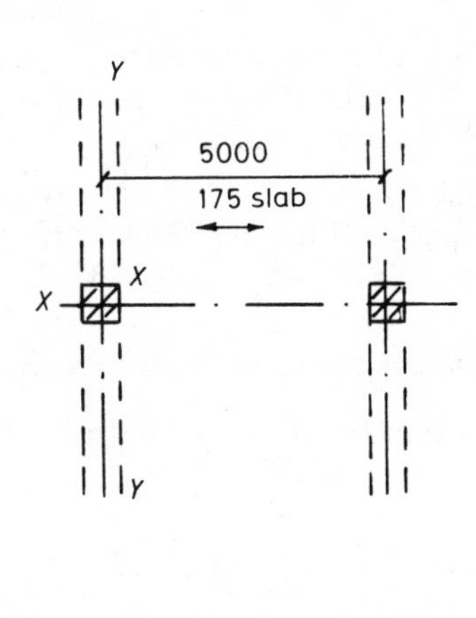


Concrete Grade C35, steel Grade 460.
Minimum cover to 10$\phi$ links = 20 mm.
Base not designed to resist moment.

### Foundation to first floor

| | Load case 1 | Load case 2 |
|---|---|---|
| Load $N$ | 2146 kN | 2006 kN |
| Moment at first floor (about $X$–$X$) | 21 kN m | 34 kN m |
| Moment at base | 0 | 0 |

Bending about $X$–$X$ axis
End conditions: top – 1, bottom – 3.
From Table 3.21 of the Code, $\beta=0.9$, $l_0=5000$,
so $l_{ex}=4500$ and $l_{ex}/h=4500/300=15$.

Bending about $Y$–$Y$ axis
End conditions: top – 2, bottom – 3.
From Table 3.21 of the Code, $\beta = 0.95$, $l_0 = 5325$,
so $l_{ey} = 5059$ and $l_{ey}/b = 5059/300 = 16.86$.

Column is therefore slender.

## Load case 1

Column will be designed for bending about $X - X$ axis, which will be referred to as major axis.

$l_e = 4500$, $b' = 300$; assume $K = 1$.

$$M_{add} = \frac{Nh}{2000}\left(\frac{l_e}{b'}\right)^2 = \frac{2146 \times 0.3}{2000}\left(\frac{4500}{300}\right)^2 = 72.4 \text{ kN m.}$$

Minimum eccentricity, $e_{min} = 0.05 \times 300 = 15$ mm.

From clause 3.8.3.2

1. $M_2 = 21$ kN m
2. $M_i = 0.4M_1 + 0.6M_2 = 0 + 0.6 \times 21 = 12.6$ kN m
   $0.4M_2 = 8.4$ kN m, so $M_i = 12.6$
   $M_i + M_{add} = 12.6 + 72.4 = 85$ kN m
3. $M_1 + M_{add}/2 = 0 + 0 = 0$ (Note: one cannot apply a moment at a pinned end)
4. $M_{min} = e_{min}N = 0.015 \times 2146 = 32.2$ kN m.

Design moment $M_t = 85$ kN m (from (2)).

$N/bh = (2146 \times 10^3)/(300 \times 300) = 23.8$
$M/bh^2 = (85 \times 10^6)/300^3 = 3.15$.

Assuming $d = 300 - 30 - 13 = 257$,

$d/h = 257/300 = 0.856$; use $d/h = 0.85$.

From chart, $100A_{sc}/bh = 4.2$ (3780 mm$^2$) and $K$ factor $= 0.35$.

Modify design moment to $12.6 + 0.35 \times 72.4 = 37.9$.

$M_t$ greater than $M_{min}$.

$N/bh = 23.8$ (as before), $M/bh^2 = 1.4$.

From chart $100A_{sc}/bh = 3$ (2700 mm$^2$) and $K = 0.2$.

Modify design moment to $12.6 + 0.2 \times 72.4 = 27.1$ kN m.

Note: this is less than $Ne_{min}$, so use minimum moment of 32.2 kN m.

$N/bh = 23.8$ (as before), $M/bh^2 = 1.2$.

From chart $100A_{sc}/bh = 2.9$ (2610 mm$^2$).

Use $4/32\phi$ (3220 mm$^2$).

We can now check some values.

$100A_{sc}/bh=3.6$, $d=250$ so $d/h=0.83$.

(a) From the design chart for $d/h=0.85$, using the values of $N/bh=23.8$ and $100A_{sc}/bh=3.6$ we can interpolate a value for $K$ as 0.28.

(b) By calculation

$$N_{uz}=\{0.45\times 35(300^2-3220)+0.87\times 460\times 3220\}\times 10^{-3}$$
$$=2655 \text{ kN}$$

$$N_{bal}=0.25\times 35\times 300\times 250\times 10^{-3}$$
$$=656 \text{ kN}$$

$$K=(2655-2146)/(2655-656)=0.255.$$

---

### 11.4.6 Biaxial bending

The Code now gives a method whereby biaxial bending is transformed into single-axis bending by increasing the moment about one axis. The definitions of the terms in clause 3.8.4.5 are a bit misleading and reference should be made to Figure 3.22 of the Code, which is reproduced here.

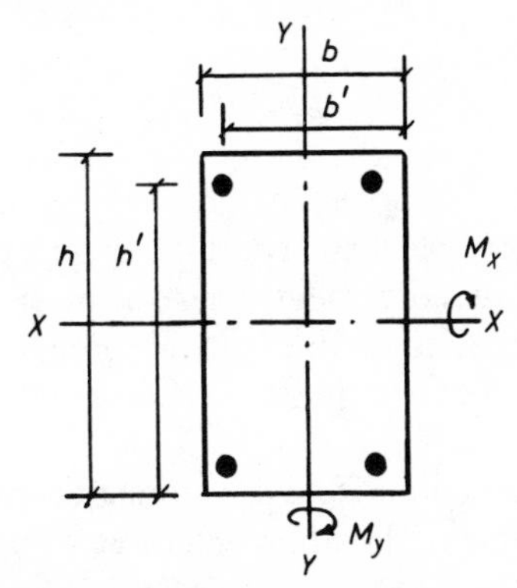


(a) If $M_x/h' \geqslant M_y/b'$ then $M_x$ is enhanced so that

$$M_x' = M_x + \beta(h'/b')M_y \qquad (11.8)$$

This is equation 40 of the Code.

(b) If $M_x/h' < M_y/b'$ then $M_y$ is enhanced so that

$$M_y' = M_y + \beta(b'/h')M_x \qquad (11.9)$$

This is equation 39 of the Code.

$\beta$ is a coefficient obtained from Table 3.24 of the Code. It can be written as

$$\beta = 1 - \tfrac{7}{6}\times N/bhf_{cu}.$$

The following example illustrates the procedure.

## EXAMPLE 11.4

A short column 250 mm square is required to carry an ultimate axial load of 600 kN together with moments of 60 kN m about one axis and 40 kN m about the other axis. Concrete is Grade C40, reinforcement Grade 460, and the cover to the main bars not less than 30.

$N=600$ kN, $M_x=60$ kN m, $M_y=40$ kN m.

Assume 250$\phi$ bars, then $b'=h'=250-30-13=207$.

$M_x/h'=60/207>M_y/b'=40/207$,

so $M_x'=M_x+\beta(h'/b')M_y$.

$N/bhf_{cu}=(600\times10^3)/(250\times250\times40)=0.24$.

From Table 3.24 of the Code, $\beta=0.72$.

$M_x'=60+0.72\times(207/207)\times40=88.8$ kN m.

$N/bh=(600\times10^3)/250^2=9.6$, $M/bh^2=(88.8\times10^6)/250^3=5.7$.
$d/h=207/250=0.828$, say 0.80.

From chart $100A_{sc}/bh=3.1$, so $A_{sc}=1938$ mm$^2$.

Use 4/25$\phi$ (1960 mm$^2$).

---

In the above example the reinforcement provided is exactly as illustrated in the diagram above – four bars, one in each corner. If this happens, then the calculation is complete, but if the area of reinforcement does not meet this requirement or, for a suitable arrangement of bars, three or four bars are provided in each face, it is not the end of the calculation. This is a very important criterion which is not made clear in the Code.

In CP110 it was common practice to find the moment capacities ($M_{ux}$ and $M_{uy}$) about each axis for a given arrangement of reinforcement by using the design charts. Where the reinforcement is distributed along the four sides this was usually done by ignoring the intermediate bars in the sides parallel to the plane of bending. Only the bars in the extreme edges at right angles to the plane of bending were used. This was because the design charts were prepared to do only this. The values for $M_{ux}$ and $M_{uy}$ were, therefore, very much underestimated. To find more accurate values for $M_{ux}$ and $M_{uy}$ means doing a rigorous analysis such as strain compatibility.

Although this method could still be used it is outside the scope of BS8110 and to avoid complications the following procedure has been found to be satisfactory. (Fuller details are given in *C & CA Development Report* No.2.)

Where there are three or more bars in a face, convert the column section into an equivalent four-bar column. This is done by concentrating all the reinforcement in a quadrant of the column into one bar at the centre of gravity of the original reinforcement. For example, an arrangement as shown in Fig. 11.6(a) becomes that shown in Fig. 11.6(b).

The bars do not all have to be the same size, but most columns will only have minor variations.

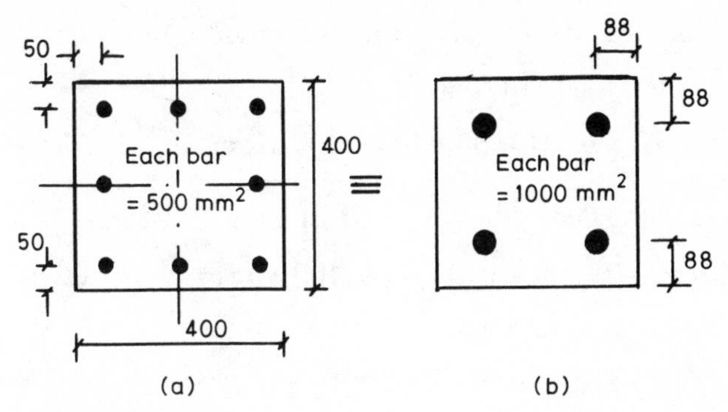


FIG. 11.6 Equivalent columns.

The following example involves the problems just discussed and illustrates a design procedure.

## EXAMPLE 11.5

A four-storey building has unbraced columns supporting beams in two directions at right angles. The columns are 400 mm square for the lower two storeys and 300 mm square for the upper two storeys. All beams are 500 deep by 300 wide and have spans of 6 m. The height from top of foundation to top of first floor is 5.5 m, and above this level the heights are 3.5 m from top of floor to top of floor. The foundation is a normal pad footing designed to resist moment. Design an internal column from foundation to first floor level for the following load case at ultimate limit state.

| | Moments about | | | |
|---|---|---|---|---|
| | *X–X* axis | | *Y–Y* axis | |
| Load | Top | Bottom | Top | Bottom |
| 3200 kN | 76 | 38 | 68 | 34 |

Concrete Grade C40, reinforcement Grade 460.
Minimum cover to main bars = 30 mm.

The end conditions of the column are both Type 1, so the effective height, $l_e = 1.2l_0$ (see Table 3.22 of the Code).

The clear height $l_0 = 5.5 - 0.5 = 5$ m.

$l_e/h = (1.2 \times 5)/0.4 = 15$, about both axes.

Column is slender and unbraced.

Note: if equations in Part 2 of the Code are used, it will be found that $l_e$ is approximately 6.5 m.

The minimum design moment $= 3200 \times 0.02 = 64$ kN m. The moments from analysis are both larger than this and additional moments will increase them, so the moments can be classed as significant about both axes and biaxial bending will be considered.

The additional moment about each axis which will be applied to the top is $N\beta_a Kh$.

$b'$ (the smaller column dimension) $= 0.4$, so $l_e/b' = 15$.

From Table 3.23 of the Code, $\beta_a = 0.11$, so, assuming $K = 1.0$,

$M_{add} = 3200 \times 0.11 \times 0.4 = 141$ kN m,

$M_x = 76 + 141 = 217$ kN m, $M_y = 68 + 141 = 209$ kN m.

From Figure 3.22 of the Code, $h' = b' = 350$ (assuming $32\phi$ bars).

$M_x/h' > M_y/b'$, $N/bhf_{cu} = (3200 \times 10^3)/(400 \times 400 \times 35) = 0.57$, so $\beta = 0.335$.

$M_x' = 217 + 0.335 \times 209 = 287$ kN m.

$N/bh = 20$, $M/bh^2 = 4.48$, $d/h = 0.85$, say.

From chart, $100A_{sc}/bh = 3.7$, $K = 0.53$.

Modify design moments to

$M_x = 76 + 0.53 \times 141 = 151$,

$M_y = 68 + 0.53 \times 141 = 143$,

$M_x' = 151 + 0.335 \times 143 = 199$ kN m.

$N/bh = 20$. $M/bh^2 = 3.10$.

From chart, $100A_{sc}/bh = 2.6$, $K = 0.43$.

Modify design moments to

$M_x = 76 + 0.43 \times 141 = 137$

$M_y = 68 + 0.43 \times 141 = 129$

$M_x' = 137 + 0.335 \times 129 = 180$

$N/bh = 20$, $M/bh^2 = 2.82$.

From chart $100A_{sc}/bh = 2.5$, $K = 0.4$.

$A_{sc} = 4000$ mm$^2$.

This requires $4/32 + 2/25$ (4198 mm$^3$) i.e. $2/32 + 1/25$ each of the two opposite faces.
By using this arrangement we must convert it to an 'equivalent' column, but as the moments about the two axes are very similar it would appear logical to start with a symmetrical arrangement of eight bars, $4/32\phi$ in the corners and $4/25\phi$ in the middle of the sides.

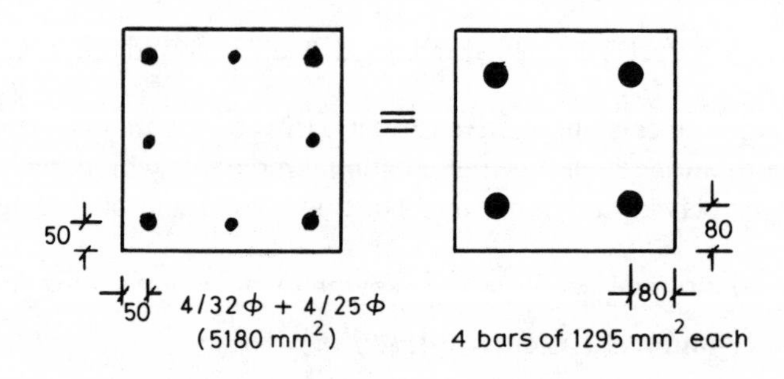

As we know the details of the column we can calculate $K$:

$$N_{uz}=0.45\times 40(400^2-5180)\times 10^{-3}+0.87\times 460\times 5180\times 10^{-3}$$
$$=4860 \text{ kN}$$

$$N_{bal}=0.25\times 40\times 400\times 320\times 10^{-3}$$
$$=1280 \text{ kN}$$

$$K=(N_{uz}-N)/(N_{uz}-N_{bal})=(4860-3200)/(4860-1280)=0.46.$$

We can find $K$ from the design chart:

$d/h=320/400=0.8$, $N/bh=20$,

$100A_{sc}/bh=(100\times 5180)/400^2=3.24$.

$K=0.48$.

Using the calculated value

$M_x=76+0.46+141=141$ kN m

$M_y=68+0.46\times 141=133$ kN m.

$\beta=0.335$.

So
$M_x'=141+0.335\times 133=186$ kN m.

From design chart, moment capacity $=3.6bh^2$
$=230$ kN m.

Section is therefore satisfactory.

---

### 11.4.7 Shear

Clause 3.8.4.6 has been introduced because there was not a method for checking shear in columns in CP110. In the majority of cases columns do not have shear problems, but in the case of high moment and small axial load there could be. The design shear strength is checked in accordance with 3.4.5.12, shear and axial compression, and is in the beam section.

For rectangular sections no check is necessary when $M/N$ is less than $0.75h$, but this is provided the shear stress does not exceed $0.8\sqrt{f_{cu}}$ or 5 N/mm$^2$, whichever is the lesser. One must still find the shear stress, whatever the value of $M/N$.

So one must calculate $V$ at the top and bottom of the column. Unless there are imposed loads up the height of the column, the shear at the top and bottom of a storey will be the sum of the moments divided by the storey height. In calculating the shear stress $v$ it would appear that the simplified approach of calculating this at $d$ from the face of the support does not apply. The enhanced shear strength $v_c$, as described in 3.4.5.8, can be applied, but as one will have a constant shear stress down the column the critical section will be at $2d$ from the face of the beam, using $v_c$, or at the face of the beam, using the maximum allowable shear stress.

The following example illustrates the procedure.

## EXAMPLE 11.6

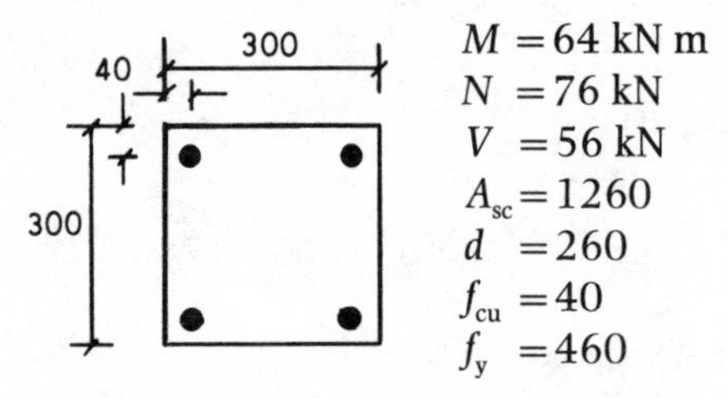


$M/N = (64 \times 10^6)/(76 \times 10^3) = 842$ mm, i.e. $> 0.75 \times 300$.

Maximum shear stress $= (56 \times 10^3)/(300 \times 260) = 0.72$ N/mm$^2$.

To obtain $v_c$, $100A_s/bd = (100 \times 630)/(300 \times 260) = 0.8$.

From formula below Table 3.9 of the Code

$v_c = (0.79/1.25)\,(0.8)^{1/3}(400/260)^{1/4} = 0.65$ N/mm$^2$.

$A_c = 300^2 - 1260 = 88\,740$.

$Vd/M = (56 \times 10^3 \times 260)/(64 \times 10^6) = 0.2275$, i.e. $< 1$.

$v_c' = 0.65 + 0.75 \times (76 \times 10^3/88\,740) \times 0.2275 = 0.8$ N/mm$^2$.

Satisfactory without shear reinforcement.

---

### 11.4.8 Areas of reinforcement

The minimum percentage of reinforcement in a column is now defined as 0.4% of the column size. The maximum percentages are 6% and 8% for vertically cast and horizontally cast columns respectively with 10% at laps in both. The links should be at least one quarter of the size of the largest compression bar or 6 mm, whichever is the greater, and at a maximum spacing of 12 times the smallest compression bar. This requirement may change at laps, which are discussed in the next section. Every corner bar, and each alternate bar, should be supported by a link passing round the bar and having an included angle of not more than 135°. No bar should be further than 150 mm from a restrained bar. So, if you have a 500 mm column face with $3/32\phi$ bars and 50 mm to the centre of the reinforcement, all three bars need holding with links.

### 11.4.9 Lapping of bars

For bars in compression, the lap length required is 1.25 times the anchorage length. In the majority of columns it can be assumed that the compressive stress in the reinforcement is at its maximum design stress. The compression lap length in terms of bar size can therefore be obtained from Table 3.29 of the Code, depending on the grade of concrete and the grade of reinforcement.

For bars in tension, the lap lengths are a factor times the anchorage length, this factor depending on the provisions as set out in clause 3.12.8.13. The provision in this clause states that if (a) the cover to either face is less than twice the size of the reinforcement, or (b) the clear distance between adjacent laps is less than 75 mm or six times the bar size, whichever is greater, then the factor is 1.4. It should be noted that even if both conditions apply the factor is still only 1.4. In many columns, particularly where larger diameter bars are used, it will be found that the 1.4 factor will apply. Assuming that it does, it can be found that up to a stress of approximately 290 N/mm$^2$ (for $f_y = 460$) in the tension steel, the compression lap will provide an adequate tension lap.

If a line is drawn on the design chart for a stress of 290 N/mm$^2$, then any value read from the chart which is above this line will require a full compression lap.

Below this line the designer has the choice of interpolation to assess the tension stress and hence determine a tension lap length, or assume a full tension lap including the 1.4 factor. The latter choice is simple to apply, does not need any calculation, and will not underestimate the lap length.

However, it can increase the length by up to 40%, and in larger diameter bars may be considered uneconomical. Although it is better to draw stress lines on the chart if a tension stress is required, it will be found that a *K*-factor line of 0.87 is very close to the stress line for 290 N/mm$^2$. The *K*-factor line of 1.0 is where the stress is 400 N/mm$^2$, the maximum design stress for Grade 460 reinforcement. Linear interpolation between these values will give a good estimate of the tensile stress in the reinforcement.

Having determined the lap lengths, there are two further points to be considered in these areas. First, clause 3.12.8.14 restricts the sum of the reinforcement sizes in a particular layer to 40% of the breadth of the section at that level. In a 300 mm wide column the maximum number of bars in a layer would be 2/25 + 2/32, i.e. 4/25 bars lapping with 4/32 bars as shown.

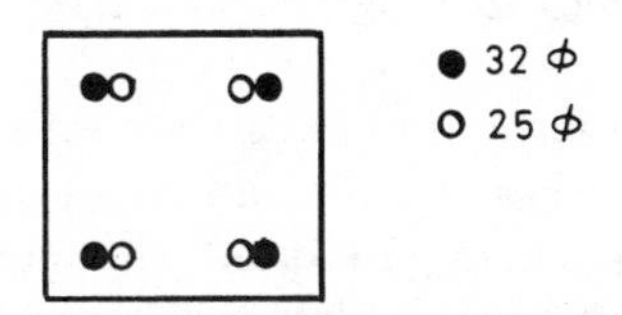


For lapping 4/32 bars with 4/32 bars in a 300 mm square column it will be necessary to choose one of the following arrangements:

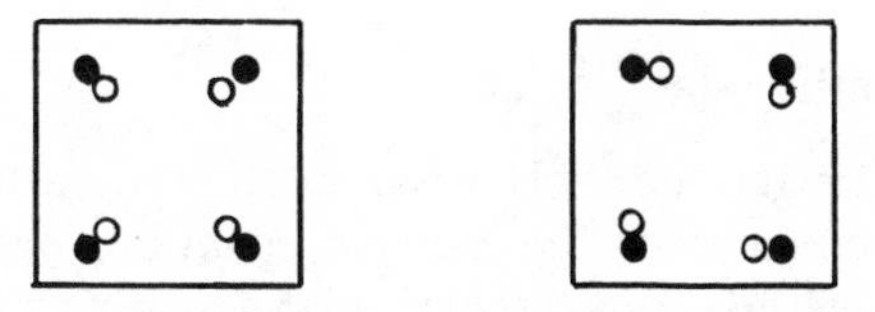

Secondly, clause 3.12.8.12 supersedes the normal requirements of clause 3.12.7.1 for size and spacing of links. Where both bars at a lap exceed size 20 and the cover is less than 1.5 times the smaller bar, the size of the links should be at least one quarter the size of the smaller bar and the spacing should not exceed 200 mm.

The normal requirement for the links in a compression zone is 6 mm or one quarter of the size of the larger bar, whichever is greater, and the maximum spacing of twelve times the size of the smallest bar.

So where a 32$\phi$ bar laps with a 40$\phi$ bar from below the arrangement in the length of column containing the 32$\phi$ bars would be:

over the lap length – 8$\phi$ links at 200 centres,
over the remainder – 8$\phi$ links at 350 centres.

If the position were reversed, so that a 40$\phi$ bar laps with a 32$\phi$ bar from below, the arrangement over the lap length would be the same, but for the remainder it would require 10$\phi$ links at 450 centres.

# 12 SOLID SLABS

In general the moments and shear forces will be found as for beams. They may also be found from an elastic analysis such as those of Pigeaud and Westergaard, or, alternatively, by Johansen's yield line method or Hillerborg's strip method provided the ratio between support and span moments are similar to those obtained by use of elastic theory. Although the analysis and design will be carried out at ultimate limit state it will be the serviceability limit state of deflection that will be the important criterion. The serviceability limit state of cracking will be controlled by bar spacing rules. In the estimation of the slab thickness the designer has to consider a span/effective depth ratio and the cover.

For the basic span/effective depth ratio, Table 3.10 of the Code (type of span and support conditions) will be used as for beams. The modification factor for tension reinforcement will be obtained from Table 3.11 of the Code and as the value of $M/bd^2$ will be less than in rectangular beams the factor will be higher. But it must be remembered that with a high loading the value of $M/bd^2$ will increase and the modification factor decrease.

Cover is also an important factor and as can be seen from Tables 3.4 and 3.5 of the Code the minimum cover will be 20 mm unless an aggregate of not greater than 15 mm is used. Any attempt at giving ratios of span to overall depth would have to be conservative, and a much better approach is to use span/effective depth ratios as given in Table 3 of the *Manual for the Design of Reinforced Concrete Building Structures* published by the Institution of Structural Engineers.

## 12.1 One-way spanning slabs

Where the length of the longer side of a slab is more than twice the length of the shorter side, it is customary to consider the slab as spanning one way only. For the arrangement of loading on slabs it was recognized that two-way spanning slabs and the empirical rules for flat slabs used the single load case of maximum design load on all spans. It was only one-way spanning slabs where patterned loading was required. Following work carried out by Dr Beeby it was agreed that the single load case could be used for one-way spanning slabs providing the following conditions are met:

1. The area of each bay exceeds 30 m$^2$ (in this context a bay is defined by a strip across the full width of the structure as illustrated in Fig. 3.7 of the Code and reproduced here);

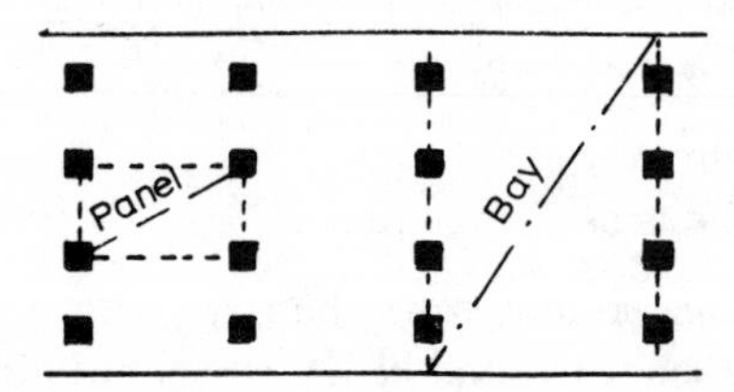

2. The ratio of the characteristic imposed load to the characteristic dead load does not exceed 1.25;
3. The characteristic imposed load does not exceed 5 kN/m$^2$ excluding partitions.

Where an analysis is carried out for the single load case, e.g. for unequal spans, the support moments should be reduced by 20% with a consequential increase in span moments. However, it is not quite as simple as that. The resulting bending moment diagram should not come within the 70% elastic bending moment diagram. As with beams this will control the length of top bars at the supports (see Fig. 12.1).

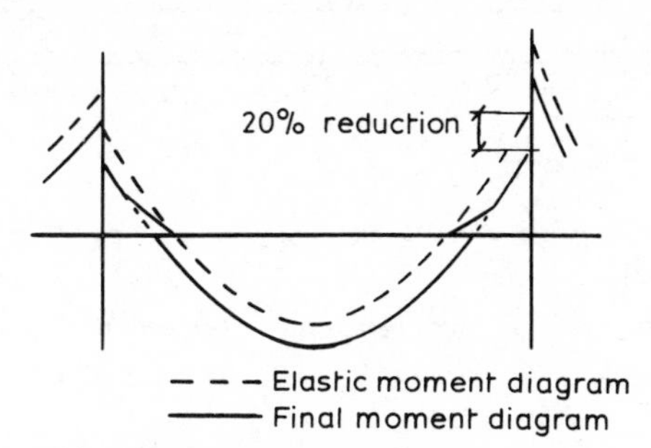


FIG. 12.1 Development of bending moment envelope.

This 20% reduction of support moments does not apply to cantilevers. Also, in the span adjacent to a cantilever of length exceeding one-third of the span of the slab, the case of the cantilever loaded with the span unloaded should also be considered.

Where the conditions stated above are met, and the spans are approximately equal, the moments and shears may be calculated using the coefficients given in Table 3.13 of the Code. 'Approximately equal' is generally taken to mean within 15% of the longest span, and although not stated the number of spans should be three or more. The coefficients in Table 3.13 of the Code include the 20% reduction mentioned above, so further redistribution is not allowed. A bending moment diagram is not required and the curtailment of reinforcement is carried out in accordance with clause 3.12.10. The relevant cases can be illustrated diagrammatically as follows:

(a) SIMPLY SUPPORTED END (FIG. 12.2)

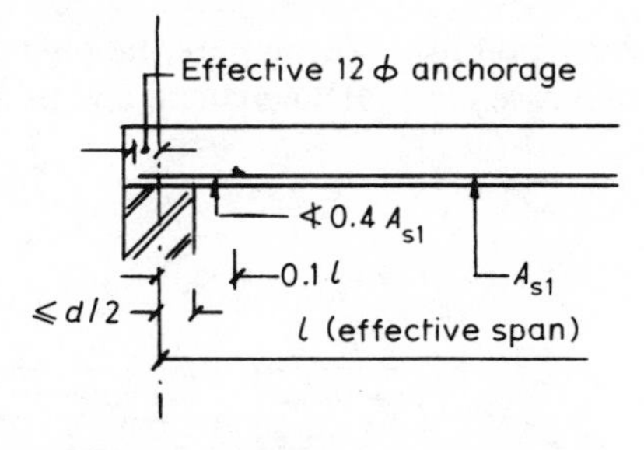


FIG. 12.2 Simply supported end.

Where there is no restraint to the free rotation at the end tensile stresses will occur in the bottom of the slab only, so bottom steel only is required.

The dimensions for curtailing the bottom reinforcement are related to the effective span. This is the distance between centre lines of supports or the clear span plus an effective depth of the slab, whichever is the lesser. So the effective line of support is never greater than $d/2$ from the face of the support.

### (b) RESTRAINED ENDS (FIG. 12.3)

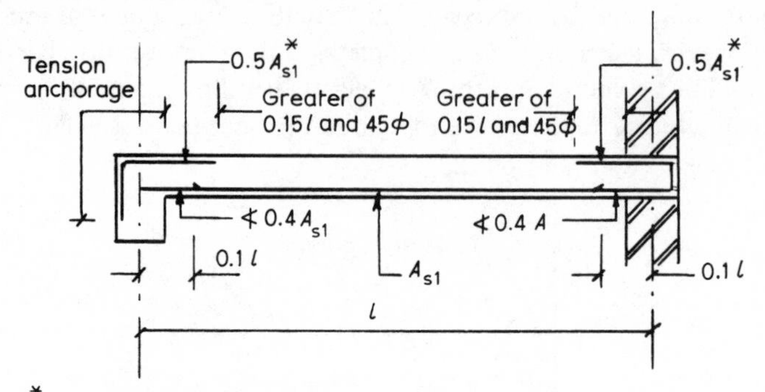


FIG. 12.3 Restrained ends.

If there is restraint at the ends, such as a monolithic connection between the slab and its supporting beams, or if the slab is built into a wall, provision should be made for the negative moment which may arise. In assessing the moments, a simple support such as in (a) will have been assumed. The Code suggests reinforcement in the top of the slab equal to half that provided at midspan, but not less than the minimum area of tension steel. It should have a full effective tensile anchorage into the support and extend not less than the greater of $0.15l$ and $45\phi$ into the span. All of these dimensions are from the face of the support.

For the bottom reinforcement the continuing reinforcement into the support need not go any further than the effective line of the support. Where restraint is provided by adjoining beams the top reinforcement can be provided by ⌐ shaped bars as shown on the left-hand side of Fig. 12.3. Where the restraint is due to the slab being held down by walls, this reinforcement can be provided as already described if the slab is of sufficient thickness. If not, the bottom bars passing through the support can be combined with the top reinforcement by utilizing bars of hairpin shape, as indicated on the right-hand side of the diagram. For shear resistance the Code gives the options of:

1. detailing as shown above and using the area of the top steel;
2. detailing the bottom reinforcement as shown in (a) and using the area of continuing bottom reinforcement.

### (c) CONTINUOUS MEMBERS (FIG. 12.4)

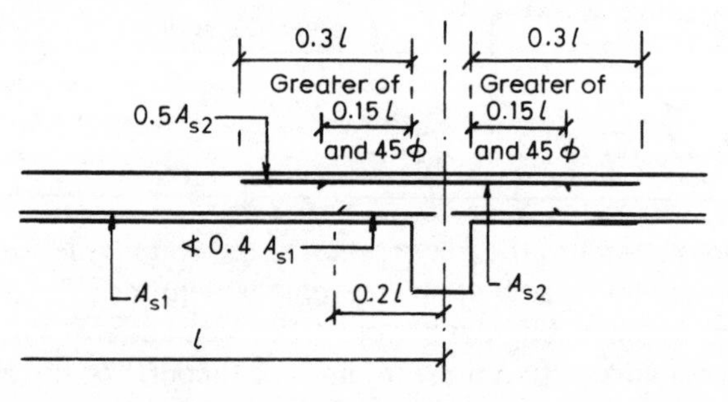


FIG. 12.4 Continuous members.

Where we have slabs continuing over supports, whether monolithic with the support (such as a beam) or not (such as a wall), the requirements are as shown.

For continuous spans it is normal to take the effective span $l$ as the distance centre to centre of supports unless they are very wide. In this case the line of effective support could be taken as $d/2$ from the face.

(d) CANTILEVER (FIG. 12.5)

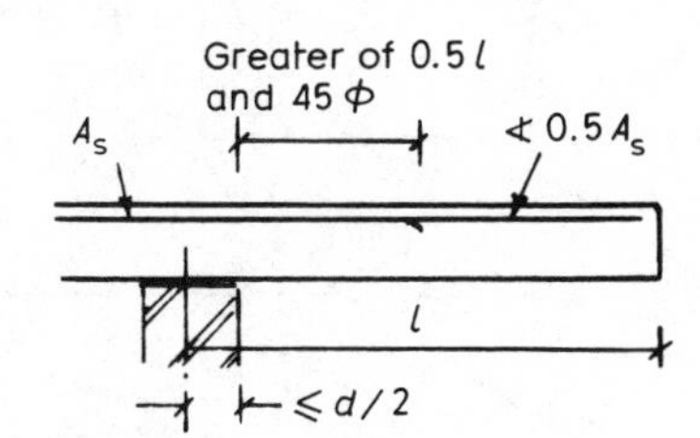


FIG. 12.5 Cantilever slabs.

Remember that the top steel should continue into the adjacent span to at least the point of contraflexure.

It should be noted that the above detailing rules apply only to where the requirements of clause 3.5.2.4 are met. In other cases, such as where the ratio of imposed load to dead load is greater than 1.25, an analysis will have to be carried out using the loading patterns as for beams. The elastic moments from this analysis can now be redistributed as for beams, i.e. a 30% reduction can be carried out if desired.

In calculating the allowable span/effective depth ratio in one-way spanning slabs, only the reinforcement at the centre of the span in the direction of span need be considered.

For crack control the Code gives bar spacing rules in clause 3.12.11.2.7. These are quite specific, but a flow chart as indicated in Fig. 12.6 can be useful. In slabs less than 200 mm thick with $f_y$ less than 460, the clear distance should not exceed three times the effective depth. The minimum area of main tension reinforcement should not be less than 0.13% $bh$ for high-yield reinforcement, or 0.24% $bh$ for mild steel reinforcement. In solid slabs which are designed per metre width, $b$ will be 1000 mm, so the minimum area will be 1.3 $h$ mm$^2$/m with high yield and 2.4 $h$ mm$^2$/m with mild steel.

In one-way spanning slabs the minimum area of secondary reinforcement (distribution steel) is the same as for main tension reinforcement. In either case the distance and the same bar spacing rules apply.

Where a slab forms the top flange of a Tee or Ell beam, the reinforcement provided in the top surface should extend across the full effective width of the flange (see Fig. 12.7). The amount should be not less than 0.15% of the longitudinal cross-sectional area of the flange, either in high yield or mild steel. In detailing this has to be watched very carefully when curtailing bars. If we have a continuous-span beam the effective width of the flange $b_e$ can be taken as $(b_w + 0.14l)$ for a Tee beam and $(b_w + 0.07l)$ for an Ell beam where $l$ is the effective span of the beam.

Where the slab is spanning at right angles to the beam, some of the reinforcement will be provided by the flexural reinforcement. Where the slab is spanning parallel to the beam the reinforcement will be provided for this requirement alone.

With a concentrated load on a simply supported span the Code suggests that the load

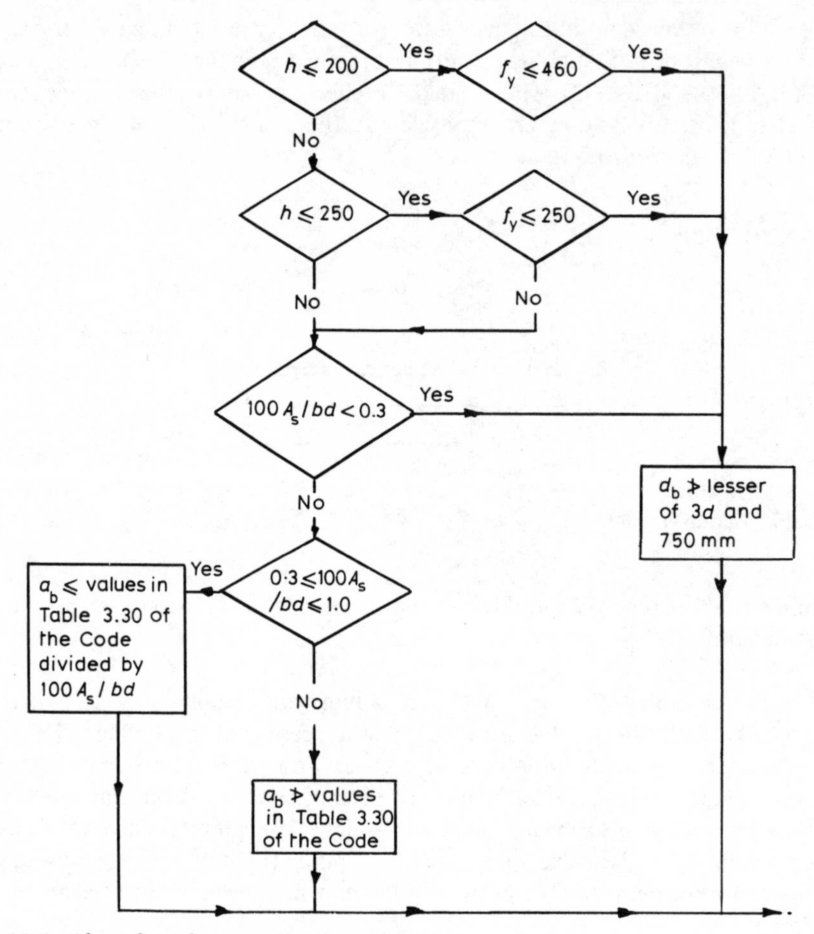


FIG. 12.6 Flow chart for spacing of bars in slab.

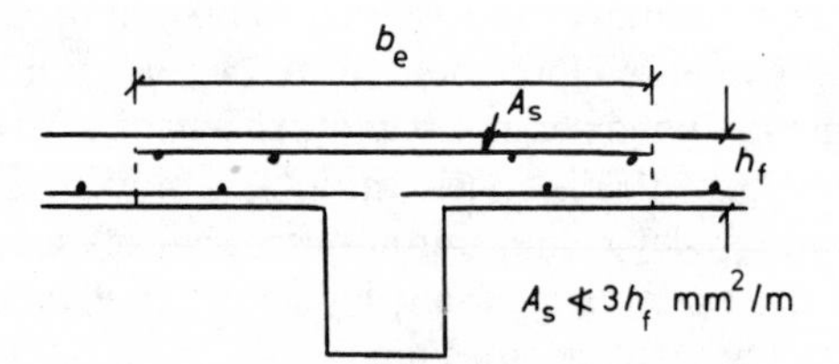


FIG. 12.7 Reinforcement in top of flanged beams.

can be spread over a distance of $2.4x(1-x/l)$ plus the load width in a direction at right angles to the span, where $x$ is the distance from the nearer support to the centre of the load. Where the load is near an unsupported edge the effective width should not exceed $1.2x(1-x/l)$ plus the distance to the unsupported edge (see Fig. 12.8).

If the load is at midspan then the distance each side of the load is $0.3l$ unless this distance is greater than the distance to the unsupported edge, in which case the smaller dimension is taken.

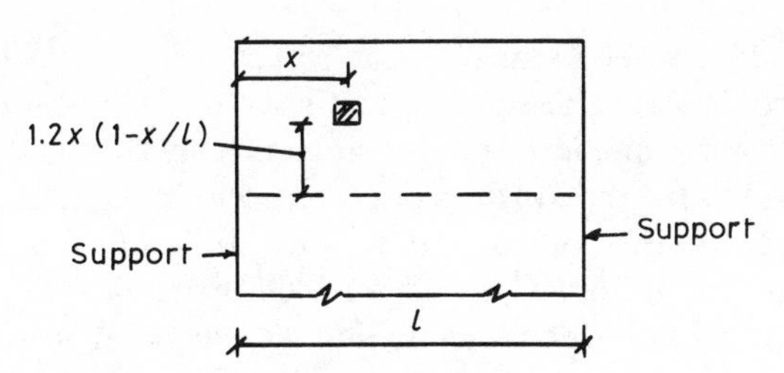


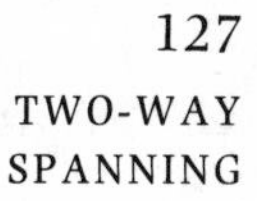

FIG. 12.8 Concentrated load on simply supported edge.

## 12.2 Two-way spanning

When slabs are carrying uniformly distributed loads and are reinforced in two directions at right angles we have the cases (a) where provision is not made to prevent the corners from lifting, i.e. simply supported, and (b) where provision is made to prevent the corners from lifting, i.e. restrained slabs.

### 12.2.1 Simply supported

Clause 3.5.3.3 gives equations (10) and (11) for the maximum moments per unit width as a factor times the total ultimate load per unit area $n$ ($n = 1.4g_k + 1.6q_k$), times the length of the shorter side $l_x$ squared, i.e.

short span: $$m_{sx} = \alpha_{sx} n\, l_x^2$$

long span: $$m_{sy} = \alpha_{sy} n\, l_x^2$$

where $\alpha_{sx}$ and $\alpha_{sy}$ are coefficients obtained from Table 3.14 of the Code which depend on the ratio of the longer side to the shorter side.

The detailing of the reinforcement is the same as for one-way spanning slabs (see Fig. 12.2).

### 12.2.2 Restrained slabs

In slabs where the corners are prevented from lifting, and provision for torsion is made, even though all four edges may be discontinuous, the moments per unit width are obtained from equations (14) and (15) of the Code. These equations are similar to those for simply supported slabs but the factors are now $\beta_{sx}$ and $\beta_{sy}$. These factors are obtained from Table 3.15 of the Code and give negative moments at continuous edges together with positive moments at midspan. The coefficient $\beta_{sx}$ varies depending on the ratio of the spans, but $\beta_{sy}$ does not.

It is important to note the conditions and rules to be observed when using these equations. These are given in clause 3.5.3.5 and are as follows.

(a) CONDITIONS

1. The characteristic dead and imposed loads on adjacent panels are approximately the same as on the panel being considered. It would appear that there is no restriction on the ratio of dead to imposed loads. This is probably because the coefficients have been obtained from a yield line analysis using the maximum design load on an individual span.

2. The span of adjacent panels in the direction perpendicular to the line of the common support is approximately the same as the span of the panel considered in that direction. This means, in effect, that rectangular panels cannot be mixed up with square panels unless the ratio of the sides comes within the term 'approximately'. It should be noted that the adjustments which can be made under clause 3.5.3.6 where support moments from adjacent panels differ significantly does not imply that adjacent spans can be outside the condition we are now considering.

An important condition which has not been stated is the edge support to these slab panels. The supports should be beams which are much stiffer in flexure than the slabs themselves. This means that the beams should project well below the slabs. A section of the slab between columns should not be considered as a hypothetical beam. If a computer analysis, such as a grillage analysis, is carried out using this hypothetical beam, it will be found that the distribution of moments is very similar to flat-slab behaviour.

### (b) RULES

1. Each panel is considered as being divided into middle strips and edge strips in each direction. The middle strip is the middle three-quarters of the width, and each edge strip is one-eighth of the width.
2. The maximum design moments calculated from equations (14) and (15) of the Code apply to the middle strips only. Redistribution cannot be carried out.
3. Reinforcement in the middle strips should be curtailed in accordance with the simplified rules in clause 3.12.10. These have been discussed earlier in this chapter and are shown in Figs 12.3 and 12.4.
4. Reinforcement in an edge strip need not exceed the minimum area of tension reinforcement, together with the recommendations given for torsion.

It should be noted that this requirement says '. . . need not exceed . . .'. In many cases the detailing of two-way spanning slabs is complicated enough, and the detailer will probably continue the middle strip reinforcement into the edge strips.

Requirements (5), (6) and (7) for torsion are set out very clearly in the Code and are not repeated here. They depend on the conditions of adjacent edges at a corner. Both edges discontinuous: full torsion reinforcement; one edge discontinuous: half the full torsion reinforcement; both edges continuous: no torsion reinforcement. Internal panels will not require torsion reinforcement, whereas external panels will.

Even when complying with conditions (1) and (2) it is often found that the support moment from a panel differs from the support moment from an adjacent panel. In the past designers have usually taken the larger moment and calculated the area of reinforcement required for that moment.

Clause 3.5.3.6 of the Code now suggests a method for adjusting these moments *if they differ significantly*. What is meant by 'significantly' is not stated, and the method described involves using coefficients obtained from a yield line analysis and distribution of moments as would be done using an elastic analysis. The moments at the support obtained from equation (14) or (15) are treated as fixed-end moments and moment distribution is then carried out according to the relative stiffnesses of the adjacent panels. If the resulting moment is significantly greater than the original value, additional steps have to be taken to ensure the top steel is extended further into the span.

## EXAMPLE 12.1

The floor slab of a building shown below is supported by beams at each grid line. The construction is monolithic and the panels are continuous at the interior supports. Floor finishes are 1.0 kN/m$^2$ and the imposed load is 5.0 kN/m$^2$. The exposure conditions are mild, period of fire resistance is 1 hour. Concrete is Grade C35 and the steel is Grade 460.

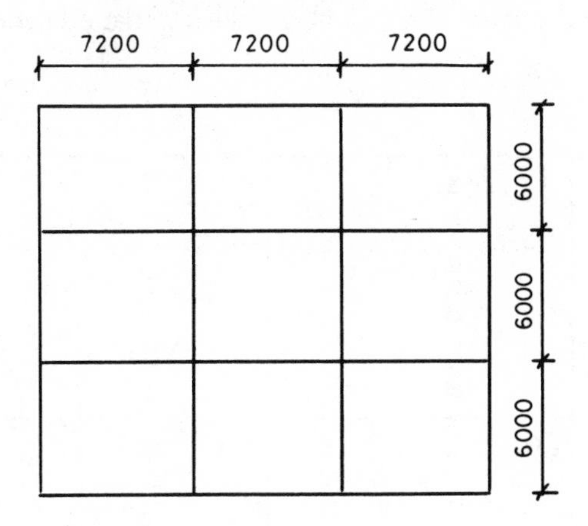


Cover. Mild exposure: 20 mm.
Fire resistance: 20 mm.

Assuming a span effective depth ratio of 36,

$d = 6000/36 = 167.$

Using 12$\phi$ bars, $h = 167 + 26 = 193$, say 200 mm.

### Loads

Dead. Slab $= 0.2 \times 24 = 4.8$ kN/m$^2$
Finishes 1.0 kN/m$^2$
5.8 kN/m$^2$.

Imposed $= 5.0$ kN/m$^2$.

Design ultimate load $= 5.8 \times 1.4 + 5 \times 1.6$
$= 16.1$ kN/m$^2$.

$l_y/l_x = 7.2/6.0 = 1.2.$

It can be seen very quickly that all of the panels come within the first four categories of Table 3.15 of the Code. The maximum span moment will be in the corner panel and is $m_{Sx} = 0.047 \times 16.1 \times 6^2 = 27.2$ kN m/m.

If the bars spanning in the shorter direction are in the bottom layer, $d = 200 - 26 = 174$ mm.

$M/bd^2 = (27.2 \times 10^6)/(1000 \times 174^2) = 0.90.$

If we assume the reinforcement provided is the same as required, then from Table 3.11 of the Code the modification factor is 1.43.

Allowable $l/d = 26 \times 1.43 = 37.2$.

Actual $l/d = 6000/174 = 34.5$.

This is satisfactory and if we do adjust the support moments will also allow for a small increase in the span moment.

The coefficients for the bending moments for the four panels in one corner are shown on the figure below.

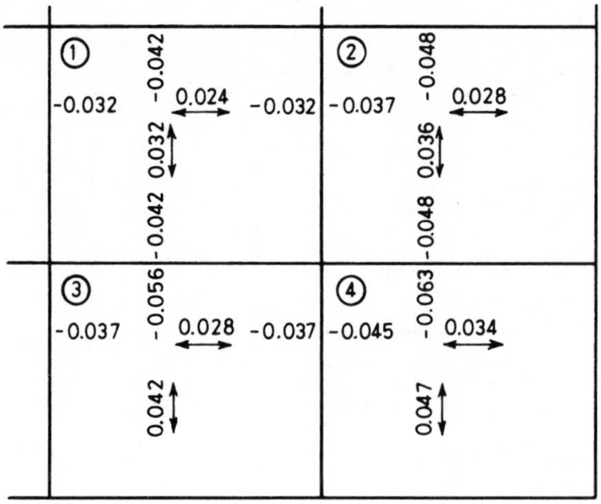

Panel 1: interior panel;
Panel 2: one short edge discontinuous;
Panel 3; one long edge discontinuous;
Panel 4; two adjacent edges discontinuous.

The moments at the junction of panels 4 and 2 show the largest difference, panel 4 being 36.5 kN m and panel 2 being 27.8 kN m.

If these are not considered as significantly different then the design would proceed using the larger moment. However, to illustrate the procedure in the Code they will be regarded as significantly different.

Spanning inthe 6.0 m direction.

1. The Code says calculate the sum of the moments at midspan and supports (neglecting signs). As we are told later to assume a parabolic bending moment we are interested in the 'overall' moment, so we must take the sum of the moment at midspan and the *average* of the support moments. The coefficients shown on the drawing will be used.
   Panel 4: sum $= 0.047 + 0.063/2 = 0.0785$.
   Panel 2: sum $= 0.036 + 0.048 = 0.084$.
2. Treat the values from Table 3.15 of the Code as fixed end moments.
   Panel 4: factor $= 0.063$.
   Panel 2: factor $= 0.048$.

3. Distribute the fixed end moments according to relative stiffnesses of adjacent spans. This is where a basic knowledge of moment distribution is useful. For a continuous slab of three equal spans, the distribution factors are 0.6 for the end span and 0.4 for the interior span. The factor for the out-of-balance moment at the supports is 0.015, so in panel 4 the distributed factor becomes $0.063-0.6\times0.015=0.054$ and in panel 2 it becomes $0.048+0.4\times0.015=0.054$.
4. Adjust midspan moments.
   Panel 4: mid-span $=0.0785-0.054/2=0.0515$, say 0.052.
   Panel 2: midspan $=0.084-0.054=0.030$.
   As the maximum span moment in panel 4 does not occur at midspan we may be slightly on the high side.

The result of the above exercise has been to increase the span moment in panel 4, reduce the support moment and reduce the span moment in panel 2.

The span moment in Panel 4 now becomes 30.1 kN m as compared with 27.2 kN m before distribution, $M/bd^2=0.99$. The modification for tension reinforcement is 1.38 so the allowable $l/d=35.9$; we are still all right. From tables the area of reinforcement in the span is given by $100\ A_s/bd=0.26$, so $A_s=452$ mm²/m. Use 12$\phi$ at 250 centres.

At the support, $M=31.3$ kN m, $M/bd^2=1.03$, $100\ A_s/bd=0.27$, $A_s=470$ mm²/m (12$\phi$ at 225 centres).

The next question is whether the increase in the support moment of panel 2 is regarded as significant. If it is then items (e) to (h) in clause 3.5.3.6 have to be carried out.

The minimum reinforcement in each direction is $0.13\%bh=260$ mm²/m, so curtailment of reinforcement will not be possible.

Torsion reinforcement in the external corner of panel 4 will be in four layers, each layer consisting of three-quarters of the area calculated above for the span moment, i.e. three-quarters of $452=339$ mm²/m. As this occurs at the junction of two edge strips where only the minimum reinforcement is required, it can be seen that the layout of the reinforcement will be quite complicated.

---

## 12.3 Loads on supporting beams

The estimation of loads on supporting beams has changed significantly from CP110. Table 3.16 of the Code gives shear force coefficients from which the loads on the beams can be calculated and takes account of the support conditions of the panel.

As with moments the shear is a factor times the shorter span, whichever direction is being considered. Figure 3.10 of the Code shows the distribution of load but it should be noted that $v_S=v_{Sx}$ when $l=l_y$ and $v_S=v_{Sy}$ when $l=l_x$.

Clause 3.5.3.7 also suggests that if the support moments used in design are substantially different from those obtained from Table 3.15, the values in Table 3.16 of the Code may need to be adjusted. Presumably this would be in a similar manner to that used for moments but, as we saw in Example 12.1, one moment is reduced and one is increased and the net result cannot be very different from the total value obtained from Table 3.16 for the two adjacent spans.

Thus, for example, on the line of common support of panels 4 and 2 in Example 12.1 the loading would be $(0.47+0.42)\times16.1\times6=86$ kN/m over the middle 5.4 m. On the line of the common support of panels 3 and 1 the load would be slightly less at $0.83\times16.1\times6=80$ kN/m.

## 12.4 Shear

As for beams, the shear stress $v$ at any cross-section is calculated from $v = V/bd$ and $b$ will usually be 1000 mm. Table 3.17 of the Code gives the form and area of shear reinforcement. The notes below the table advise that shear reinforcement should not be used in slabs less than 200 mm deep.

Shear in solid slabs carrying uniformly distributed loads is seldom a problem and although the value can be calculated at distance $d$ from the face of the support this is not usually done.

As we saw when finding the loads on supporting beams the maximum shear force in Example 12.1 will be from panel 4 and is $0.47 \times 16.1 \times 6 = 45.5$ kN/m. Here $v = (45.5 \times 10^3/(1000 \times 174) = 0.26$ N/mm$^2$. This is less than the minimum value in Table 3.9 of the Code, so there is no need to calculate $v_c$.

Shear under concentrated loads is covered in the section on flat slabs.

## 12.5 Deflection

This was covered in the example and is based on the shorter span and the amount of reinforcement in that direction.

# RIBBED SLABS (SOLID OR HOLLOW BLOCKS OR VOIDS)

# 13

## 13.1 Definitions

The term 'ribbed slab' in clause 3.6.1.1 refers to *in situ* slabs constructed in one of the following ways.

1. Where topping is considered to contribute to structural strength:
   (a) as a series of concrete ribs cast *in situ* between blocks which remain part of the completed structure; the tops of the ribs are connected by a topping of concrete of the same strength as that used in the ribs;

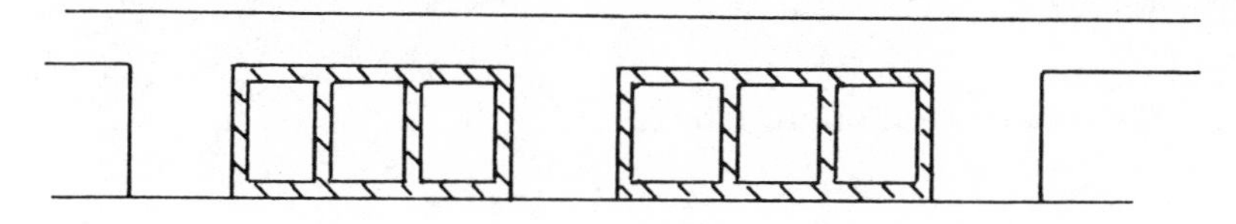

   (b) a series of concrete ribs with topping cast on forms which may be removed after the concrete has set;

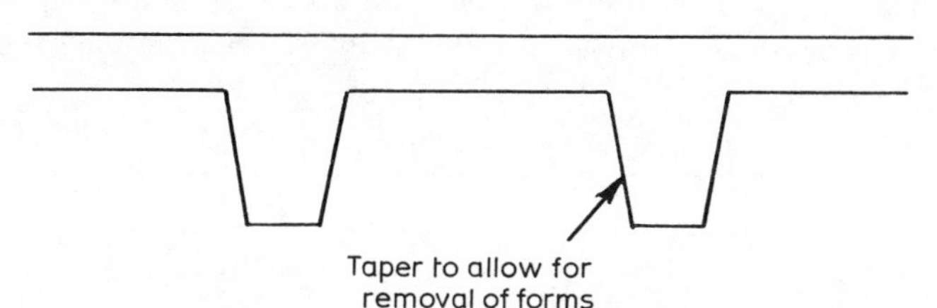


   (c) with a continuous top and bottom face but containing voids of rectangular, oval or other shapes. The formers for the voids may be permanent or removable.

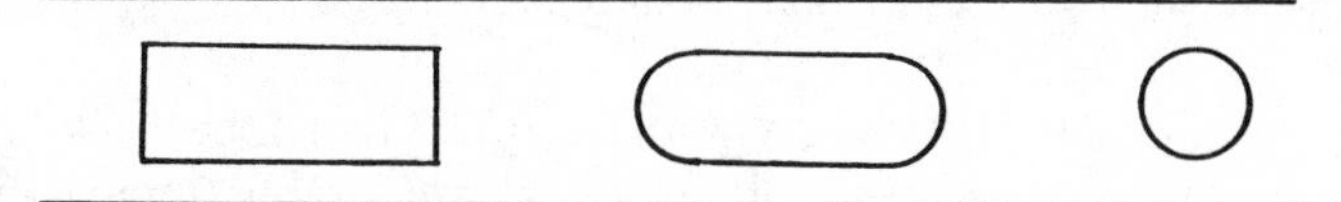

2. Where topping is not considered to contribute to structural strength:
   as a series or concrete ribs cast *in situ* between blocks which remain part of the completed structure; the tops of the ribs may be connected by a topping of concrete (not necessarily of the same strength as that used in the ribs).

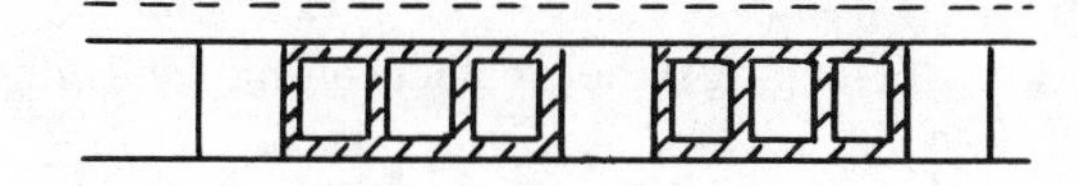

## 13.2 Hollow or solid blocks and formers

In items (1a) and (1b) we have blocks which remain part of the completed structure. These blocks themselves may or may not contribute to the structural strength of the slab. If not, they can be regarded as non-removable formers. It should be remembered that we are talking about *in situ* concrete slabs, not slabs consisting of precast concrete ribs with infill blocks between them, on top of which is cast a concrete topping.

Where the blocks do contribute to the structural strength they will be referred to as structural-type blocks which comply with clause 3.6.1.2. Although these blocks may contribute to the flexural strength, their main contribution is regarding shear and deflection.

## 13.3 Thickness of topping

We have already categorized the slabs as (1) structural topping, where topping does contribute, and (2) non-structural topping, where topping does not contribute to the structural strength.

Table 3.18 of the Code divides the slabs into those with permanent blocks and those without permanent blocks for the purpose of thickness of topping required. We shall use this classification.

### 13.3.1 Slabs with permanent blocks

#### (a) STRUCTURAL TOPPING AND STRUCTURAL-TYPE BLOCKS

The clear distance between ribs should be not more than 500 mm. The width of the rib will be determined by considerations of cover, bar spacing and fire requirements, but the depth of the rib excluding the topping should not exceed four times the width – see Fig. 13.1.

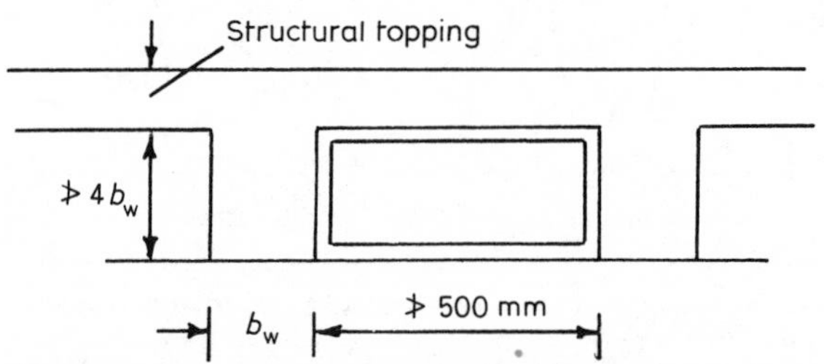


FIG. 13.1 Permanent blocks contributing to structural strength.

The minimum thickness of topping required is: (a) 25 mm if the blocks are jointed in cement:sand mortar not weaker than 1:3 or 11 N/mm$^2$; or (b) 30 mm if the blocks are not jointed.

#### (b) STRUCTURAL TOPPING AND NON-STRUCTURAL-TYPE BLOCKS

In this case the spacing of the ribs can be increased above 500 mm clear but the centres of the ribs must not exceed 1500 mm. As in (a) the depth of the rib should not exceed four times the width.

The minimum thickness of topping is the greater of 40 mm and one-tenth of the clear distance between ribs (see Fig. 13.2).

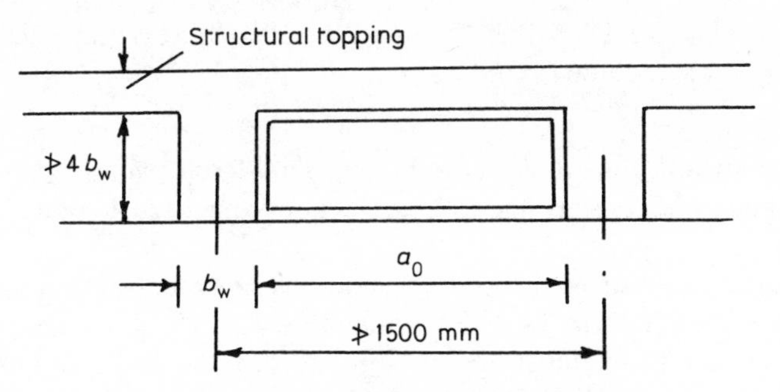


FIG. 13.2 Permanent blocks not contributing to structural strength.

(c) NON-STRUCTURAL TOPPING AND STRUCTURAL-TYPE BLOCKS

If the topping does not contribute to the strength, then the blocks must be structural type (see clause 3.6.1.6). The blocks do not need to be jointed in cement: sand mortar. There are further requirements as shown in Fig. 13.3, and it should be noted that there is a small difference from CP110. The new requirement concerns the thickness of the block material above its void, not block material plus topping as previously required.

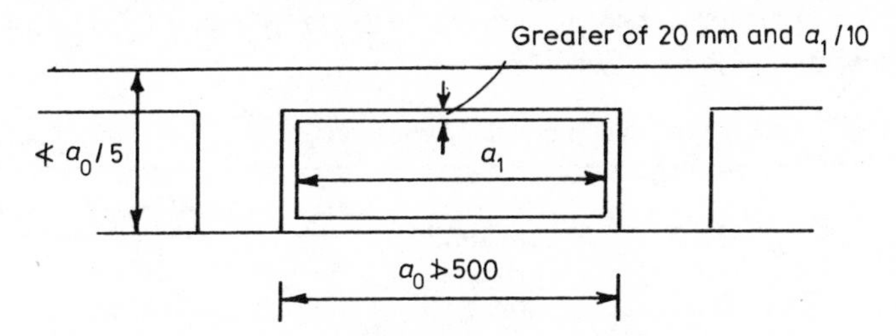


FIG. 13.3 Blocks contribute to structural strength but topping does not.

### 13.3.2 Slabs without permanent blocks

For all these slabs such as shown in (1b) and (1c) under *Definitions* (§13.1), the minimum thickness of topping is the greater of 50 mm and one-tenth of the clear distance between ribs. In (1b) this is relatively easy to determine, but in (1c) it is suggested that the 'rib' width is determined as in clause 3.6.5.2 for deflection.

## 13.4 Fire requirements

For complying with the fire requirements for cover and thickness reference should be made to Figure 4.2 of Part 2 of the Code. From this it can be seen that slabs with permanent blocks are classed as plain soffit floors whereas slabs without permanent blocks are classed as ribbed soffit floors.

For plain soffit floors the bottom cover to the reinforcement is obtained from Table 3.5 of Part 1. In complying with this requirement, plaster or sprayed fibre can be taken into account. The equivalent thickness of concrete can be obtained from clause 4.2.4 of Part 2. For example, a mortar thickness of 12 mm is equivalent to $12/0.6=20$ mm concrete. The width of the rib and the side cover to the reinforcement in the rib are not controlled by fire requirements. The floor thickness required is obtained from Figure 3.2 of Part 1, and with hollow slabs the effective thickness is calculated from clause 4.2.5 of Part 2. Equation (16) of the original version of BS8110 was incorrect and should be

$$t_e = h \times \sqrt{\xi} + t_f$$

where $h$ is the overall thickness of the hollow slab floor, $\xi$ is the proportion of solid material per unit width of slab, and $t_f$ is the thickness of any non-combustible finish on top.

For ribbed soffit floors the width of the rib and side cover to the reinforcement have to meet the fire requirements in Table 3.5 and Figure 3.2 of Part 1. The thickness of the concrete joining the ribs is illustrated in Figure 4.2 of the Code. Values are obtained from Table 4.5 in Part 2.

## 13.5 Hollow clay floor blocks

Sizes and weights of clay blocks vary with different manufacturers and for an accurate weight of the floor this should be worked out from the manufacturer's literature. Generally they are 300 mm × 300 mm in plan and the depth varies from 75 to 250 mm in 25 mm intervals. Table 13.1 gives typical weights of floor construction for three different widths of rib, and for a structural topping 40 mm thick. For different thicknesses of topping adjust the weights by 0.24 kN/m$^2$ for each 10 mm thickness; e.g. the overall weight for a structural depth of 200 mm using a 50 mm topping and 100 mm ribs would be 2.78 kN/m$^2$.

**Table 13.1** Weights of hollow clay block floor construction with 40 mm topping

| *Block size* (mm) | *Wall thickness* (mm) | *Structural depth* (mm) | *Overall weight* (kN/m$^2$) 75 mm *rib* | 100 mm *rib* | 125 mm *rib* |
|---|---|---|---|---|---|
| 300 × 300 × 75 | 15 | 115 | 1.75 | 1.80 | |
| 300 × 300 × 100 | 15 | 140 | 1.95 | 2.00 | |
| 300 × 300 × 125 | 15 | 165 | 2.15 | 2.25 | |
| 300 × 300 × 150 | 18 | 190 | 2.39 | 2.54 | |
| 300 × 300 × 175 | 18 | 215 | 2.58 | 2.73 | 2.88 |
| 300 × 300 × 200 | 18 | 240 | 2.82 | 2.98 | 3.13 |
| 300 × 300 × 225 | 18 | 265 | | 3.22 | 3.42 |
| 300 × 300 × 250 | 20 | 290 | | 3.51 | 3.70 |
| 300 × 300 × 250* | 20 | 340 | | | 4.49 |

*250 mm blocks on their sides to give 300 mm deep block.

## 13.6 Edges

Where ribs are running parallel to a beam or wall and bear on the beam or wall, the rib must be at least as wide as the bearing. This means the block or void must not be on the bearing.

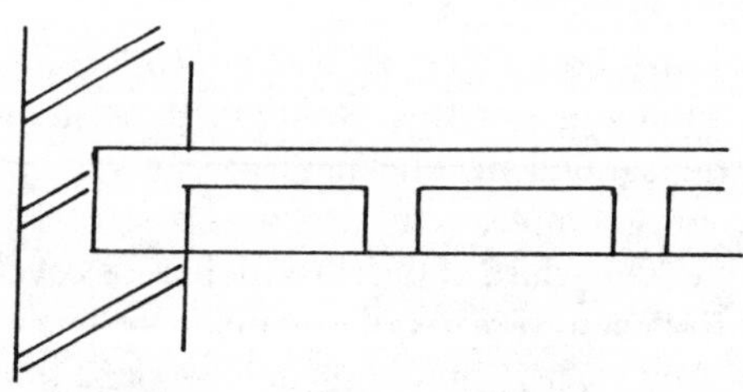

## 13.7 Analysis

Moments and forces due to ultimate design loads in continuous slabs may be found as for solid slabs. Alternatively, provided they are not exposed to weather or corrosive conditions, they may be designed as a series of simply supported slabs. If this is done, sufficient reinforcement should be provided over the support to control cracking. The Code recommends that such reinforcement should have an area of not less than 25% of that in the middle of the adjoining spans and should extend at least 15% of the spans into the adjoining spans. Although reinforcement has to be provided in the top of the slab over the supports it may not be sufficient to resist moments that could develop at this point and the steel may yield with cracks developing in the top surface of the concrete. These cracks may be covered with floor finishes, but the engineer should be aware that this method of design does have risks of cracking associated with it.

### 13.7.1 Resistance moments

The same methods as for beams apply, and where the topping contributes to the structural strength we shall have a Tee section in the span where the width of the flange is the lesser of the distance centre to centre of ribs and the effective flange width as determined from clause 3.4.1.5 (i.e. $b_w + kl_z/5$) where $k$ is 1.0 for simply supported spans and 0.7 for continuous spans. In determining the flange width $b_w$ is taken as the width of the concrete rib only. When analysing the section the stresses in burnt clay blocks or solid blocks in the compression zone may be taken as 0.25 times the crushing strength as determined in clause 3.6.1.2(b). In most cases designers ignore this in strength calculations and in examples in this chapter it will also be ignored. However, if the designer wishes to take advantage of this item it is suggested that flange depth in the span be increased by an amount equal to

$$\frac{\text{Allowable stress in block}}{\text{Design stress in concrete}} \times \text{thickness of block wall.}$$

For example, if we have a block of crushing strength 16 N/mm$^2$, thickness of block wall at underside of topping is 18 mm, concrete in topping is grade 35, thus topping thickness can be increased by $(0.25 \times 16)/(0.45 \times 35) \times 18 = 4.6$, say 4 mm. As can be seen, this is very small and the complete 18 mm of the block wall thickness must come with the distance of 0.9 times the neutral axis depth.

Where the slab is continuous over supports the thickness of the concrete rib in the compression zone could be increased in the same way.

### 13.7.2 Shear

The shear stress $v$ should be calculated as

$$v = V/b_v d$$

where $V$ is the shear force due to ultimate loads on a width of slab equal to the distance centre to centre of ribs, and $b_v$ is the average width of the rib.

If hollow blocks of the structural type are used the effective rib width is the width of the concrete rib plus one wall thickness of the block.

Here again many designers prefer to take the shear on the concrete rib only. The

Code, in fact, does not state that the blocks have to be the structural type, but in the past it has been restricted to this type and it is assumed that this is still meant.

If solid blocks are used the effective rib width may be increased by one-half of the rib depth on each side of the rib.

For voided slabs the Code does not say how to calculate $b_v$ but it would seem appropriate to use the same method as used for deflection.

For the allowable shear stress we have the same requirement as for solid slabs. No shear reinforcement is required for $v$ less than $v_c$. If $v$ is greater than $v_c$ reinforcement to Table 3.17 of the Code should be provided.

### 13.7.3 Deflection

The span/effective depth ratios as for flanged sections should be used, but when considering the final modification factor for a flanged beam, the rib width may be increased to take account of the thickness of the walls of the blocks on both sides of the rib. For voided slabs and slabs constructed of box or I-section units, an effective rib width should be calculated assuming all material below the upper flange to be concentrated in a rectangular rib having the same cross-sectional area and depth (see, for example, Fig. 13.4).

For two-way spanning slabs the check should be carried out for the shorter span.

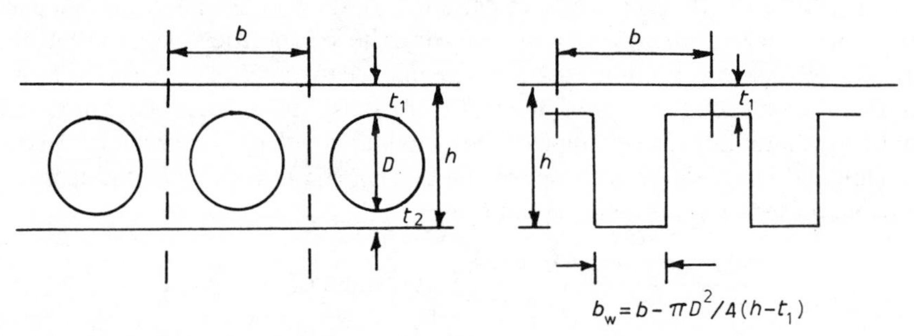


FIG. 13.4 Equivalent rib widths for voided slabs.

### 13.7.4 Arrangement of reinforcement

The curtailment of the reinforcement will depend on how the moments have been determined, i.e. by analysis or simplified rules.

In ribbed or hollow block slabs the Code recommends providing a single layer of mesh, with a cross-sectional area of not less than 0.12% of the topping, in each direction. The spacing of the wires should not be greater than half the centre-to-centre distance between the ribs. For the reinforcement in the ribs themselves, if only a single bar is needed, links are not necessary unless shear or fire resistance requirements so dictate. However, consideration should be given to the use of purpose-made spacers occupying the full width of the rib to ensure correct cover to the bar. Where two or more bars are used in a rib, the use of link reinforcement is recommended to ensure correct cover to the main bars. These links are in addition to normal spacers and will generally be at a spacing of 1.0 to 1.5 m, depending on the size of the main bars. The cover to the link should satisfy the durability requirement but need not satisfy the fire requirement provided the cover to the main bars does so.

## EXAMPLE 13.1

Calculate a suitable section for a 6.5 m span, continuous but treated as simply supported, using the following data: Imposed load is 4.0 kN/m$^2$, 12 mm plaster ceiling, structural-type hollow clay blocks, mild exposure, 2 h fire resistance. The concrete is Grade C35, and reinforcement Grade 460.

Assuming a span/effective depth ratio of 25, then $d=260$ mm.

Durability: cover = 20 mm (Table 3.4 of the Code).

Fire resistance: cover = 35 mm (Table 3.4 of the Code)
thickness of floor = 125 mm (Figure 3.2 of the Code).

With 20 mm concrete cover in the rib, the effective cover using clause 4.2.4 of Part 2 = 20 + 12/0.6 = 40 mm.

From Table 13.1 try a structural depth of 290 mm using 300 × 300 × 250 blocks and 100 mm wide ribs as shown below:

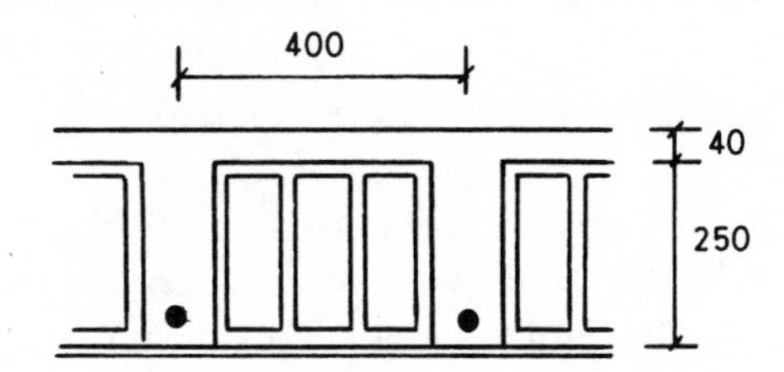


The walls of the blocks are 20 mm thick so

$\xi = (400 \times 290 - 210 \times 220)/(400 \times 290) = 0.6.$

Effective thickness, $t_e = 290 \times \sqrt{0.6} = 223$ mm. Satisfactory.

Loads

Dead: self weight = 3.51 kN/m$^2$
plaster = 0.25 kN/m$^2$

3.76 kN/m$^2$,

Imposed = 4.0 kN/m$^2$.

Ultimate design load per rib = $(3.76 \times 1.4 + 4 \times 1.6) \times 0.4$
= 4.67 kN/m.

Assuming 20$\phi$ bar, $d = 290 - 20 - 20/2 = 260$ mm.

$M = 4.67 \times 6.5^2/8 = 24.7$ kN m.

$M/bd^2 = (24.7 \times 10^6/(400 \times 260^2) = 0.91.$

From tables, neutral axis depth is $0.111 \times 260 = 29$ mm, so neutral axis is in flange, and $100A_s/bd = 0.24$.

$A_s = 0.0024 \times 400 \times 260 = 250$ mm$^2$. Use 1/20$\phi$ (314 mm$^2$).

This is more than minimum requirement in Table 3.27 of the Code.

$f_s = 288 \times 250/314 = 230$ N/mm$^2$.

Modification for tension reinforcement (Table 3.11 of the Code) $= 1.69$.

$b_w/b = 140/400 = 0.35$, i.e. $> 0.3$, but take as 0.3.

Allowable $l/d$ ratio $= 16 \times 1.69 = 27.0$.

Actual ratio $= 6500/260 = 25$. Satisfactory.

$V = 4.67 \times 6.5/2 = 15.2$ kN.

$v = (15.2 \times 10^3)/(120 \times 260) = 0.49$ N/mm$^2$.

$100A_s/b_vd = (100 \times 314)/(120 \times 260) = 1.0$.

$v_c$ from Table 3.9 of the Code $= 0.71 \times 1.12 = 0.79$ N/mm$^2$. Satisfactory.

In assessing the shear resistance we have assumed that the bottom bar in each rib is effectively anchored. In this type of floor, however, it is usual to make the slab solid near the bearings.

For this area $v = (15.2 \times 10^3)/(400 \times 260) = 0.15$ N/mm$^2$.

This is less than half the allowable $v_c$ which can be obtained from Table 3.9 of the Code, so an effective anchorage can be obtained by taking the bars a distance equal to the greater of 30 mm and one third of the support width beyond the centre line of the support.

For the top steel over the interior support we need $250/4 = 63$ mm$^2$ per rib, so suggest $1/10\phi$. This should extend 15% of $6500 = 975$ mm into the span from the face of the support.

---

# FLAT SLAB CONSTRUCTION 14

Flat slabs are dealt with in Clause 3.7 of the Code.

## 14.1 Definitions

The term 'flat slab' means a reinforced concrete slab with or without drops and supported, generally without beams, by columns with or without column heads. The slab may be solid or may have recesses formed on the soffit so that the soffit comprises a series of ribs in two directions. Where recesses are formed on the soffit, it will usually be necessary to make the slab solid in the region surrounding the column heads in order to provide adequate shear strength.

A 'drop' is a local thickening of the slab in the region of a column.

A 'column head' is a local enlargement of the top of the column providing support to the slab over a larger area than the column section alone. The head may be of uniform cross-section or may be tapered (generally referred to as flared).

The different types of column head are illustrated in Fig. 14.1.

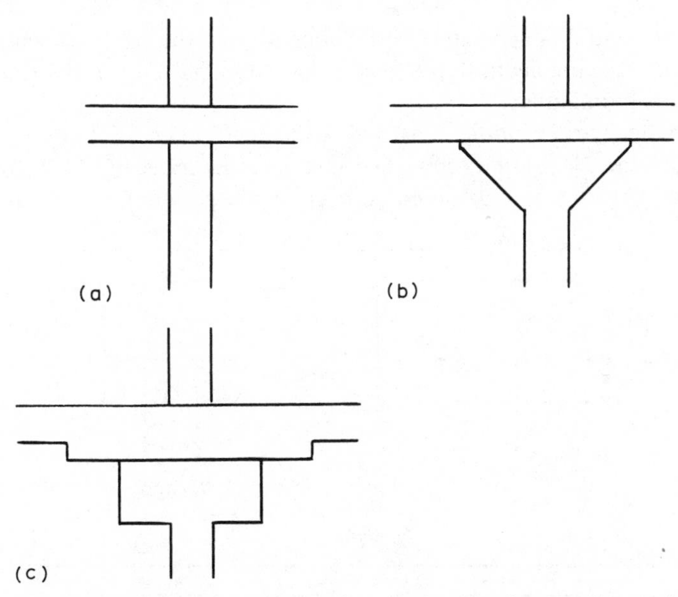


FIG. 14.1 Types of column head. (a) Slab without drop and column without column head. (b) Slab without drop and column with flared head. (c) Slab with drop and column with column head of uniform cross-section.

## 14.2 Notation

A flat slab will generally span in two directions at right angles, and as each direction is considered separately it is important to get the notation clear to avoid confusion with two-way spanning slabs on to beams.

$l_1$ is the length of the panel, centre to centre of columns, in the direction of the span being considered.
$l_2$ is the width of the panel, centre to centre of columns, at right angles to the direction of the span being considered.
$l_y$ is the longer span of a panel.
$l_x$ is the shorter span of a panel.
$l$ is the effective span of a panel ($=l_1-2h_c/3$).
$h_c$ is the effective diameter of a column or column head, which shall be taken as the diameter of a circle of the same area as the cross-section of the head based on the effective dimensions as described below, but in no case greater than one quarter of the shortest span framing into the column.
$n$ is the total ultimate design load per unit area ($=1.4g_k+1.6q_k$).

## 14.3 Column heads

The dimensions of a column head which may be considered as effective depend on the depth of the head. The angle of slope of the head, if flared, or theoretical slope if uniform, should not be less than 45° from the horizontal. The dimension should be measured at a distance of 40 mm below the underside of the slab or drop where provided. If the actual dimensions of the head are less than those obtained from the 45° requirement then those dimensions should be used. These requirements can be written mathematically as:

$$l_h = \text{the lesser of } l_{ho} \text{ and } l_{h,max} = l_c + 2(d_h - 40) \text{ mm}$$

where $l_{ho}$ is the actual dimension, $l_c$ is the dimension of the column measured in the same direction, $d_h$ is the depth of the head below the soffit of the slab or drop, and all dimensions are in millimetres.

Diagrammatically they can be illustrated as shown in Fig. 14.2.

If the column head is circular then $l_h$ becomes $h_c$, as described in §14.2. For any other shape a value of $h_c$ has to be calculated. As $h_c$ cannot be greater than one-quarter of the

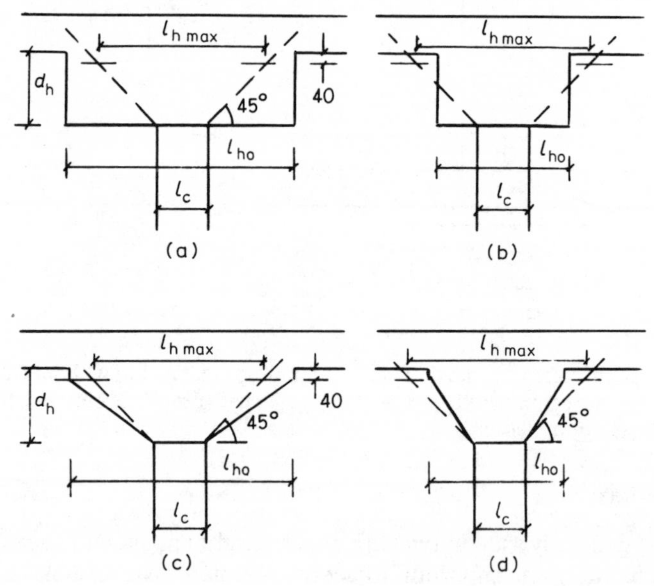


FIG. 14.2 Dimensions for column head. (a) $l_h=l_{h,max}$. (b) $l_h=l_{h0}$. (c) $l_h=l_{h,max}$. (d) $l_h=l_{h0}$.

shortest span framing into the column it is usual to start with this dimension and then calculate the largest size of head that can be used. For instance, if we have a square column and square head, then the size of the head will be $0.886h_c$, which is approximately $0.22l_{min}$. Having decided on the size of a suitable square head, the value of $h_c$ is then determined and it is this value which is used in all analyses to calculate bending moments.

## 14.4 Division of panels

It must be remembered that a panel is the area within the lines joining the centres of the columns.

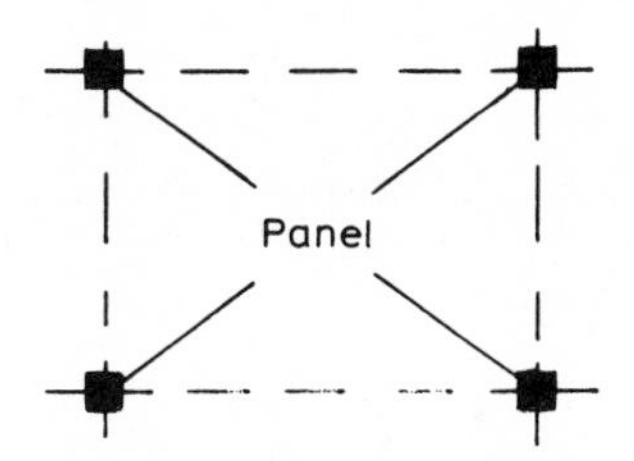


A panel should be assumed to be divided into column strips and middle strips. These are illustrated in Fig. 14.3 for panels with and without drops.

For panels with drops, the total width of the column strip should be taken as the width of the drop unless this is less than one-third of the smaller dimension of the panel. In this case the panel is treated as for a slab without drops.

Where there is a support common to two panels of such dimensions that the strips in one panel do not match those in the other, the division of the panels over the region of the common support should be taken as that calculated for the panel giving the wider column strip. In the span, the division of the panels will remain unaltered. This will be illustrated in the design example which will follow later. No requirements are given for the span dimensions of drops, except that the following points have to be taken into account:

1. Unless the smaller dimension of the drop is at least one-third of the smaller dimension of the surrounding panels it will not influence the distribution of moments within the slab (see previous comments concerning width of column strip). Even if it does extend the necessary distance the Code says that one can ignore the stiffening effect of the drops when calculating the stiffness of the slab.
2. For deflection control, the total width of the drops in each direction should be at least equal to one-third of the respective span.
3. When it comes to apportioning moments within the panels a further calculation is introduced if the column strip is taken as the width of the drop and this is less than half the smaller dimension of the panel ($l_x/2$).

## 14.5 Thickness of panels

The Code states that the thickness of the slab will generally be controlled by deflection considerations, but the minimum thickness is 125 mm. It should be pointed out that this statement is true only in flat slabs without drops if shear considerations are ignored and the designer is prepared to put in shear reinforcement around the column head, if the occasion arises. As will be seen later, if the designer wants to avoid shear

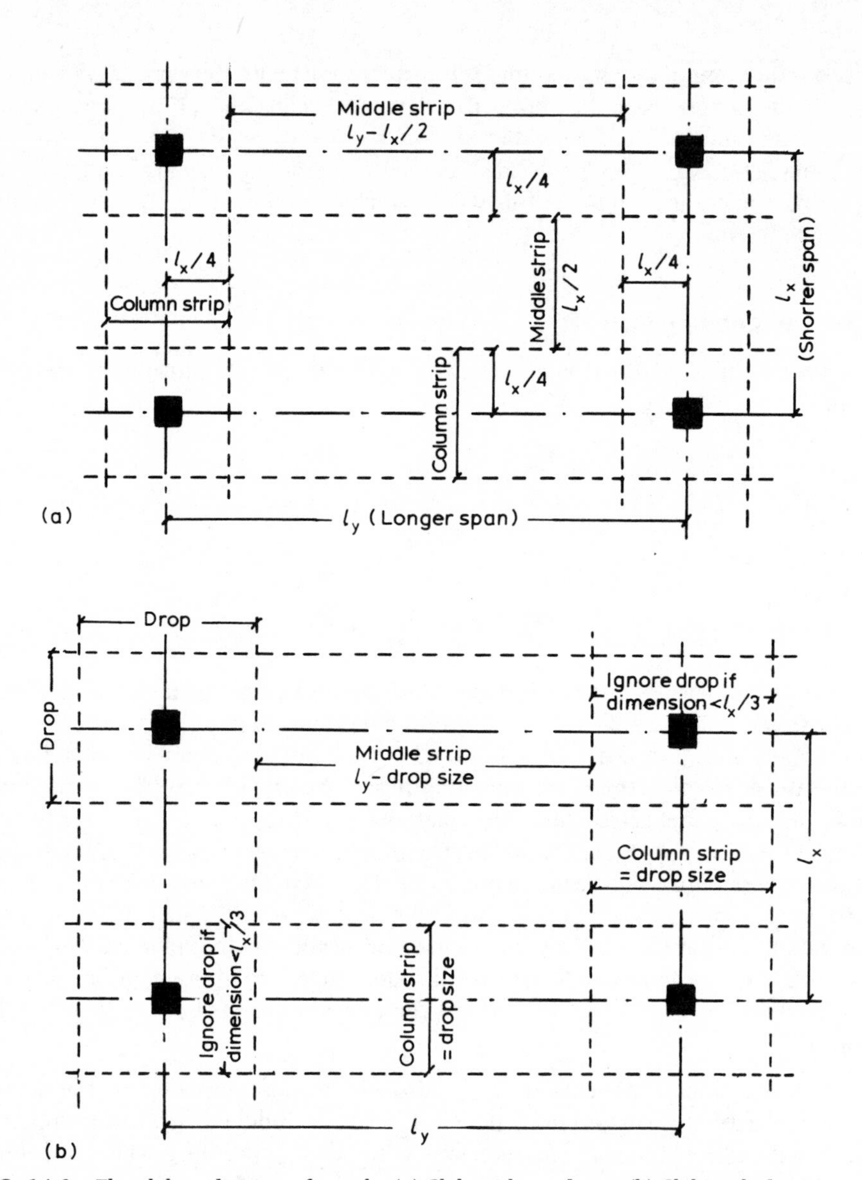


FIG. 14.3 Flat slabs – division of panels. (a) Slab without drops. (b) Slab with drops.

reinforcement he may have to introduce additional flexural reinforcement or thicken the slab.

Where drops are used it is usual for the overall thickness of the drop to be 1.50 times the thickness of the slab. An initial assessment of a slab thickness can be made based on solid slabs, modified as necessary by the deflection criteria for flat slabs as described below.

## 14.6 Deflection

For deflection criteria, the span/effective depth ratios as for slabs will be used with modifications as follows:

### 14.6.1 Slabs without drops

For solid slabs it will be the factor as obtained for solid rectangular sections and then multiplied by 0.9. To obtain the modification factor for tension reinforcement from Table 11 of the Code the average value of $M/bd^2$ across the width of the sum of the column strip plus middle strip at midspan should be used.

For slabs with recesses on the underside there is the further reduction factor for Tee sections which could be 0.8. So the overall reduction factor could be 0.72 as compared with a solid rectangular section. The check should be carried out for the more critical direction, which is probably the longer span.

### 14.6.2 Slabs with drops

Provided the dimensions of the drops are as stated earlier, the factors will be as for rectangular sections when using solid slabs and the section considered will be at midspan of the slab. So it will be the effective depth of the slab between the drops which will control.

It is unlikely that this type of slab will have recesses on the underside, but if it does, then the modification factor for Tee-sections will apply. Again, the check should be carried out for the more critical direction.

### 14.6.3 Effective depth

An effective depth is required for

1. Strength of sections. Some designers use the average effective depth of the two directions, but with the methods of analysis now used in the Code it would appear that the actual effective depth in each direction will be used.
2. Shear. The average effective depth is required.
3. Deflection. The actual effective depth in each direction is required.

With rectangular panels the bending moment will be greater in the longer direction and the effective depth will usually be taken as large as possible. To avoid having different effective depths for span and support positions the arrangement shown in Fig. 14.4 would appear to cause the least confusion.

FIG. 14.4 Positioning of bars in flat slabs.

## 14.7 Crack control

In general, the bar spacing rules described for solid slabs will be used, although calculations can be carried out if desired.

## 14.8 Analysis

In the first edition of CP110 there were two quite independent methods of analysing the structure to obtain the bending moments and shear forces. One method was called the equivalent frame method, where the structure was divided into frames in two directions. The other method was called the empirical method and considered one panel at a time and the moments in the various strips obtained as a proportion of the total moment in the panel.

In the revision to the Code, the empirical method has been dropped, and although there are still two methods, both are equivalent frame methods. The first method is the true equivalent frame, where a frame analysis is carried out. The second method is a simplified version of this, where coefficients are used to determine the moments and shears without carrying out an analysis. Quite obviously there are restraints to using the simplified method.

### 14.8.1 Frame analysis

In dividing the structure into frames the designer has to take into account whether the frame is being subjected to vertical loads only or vertical and horizontal loads, as with a framed structure of beams and columns.

For vertical loads only, i.e. load combination (1), the equivalent beam for stiffness purposes will be based on a strip of slab of width equal to the distance between centre lines of panels on each side of the columns. With vertical and horizontal loads, i.e. load combinations (2) and (3), the equivalent beam will be half the value. The Code actually says that for 'horizontal loading' one takes half the value, but it would not be reasonable to carry out an analysis changing the frame properties in this manner.

In calculating the relative stiffnesses of the columns and slabs, it would also seem appropriate to consider the concrete section only. The stiffening effect of drops and column heads may be ignored, as illustrated in Fig. 14.5.

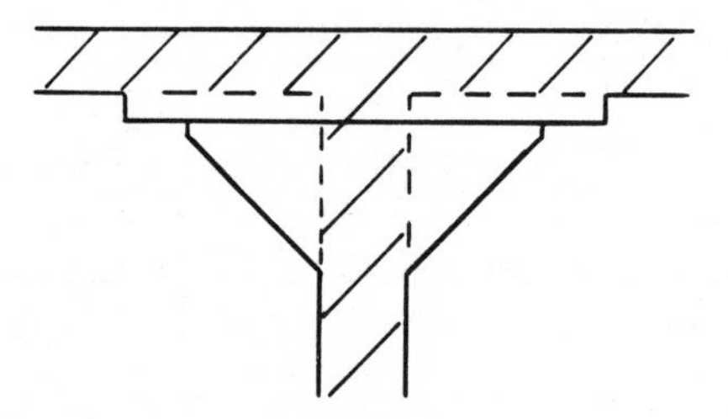

FIG. 14.5 Sections for calculating relative stiffnesses.

Where recessed or coffered slabs are used and these are made solid in the region of the columns, usually for shear purposes, the stiffening effect may be ignored provided the solid portion does not extend more than 0.15 of the span into the span, measured from the centre line of the columns. Irrespective of which section is taken for the stiffness of the slabs, the loads to be carried are the appropriate design load for a strip of slab of width equal to the distance between centre lines of panels on each side of the columns.

At this stage it might be useful to try to clear up the confusion that often arises when analysing frames in two directions. Some designers feel that the load is being carried twice, as compared with a two-way spanning slab carried on beams where the load is considered as being carried in two directions simultaneously. What is generally forgotten in a two-way spanning slab is that the load from the slab is carried to the beams, and the beams then carry the load to the columns. Consider such a slab carried on perimeter beams:

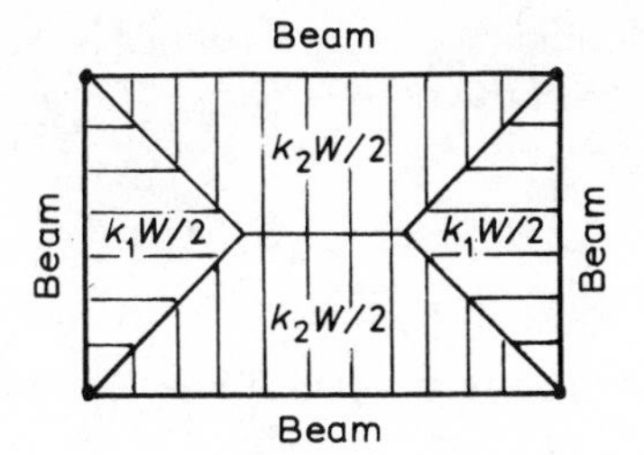


In the short direction:

$$\begin{aligned}\text{Load carried by slab} &= k_2W\\ \text{Load carried by beams} &= k_1W\\ \text{Total load carried} &= (k_1+k_2)W = W.\end{aligned}$$

Similarly, in the long direction the total load carried is $W$. Therefore, in each direction considered separately, it is the total load that is carried. This means that in flat slabs the full load must be carried in both directions by the slab alone.

When considering vertical loads for load combination (1), the general cases for load patterning will be as for beams, i.e. maximum design load on all spans and also for alternate spans having maximum and minimum design load. However, if the conditions as set out for simplification of load arrangements for solid slabs are satisfied, then one load case of maximum design load on all panels is all that is necessary.

Having used the single load case we now have to comply with clause 3.5.2.3 whereby the support moments are reduced by 20%, with a corresponding increase in the span moments. The resulting envelope should satisfy the provisions of redistribution as already discussed for slabs, i.e. we do not come within 70% of the ultimate envelope.

Having found the moments in the slabs and columns for the equivalent frame either by a full-frame analysis or series of subframe analyses we now modify these moments as follows:

1. The moment transferred between a slab and an edge column will generally be much less than indicated by the analysis unless there is an edge beam, or strip of slab along the free edge, specially designed and detailed to transfer the extra moment into the column by torsion. The moment transferred direct to the column is given as $M_{t,max} = 0.15b_ed^2f_{cu}$, where $b_e$ is a breadth of strip dependent on the relative position of the column and the free edge of the slab, but should not be taken as greater than the column strip width appropriate for an interior column. In this case $d$ is the effective depth for the top steel in the column strip. If the negative moment at the edge column is reduced from that obtained from analysis then the positive moment in the end span should be increased accordingly.
2. Negative moments greater than those at a distance of $h_c/2$ from the centre line of the column may be ignored provided the sum of the maximum positive moment and the average of the negative moments in any one span for the whole panel width is not less than

$$(n/8)l_2(l_1 - 2h_c/3)^2,$$

where $n$ is the maximum design load per unit area of the span under consideration.

When this condition is not satisfied the negative moments at both supports should be increased so that the above requirement is met.

The modified moments should now be apportioned between column strip and inner strip as shown in Table 3.20 of the of the Code.

### 14.8.2 Simplified method

Where the following conditions are satisfied, Table 3.19 in the Code gives coefficients for determining the bending moments and shear forces in the frame:

1. Lateral stability is not dependent on slab column connections, i.e. it is a braced frame structure.
2. The conditions of clause 3.5.2.3 are satisfied. The area of a bay will usually be satisfied, so this in effect means that the characteristic imposed load does not exceed 1.25 times the characteristic dead load, and the characteristic imposed load, excluding partitions, does not exceed 5 $kN/m^2$.
3. There are at least three rows of panels in the direction being considered and the spans are approximately equal.

The values in Table 3.19 of the Code are based on the analysis of a frame of equal spans considering the single load case of maximum design load on all spans. Reduction of the support moments with an increase in span moments has also been carried out. Modifications to allow for the size of a column head have not been carried out and the moments over the supports will therefore generally be larger than those from a proper frame analysis. It should also be noted that the total design load $F$ is based on the complete span, centre to centre of columns, whereas the moment is based on $Fl$, where $l$ is the effective span $(l_1 - 2h_c/3)$. The moments at the outer support are given for both a column and a wall, the wall being a simple support providing no restraint to rotation.

For the external column we again have to consider the moment which can be transferred from the slab, as with the frame analysis. The value given in the table may have to be adjusted; this will also affect the moment near the centre of the first span. If the moment at the outer support is reduced by a certain numerical amount then the span moment will be increased by half this numerical amount. The moments obtained from the table can now be apportioned into column strip and inner strip as before.

## 14.9 Shear

One of the major problems in assessing the thickness of a flat slab is that of shear, particularly if the provision of shear reinforcement is to be avoided. Shear around column heads is dealt with in clause 3.7.7 as shear under concentrated loads. With column heads, however, the shear forces are enhanced to allow for non-symmetrical distribution of shear around the column. The Code calls this the effects of moment transfer.

For concentrated loads supported by slabs this enhancement is not required but the procedures are the same. An important requirement is the understanding of what is a shear perimeter and failure zone. The Code refers the reader back to clause 1.2.3 for definitions of terms specific to perimeters, but the diagrams which illustrate some of the terms are given in Figures 3.16 and 3.17 of the Code.

For the increases in shear forces, and hence shear stresses, the following cases have to be considered.

### 14.9.1 Internal columns

The effective shear force, $V_{eff} = V_t(1 + 1.5M_t/V_t x)$, where $V_t$ is the support reaction obtained from the sum of the beam shears. If the single load case has been used these values will be after the 20% reduction at the supports. $M_t$ is the moment transmitted from the slab to the connection, i.e. the sum of the column moments above and below the slab. The diagram in Figure 3.14 of the Code shows this as the difference in beam moments at the connection. If the single load case has been used these moments will be before the 20% reduction at the supports. $x$ is the length of the side of the shear perimeter under consideration parallel to the axis of bending (see Fig. 14.6).

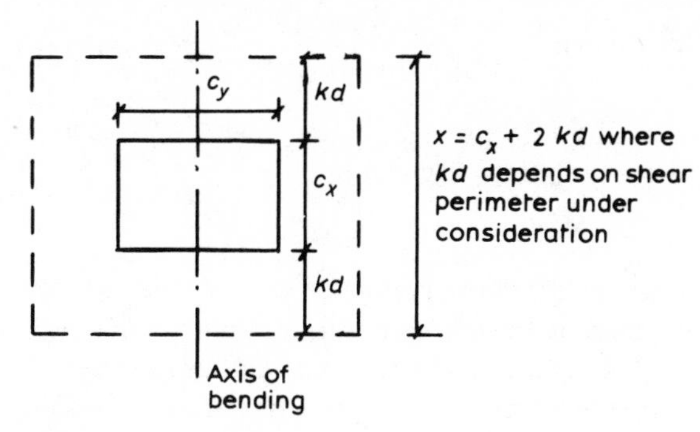


FIG. 14.6 Shear perimeter.

The Code says that in the absence of calculation it will be satisfactory to take a value of $V_{eff} = 1.15V_t$ for internal columns in a braced structure with approximately equal spans. $V_t$ is calculated on the assumption that the maximum design load is applied to all panels adjacent to the column itself.

Where patterned loading has been used in analysing the frame the value of $M_t$ may be reduced by 30%. The calculation for $V_{eff}$ should be applied independently for the moments and shears about both axes of the columns and the design checked for the worst case.

### 14.9.2 External columns

Here we have corner columns and edge columns.

#### (a) CORNER COLUMNS

$V_{eff} = 1.25V_t$ and applies to both axes.

#### (b) EDGE COLUMNS

When considering bending about an axis parallel to the free edge, $V_{eff} = 1.25V_t$.

When considering bending about an axis perpendicular to the free edge, $V_{eff} = V_t(1.25 + 1.5M_t/V_t x)$. If spans are approximately equal $V_{eff}$ can be taken as $1.4V_t$. As for internal columns $M_t$ can be reduced by 30% if patterned loading is used.

From Figure 3.16 of the Code it can be seen that the shear perimeters are rectangular whatever the actual shape of the column.

It is fairly obvious that the shear perimeter under consideration is extremely important. Clause 3.7.7.6 gives the design procedure which says that the first critical perimeter is at 1.5$d$ from the loaded area. If the shear stress in this perimeter does not exceed the allowable $v_c$ then no further check is required. If the shear stress exceeds $v_c$ then the shear reinforcement will be provided, but successive perimeters will be checked until $v$ does not exceed $v_c$. The enhancement factor will therefore vary from perimeter to perimeter, but the largest calculated factor will be at the face of the loaded area itself.

For example, in Fig. 14.6, assume $c_y = 400$ mm, $c_x = 300$ mm, $V_t = 700$ kN, $M_t = 25$ kN m and $d = 250$ mm. At the first critical perimeter $x = 300 + 3 \times 250 = 1050$ mm. So

$$V_{eff} = 700\left(1 + \frac{1.5 \times 25 \times 10^6}{700 \times 10^3 \times 1050}\right) = 700 \times 1.05.$$

On the column face itself $x = 300$ so $V_{eff} = 700 \times 1.18$.

The shear stress on a perimeter is given by

$$v = V_{eff}/ud,$$

Where $u$ is the effective length of the perimeter, taking into account holes or adjacent edges. On the column face itself, $u$ becomes $u_0$ and the value of $v_{max}$ must not exceed $0.8\sqrt{f_{cu}}$ or 5 N/mm$^2$ if less. The reinforcement percentage used to calculate $v_c$ is given by $100 \times$ (effective reinforcement area)/$ud$, where the effective reinforcement is the total area of all tension reinforcement that passes through a zone and extends at least an effective depth or twelve times the bar size beyond the zone on either side. As stated earlier, the effective depth for shear is the average effective depth for all effective reinforcement passing through a perimeter.

A zone is defined in the Code as an area of slab bounded by two perimeters 1.5 d apart. The length of a perimeter will be modified if there are holes within a distance of 6$d$ from the edge of the load or if the load is located close to a free edge. Figure 3.18 of the Code illustrates the reduction in perimeter for holes and Figure 3.19 of the Code shows alternative perimeters for a load close to a free edge.

Holes are sometimes left for a service pipe and these are adjacent to the column. Provided the greatest width is less than one-quarter of the column side or one-half the slab depth, whichever is the lesser, its presence may be ignored.

If shear reinforcement is required, and the slab must be more than 200 mm deep to do this, then this is to be in accordance with equation (29) of the Code, which is

$$\sum A_{sv} \sin \alpha \geqslant (v - v_c)ud/0.87f_{yv},$$

where $f_{yv}$ is the characteristic strength of the shear reinforcement, $A_{sv}$ is the area of the shear reinforcement, and $\alpha$ is the angle between the shear reinforcement and the plane of the slab.

The minimum value to be taken for $(v - v_c)$ in the above equation is 0.4 N/mm$^2$.

As the shear reinforcement will normally take the form of vertical links, the angle $\alpha$ will be 90° and hence sin $\alpha$ will be 1.

The shear reinforcement itself will be distributed evenly around the zone on at least two perimeters and spacing around a perimeter should not exceed 1.5$d$. Figure 3.17 of the Code illustrates the punching shear zones and where the reinforcement should be positioned. If ribbed slabs, as described in the previous chapter, are used and a critical perimeter cuts ribs, then each rib should be designed to resist an equal proprtion of the applied effective design shear force (see clause 3.6.4.1).

The following design example illustrates the procedures to be adopted in analysing a braced frame structure to determine the bending moments and shear forces. Calculations to find the areas of reinforcement required are not shown as these procedures are described elsewhere. Some of the results, however, will be used to illustrate the deflection and shear calculations.

## EXAMPLE 14.1

A floor slab in a building where stability is provided by shear walls in one direction (N–S) and by stairs and lift well in the other direction (E–W) is divided into bays as shown in Fig. 14.7. The slab is to be without drops and is supported internally and on the external long sides by square columns with heads. The imposed loading on the floor is 5.0 kN/m$^2$ and an allowance of 2.5 kN/m$^2$ should be made for finishes, etc. The exposure conditions will be mild and the fire resistance period is half an hour. Use Grade 40 concrete and reinforcement Grade 460, Type 2.

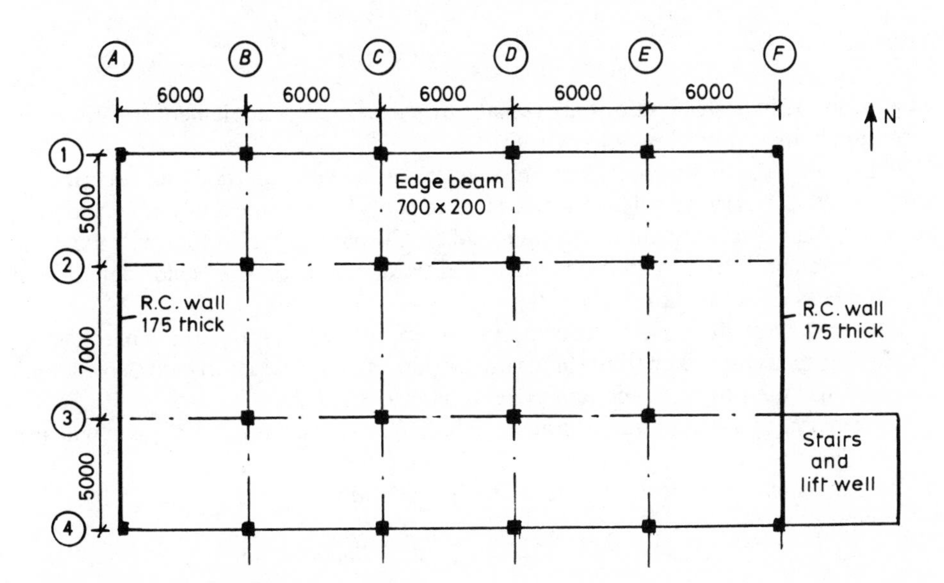


FIG. 14.7 Layout of building.

For the slab thickness assume a span/effective depth ratio of 33. The longest span is 7.0 m so $d$ would be 212 mm. Assuming 20 mm cover and 12 mm bars the overall depth would be 238 mm. Assume a 250 mm slab.

As the columns are square, the heads will also be made square. The maximum value of $h_c$, the effective diameter, is one-quarter of the smallest span, so $h_c = 5.0/4 = 1.25$ m. If the column head is 1.1 m square then $h_c = 1.24$ m. Now decide on the depth of the head.

From clause 3.7.1.3, $l_{hmax} = l_c + 2(d_h - 40)$.
So $1100 = 300 + 2(d_h - 40)$, i.e. $d_h = 440$ mm.

Make head 450 mm deep overall and make edge beam 700 mm deep overall.

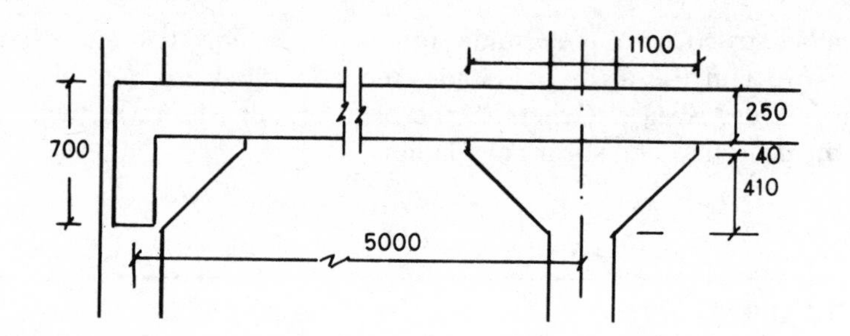


Characteristic loads:

Dead, slab $=6.0\ \text{kN/m}^2$

finishes $=2.5\ \text{kN/m}^2$

Total $=8.5\ \text{kN/m}^2$

Imposed $=5.0\ \text{kN/m}^2$.

Maximum design load $n=8.5\times1.4+5\times1.6=19.9\ \text{kN/m}^2$.

## Analysis

As wind forces in both directions are resisted by shear walls or equivalent the frames in both directions can be classes as braced.

In the north–south direction, as the characteristic imposed load does not exceed 1.25 times the characteristic dead load we can use an equivalent frame analysis considering the single load case of maximum design load on all spans provided that we adjust the support moments and span moments in accordance with the solid slab design requirements given in clause 3.5.2.3.

In the east–west direction, we can analyse as for the north–south direction, but as we have equal spans and more than three rows of panels in the direction being considered we could use the simplified method given in clause 3.7.2.7.

It is proposed to use a subframe analysis in both directions, but a comparison will also be given for the east–west direction using the simplified method.

The panels will be divided into strips as shown below.

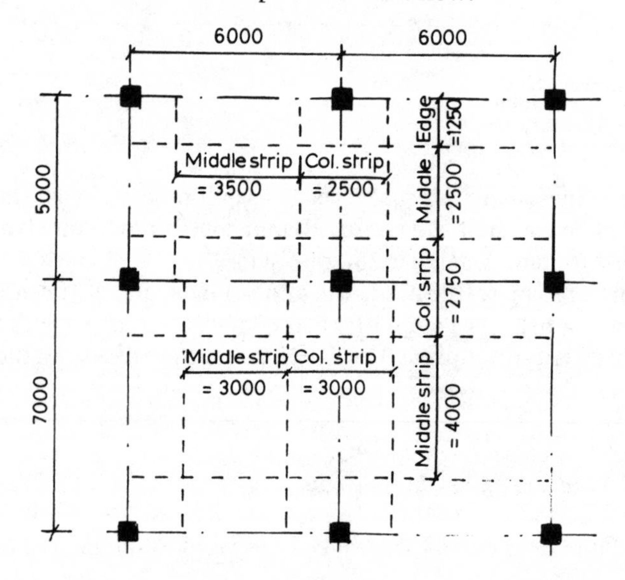

## North–south direction

On the line of the internal columns the width of the strip for loading and analysis is 6.0 m, so the maximum design load per unit length = 19.9 × 6 = 119.4 kN/m.

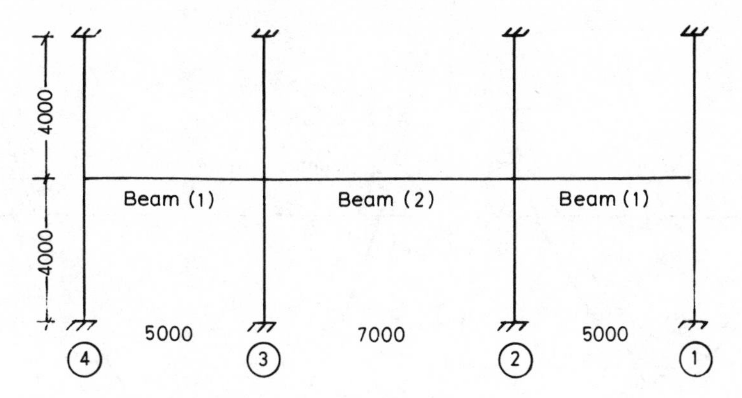


Stiffness:

Columns $I = 300 \times 300^3/12 = 0.675 \times 10^9\ \text{mm}^4$,

$k_c = (0.675 \times 10^3)/(4 \times 10^3) = 0.169 \times 10^6\ \text{mm}^3$.

Equivalent beam:

$I = 6000 \times 250^3/12 = 7.81 \times 10^9\ \text{mm}^4$.

Beam (1)

$k_{b1} = (7.81 \times 10^9)/(5 \times 10^3) = 1.56 \times 10^6\ \text{mm}^3$

Beam (2)

$k_{b2} = (7.81 \times 10^3)/(7 \times 10^3) = 1.116 \times 10^6\ \text{mm}^3$.

From an analysis of the single load case of all spans loaded with the maximum design load we get the following results:

| 4 | 3 | 3 | 2 | 2 | 1 | | |
|---|---|---|---|---|---|---|---|
| −35 | −430 | −452 | −452 | −430 | −35 | Support | Beam moments |
| +167 | | +279 | | +167 | | Span | |
| 17.5 | | 11 | | 11 | 17.5 | Upper | Column moments |
| 17.5 | | 11 | | 11 | 17.5 | Lower | |
| 220 | 377 | 418 | 418 | 377 | 220 | Shear | |

The bending moment diagram is shown in Fig. 14.8.

We can now reduce the support moments by 20%, but as the exterior support is not very large, we will reduce the internal support only. The span moments will increase accordingly. We also have to comply with the 70% maximum moment requirement.

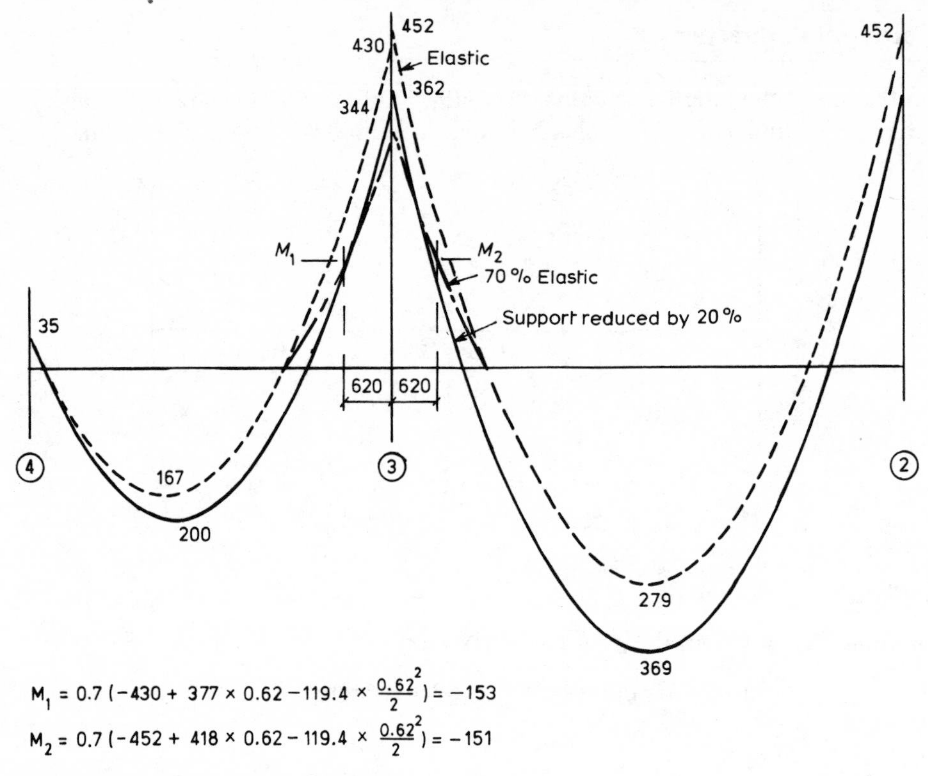


FIG. 14.8 Bending moment diagram.

From the bending moment diagram it can be seen that this 70% line controls part of the new diagram. We are now allowed to reduce the support moments even further by taking the moment at $h_c/2$ from the centre line of the column, provided the sum of the positive moment and average of the negative moments is not less than $19.9 \times \frac{6}{8}(l-2 \times 1.24/3)^2$ where $l$ is 5 m and 7 m respectively for the short and long span. So short-span value = 260 kN m and long-span value = 569 kN m. In the short span at the external column the moment will be positive if we take the value at $h_c/2$ from the centre line of the support. This will be ignored and the value of 35 will be used. The average of the negative moments is therefore $(35+153)/2=94$, which added to the span moment of 200 gives a total of 294; this is in excess of the required moment.

In the long span the sum of the moments is $151+369=520$, which is less than the minimum value. The negative moments must be increased by 49 to 200 to satisfy the minimum requirement.

For the external column the maximum moment which can be transferred is given by

$M_{t,max}=0.15b_e d^2 f_{cu}$
$b_e=300+250=550$
$d=250-20-8=222$ mm (assuming $16\phi$ bars which will be in outer layer).

So
$M_{t,max}=0.15 \times 550 \times 222^2 \times 40 \times 10^{-6}=163$ kN m
which is well in excess of actual moment.

The design moments can be shown diagrammatically as follows:

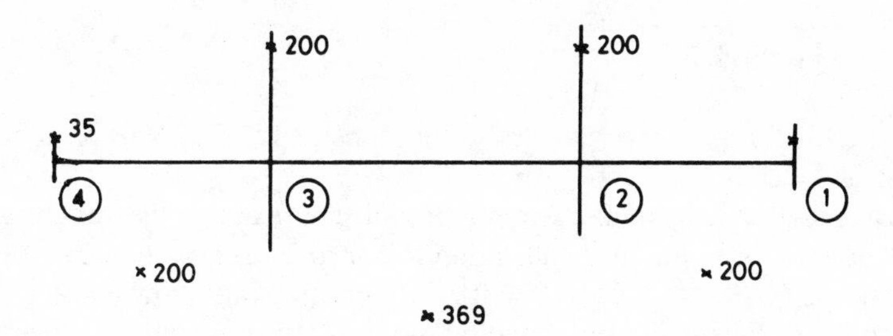


At the first interior support, i.e. at the junction of the 5 m and 7 m spans, we have the problem that the column strip in the 5 m span is 2.5 m while in the 7 m span it is 3 m, so the wider column strip will be used.

The moments from analysis can now be apportioned in accordance with Table 3.20 of the Code.

## Column strip

Exterior support

$=0.75 \times 35 = 26.3$ kN m on strip 2.5 m wide $= 10.5$ kN m/m.

Centre of first span

$=0.55 \times 200 = 110$ kN m on strip 2.5 m wide $= 44.0$ kN m/m.

First interior support

$=0.75 \times 200 = 150$ kN m on strip 3.0 m wide $= 50.0$ kN m/m.

Centre of interior span

$=0.55 \times 369 = 203$ kN m on strip 3.0 m wide $= 67.7$ kN m/m.

## Middle strip

Exterior support

$=0.25 \times 35 = 8.8$ kN m on strip 2.5 m wide $= 3.5$ kN m/m.

Centre of first span

$=0.45 \times 200 = 90$ kN m on strip 2.5 m wide $= 36.0$ kN m/m.

First interior support

$=0.25 \times 200 = 50$ kN m on strip 3.0 m wide $= 16.7$ kN m/m.

Centre of interior span

$=0.45 \times 369 = 166$ kN m on strip 3.0 m wide $= 55.4$ kN m/m.

For positioning the reinforcement, the bars in this direction will be in the outer layers, so assuming $12\phi$ bars, $d = 250 - 20 - 6 = 224$ mm.

Minimum area $= (0.13/100) \times 1000 \times 250 = 325$ mm$^2$/m.

Use $12\phi$ at 300 centres (377 mm$^2$/m).

With this area of reinforcement the moment of resistance can be calculated as 32.1 kN m/m. Any moment less than this value will be covered by the minimum reinforcement. It can also be found that $12\phi$ bars will be satisfactory in all positions.

## Deflection – clause 3.7.8

Consider interior span. In the span the total moment is 369 kN m on a band 6.0 m wide, so $M/bd^2 = 1.23$.

If we assume that the reinforcement provided is exactly the amount required, Table 3.11 of the Code gives a modification factor for tension reinforcement as 1.30.

The allowable $l/d = 26 \times 1.3 \times 0.9 = 30.4$ which gives an allowable span of 6.8 m. This is not quite sufficient, but by providing more reinforcement than is required it will be found to be satisfactory.

## Shear

Internal column using equation (25) of the Code:

$V_t = 795$ kN

$d = (224 + 212)/2 = 218$ average

$x = 1100 + 3 \times 218 = 1754$ mm

$u = 4(1100 + 3 \times 218) = 7016$ mm

$M_t = 11 + 11 = 22$ kN m

$$V_{eff} = 795\left(1 + \frac{1.5 \times 22 \times 10^6}{795 \times 10^3 \times 1754}\right)$$

$$= 795 \times 1.024 = 814 \text{ kN.}$$

Note that the enhancement factor of 1.15 as suggested in the Code appears to be conservative in this case.

$v = (814 \times 10^3)/(7016 \times 218) = 0.53$ N/mm$^2$.

This cannot be checked completely until the reinforcement in the direction at right angles has been calculated, but working the reverse way from Table 3.9 of the Code we shall need an average percentage of reinforcement of approximately 0.25%. At the column head itself it can be found that $V_{eff} = 825$ kN and $v_{max} = 0.86$ N/mm$^2$.

## East–west direction: frame analysis

The strip for loading and for analysis will be 6.0 m wide. The exterior support is a continuous wall 175 mm thick, and we shall take a 6.0 m length, the same as the beam strip.

The maximum design load per metre run will be the same as in the north–south direction. By carrying out the same procedures as before we can arrive at design moments which can be shown diagrammatically as follows:

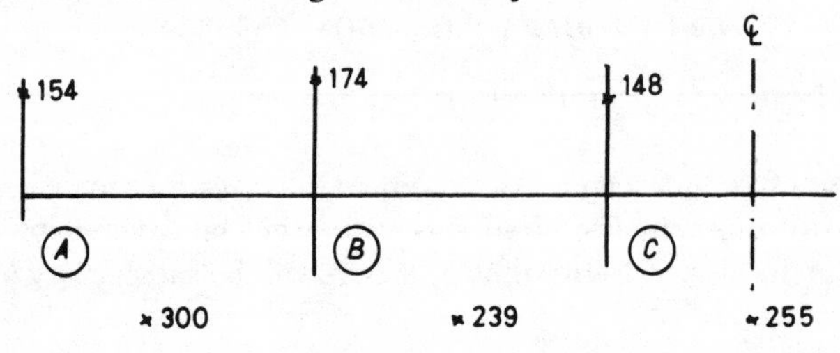

From the layout of the division of panels it can be seen that the column strip is 2.75 m wide, and the middle strip is $6.0-2.75=3.25$ m, which is made up of 2.0 m from 7 m wide panel and 1.25 m from 5 m wide panel.

The moments in the strips would now be apportioned in accordance with Table 3.20 of the Code, and in calculating the areas of reinforcement it must be remembered that $d$ will now be 212 mm. The minimum reinforcement will be as before, but due to the slightly reduced effective depth will provide a smaller moment of resistance.

The deflection will be checked on the exterior span. As the reinforcement in both directions is now known the shear around the column heads can also be checked.

## Simplified method

Maximum design load per unit length $=19.9\times 6=119.4$ kN/m.
From Table 3.19 of the Code, $F=119.4\times 6=716.4$ kN.

Effective span, $l=6-2\times 1.24/3=5.17$ m for internal panels

and

$l=6-(1.24+0.175)/3=5.53$ m for external panels.

Bending moments:

Outer support (classed as column) $=-0.04\times 716.4\times 5.53$
$=-158.5$ kN m.

Near centre of first span $=0.083\times 716.4\times 5.53$
$=329$ kN m.

First interior support $=-0.063\times 716.4\times 5.53$
$=-250$ kN m

or $-0.063\times 716.4\times 5.17$
$=-233$ kN m

(The larger value would be used).

Centre of interior span $=0.071\times 716.4\times 5.17$
$=263$ kN m.

Interior support $=-0.055\times 716.4\times 5.17$
$=-204$ kN m.

If one compares these moments with those from the frame analysis it can be seen that the simplified method gives higher values in all cases. These are particularly noticeable at the interior supports; this is because the simplified method does not make allowance for the flared column head.

For shear we obtain

| | Frame analysis | Simplified |
|---|---|---|
| (i) Exterior support | 322 kN | 322 kN |
| (ii) First interior support | 760 kN | 788 kN |
| (iii) Interior support | 708 kN | 716 kN |

which are very similar.

The simplified method, however, will give considerably higher column moments.

## 14.10 Arrangement of reinforcement

Where a frame analysis has been carried out and a bending moment envelope derived, the reinforcement will be curtailed in accordance with the normal detailing rules of clause 3.12.9. For the simplified method of analysis the simplified rules of clause 3.12.10 may be used.

Whichever method is used, the column strip reinforcement over the column head should be arranged so that two-thirds of the amount of reinforcement required should be placed in a width equal to half that of the column strip and central with the column.

The detailing for the simplified method for slabs as shown in Figure 3.25 of the Code gives dimensions from the face of the support as a proportion of the effective span. For columns with heads the face of the support will be taken as the edge of the head. As these heads can be classed as wide supports, the effective span will be the clear distance between heads plus the effective depth of the slab. The dimensions for the middle strip reinforcement will be the same as for the column strip.

# STAIRCASES 15

## 15.1 General requirements

The generally accepted definitions of parts of a flight of stairs are given in Fig. 15.1.

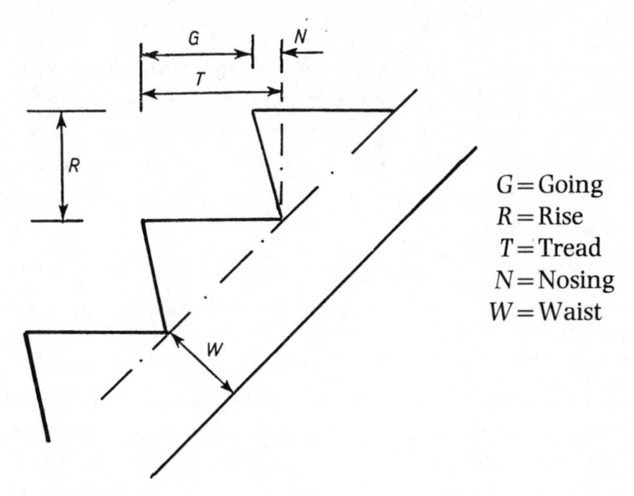


FIG. 15.1 Notation for stairs.

Although there are different requirements for the maximum rise and going for different occupations, the general rules for offices are:

Rise: not greater than 190 mm.
Going: not less than 250 mm.
Number of risers in a flight: not more than 16.

Another value in common usage is that twice the rise plus the going shall be between 550 mm and 700 mm, i.e. $550 < (2R + G) < 700$.

Another requirement is that for headroom the vertical distance from the pitch line should not be less than 2.0 m, where the pitch line is a notional line connecting the nosings of all treads in a flight.

The dimensions given above apply to the finished stairs, not to the reinforced concrete part of the stairs. As each staircase must have an equal rise and an equal going for every step between consecutive floors, it frequently means that with different finishes on the floors and stairs the risers at the top and bottom of the concrete stairs have to be adjusted to suit. For example, if we have 2.85 m between finished floors we could have 15 rises of 190 mm each (finish). If we have 50 mm finish on the lower floor, 40 mm finish on the upper floor and 20 mm finish on the stairs, the height of the bottom concrete riser would be $190 + 50 - 20 = 220$ mm, and the top concrete riser would be $190 - 40 + 20 = 170$ mm. This means that the junctions of treads and risers will not lie on a straight line and will have to be taken into account when determining the thickness of the waist. If the staircase cannot be accommodated in a single flight, then we have to provide two flights, with a landing, generally at mid height.

Stair spans may be divided broadly into two types, spanning transversely and longitudinally.

## 15.2 Transverse spans

In this category we have:

1. steps cantilever from wall at one side
2. steps span between supports at each side – e.g. wall, stringer beam
3. steps cantilever across a central spine beam.

In the first two cases the waist need only be thick enough to accommodate the distribution reinforcement with the necessary cover, and thicknesses between 50 and 75 mm are adequate. This thickness is considered adequate to provide effective lateral distribution of load so that we need only consider a uniformly distributed load. In case (3) there is no lateral distribution between adjacent treads so that each tread must be designed for concentrated load.

### EXAMPLE 15.1

Consider a case (2) staircase with a 2.0 m effective span between centres of supporting walls. The staircase is internal (i.e. mild exposure) without any additional finishes and is to carry an imposed load of 3 kN/m$^2$. $R=175$ mm, $G=250$ mm, $N=25$ mm, $W=75$ mm. Use concrete Grade C35 and reinforcement Grade 460.

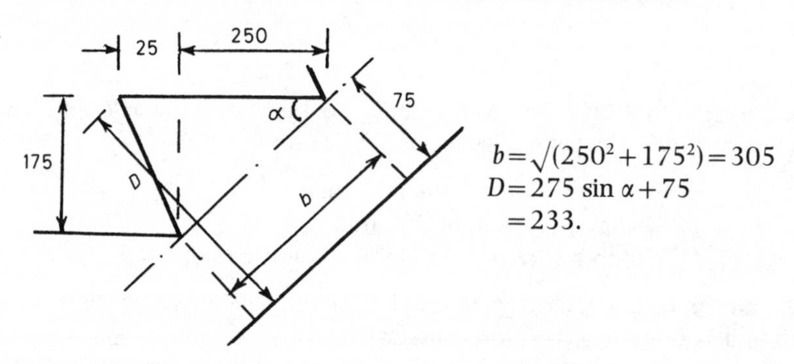


There is no universal agreement as to how these steps should be designed, but one method is to assume the lever arm as half the maximum thickness from the nose to the soffit measured normal to the soffit.

#### Loading

Dead:

One step $=\frac{1}{2}\times 0.275\times 0.175\times 24=0.58$ kN/m run of step

Waist $\quad =0.305\times 0.075\times 24 \quad =0.55$ kN/m run of step

1.13 kN/m run of step.

Imposed $= 0.25 \times 3.0 = 0.75$ kN/m run of step.

Design load $= (1.13 \times 1.4 + 0.75 \times 1.6) \times 2 = 5.6$ kN.

$M = (5.6 \times 2^2)/8 = 2.80$ kN m.

Effective lever arm $= 233/2 = 116$ mm.

$$A_s = \frac{2.8 \times 10^6}{116 \times 0.87 \times 460} = 60 \text{ mm}^2. \text{ Use } 1\text{–}10\phi.$$

For distribution use

$(0.13 \times 1000 \times 75)/100 = 98$ mm$^2$/m. Suggest $8\phi$ at 300 mm centres.

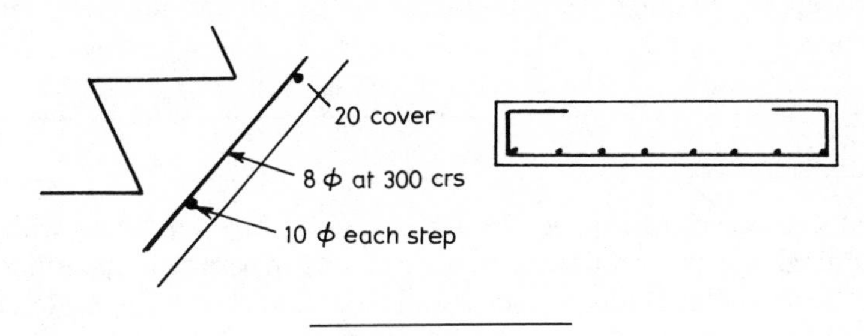


## 15.3 Longitudinal spans

These stairs span between supports at the top and bottom of the flight and are unsupported at the sides. The supports themselves may be (a) beams which are cast monolithic with the stairs at the top and bottom of the actual stairs; or (b) beams or walls at the outside edges of the landing; or (c) the landings themselves spanning at right angles to the stairs. Cases (a) and (b) only are shown in Fig. 15.2, case (c) will be dealt with later.

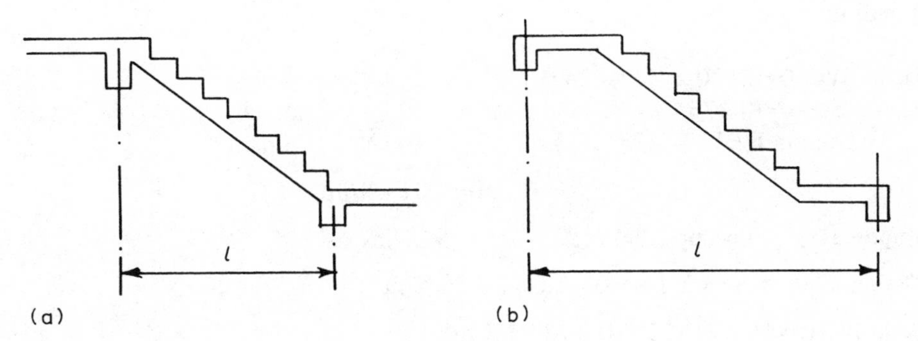


FIG. 15.2 Supports for longitudinal spanning stairs.

In both cases shown, the span of the stairs will be taken as the horizontal distance $l$, the stairs designed as a slab with a thickness equal to the waist thickness and the self weight of the slab as equivalent to the weight of the stairs on the slope.

To find the weight of the waist on plan we must increase the weight on the slope by the ratio $\sqrt{(R^2 + G^2)}/G$. So if $R = 175$, $G = 250$ the ratio is $305/250 = 1.22$. Therefore the weight on plan $=$ weight on slope $\times$ 1.22. For estimating their weight, the steps can be regarded as a slab thickness equal to half the rise and no increase is necessary to obtain the weight on plan.

The imposed load, being on the treads, is the actual load on plan and need not be increased.

In case (a) where the stairs are continuous over the beams it is usual to take the maximum bending moment in the span as $Fl/10$ and allow for continuity in the top over the beams. As we are taking some continuity at the ends of the stairs we must treat it as such when we come to detailing.

For deflection, the Code now recognizes that a flight of stairs is stiffer than a slab of thickness equal to the waist of the stairs. Where the stair flight occupies at least 60% of the span, the allowable span/effective depth ratio can be increased by 15%. This, of course, only applies to staircases spanning in the direction of the flight and without stringer beams.

## EXAMPLE 15.2

An internal staircase consisting of 175 mm risers and 275 mm treads with 250 mm going is required to span a horizontal distance of 3.35 m between supporting beams at the top and bottom of the flight, the total rise of the stairs being 2.275 m (13 steps). The treads have 15 mm granolithic finish. Design the stairs assuming they lead to a place of public assembly without fixed seating. Concrete Grade 35 and high yield reinforcement.

For a one-way spanning slab, with continuity, try a span/effective depth ratio of 30; we can increase this by 15%, so try 34.5.

So $d = 3350/34.5 = 97$ mm. For Grade 35 concrete minimum cover = 20 mm, and this will satisfy 1 hour fire resistance. If we have 12$\phi$ bars then $h = 123$ mm, so try a waist thickness of 130 mm, $d = 104$ mm.

Since $R = 175$ mm and $G = 250$ mm, ratio for weight increase = 1.22.

### Loading

Dead: Waist = 0.130 × 1.22 × 24.0 = 3.81 kN/m$^2$ on plan
Steps = (0.175/2) × 24.0 = 2.10 kN/m$^2$ on plan
Grano = 0.015 × (275/250) × 24.0 = 0.40 kN/m$^2$ on plan

Total = 6.31 kN/m$^2$

Imposed (due to occupancy) = 5.00 kN/m$^2$

Design load = 6.31 × 1.4 + 5.0 × 1.6 = 16.8 kN/m$^2$

$M$ at $Fl/10 = 16.8 \times 3.35^2/10 = 18.9$ kN m

$$M/bd^2 = \frac{18.9 \times 10^6}{10^3 \times 104^2} = 1.75$$

and from tables

$100A_s/bd = 0.46$.

So $A_s = (0.46/100) \times 1000 \times 104 = 478$ mm$^2$/m.
Use 12$\phi$ at 225 crs (503 mm$^2$).

For deflection, modification factor for tension reinforcement is 1.19, so allowable ratio $= 26 \times 1.19 \times 1.15 = 35.6$
Actual ratio $= 32.2$.

Distribution reinforcement $= (0.13 \times 1000 \times 130)/100 = 169\ \text{mm}^2/\text{m}$.

Use $8\phi$ at 300 centres.

The arrangement of reinforcement is shown in Fig. 15.3.

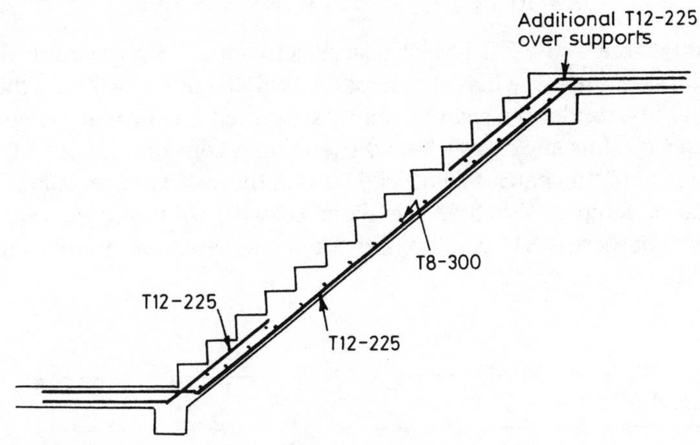


FIG. 15.3 Arrangement of reinforcement in Example 15.2.

## 15.4 Flights or landings built into walls

Where flights or landings which span in the direction of the flight are built at least 110 mm into the walls, a 150 mm strip adjacent to the wall may be deducted from the loaded area. The effective breadth of the staircase will be the clear distance projecting from the wall plus two-thirds of the embedded distance up to a maximum of 80 mm. This is illustrated in Fig. 15.4 and has been taken from CP110 as it appears to have been omitted from BS8110.

For example, assume that the clear breadth of the stair in Example 15.2 is 1.5 m and that there is an open well on one side, the stair flight being built 110 mm into a brick wall on the other side. The loaded area may be taken on a 1.35 mm wide strip and the effective breadth as $1.50 + \frac{2}{3} \times 0.110 = 1.573$ m.

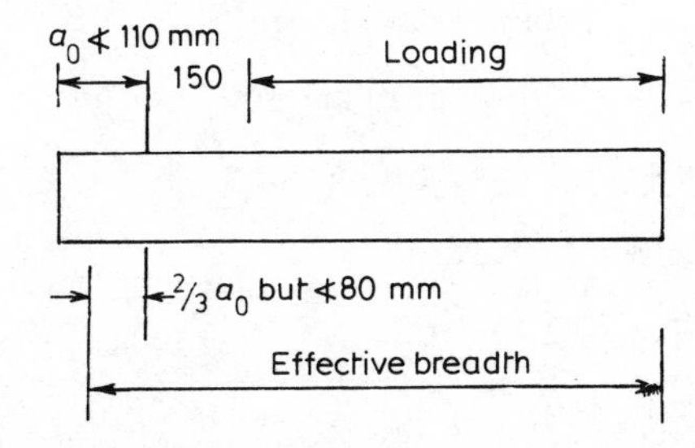


FIG. 15.4 Flights or landings built into walls.

The total design load on the flight $= 17.4 \times 3.35 \times 1.35 = 78.7$ kN.

M at $Fl/10 = (78.7 \times 3.35)/10 = 26.4$ kN m.

This is on a breadth of 1.573 m, so

$$\frac{M}{bd^2} = \frac{26.4 \times 10^6}{1.573 \times 10^3 \times 104^2} = 1.6.$$

$100A_s/bd = 0.42$, so

$$A_s = (0.42/100) \times 1573 \times 104 = 687 \text{ mm}^2.$$

This is the total area so use $7/12\phi$ bars spaced across the 1.5 m clear breadth of stairs.

For case (b) stairs, that is where the supports are at the outside edges of the landing, it is more usual to consider the span as simply supported, the distance $l$ being centre to centre of bearings. In some cases where the landing is continuous with a floor slab the bending moment in the span is taken as $Fl/10$ as in the last example, but this is a matter of individual preference. Whether it is designed with continuity or not, it must be remembered that there will be a monolithic connection and some reinforcement should be provided to prevent cracking.

## EXAMPLE 15.3

Use the information given in Example 15.2 but now omit the beams at the top and bottom of the flight and take the bearings at the extreme ends of the landings, which increases the span by 1.5 m at each end.

The span is now 6.35 m and designing the span as simply supported try a span/effective depth ratio of 24. The stair flight does not occupy 60% of the span so no enhancement factor.

So $d = 6350/24 = 265$. Assuming $16\phi$ bars plus 20 cover, $h = 293$ mm.

Try waist of 300 mm, $d = 272$ mm.

### Loading

Dead: Waist $= 0.30 \times 1.22 \times 24.0 =$ 8.80 kN/m$^2$
Steps and finish $=$ 2.50 kN/m$^2$

Total $=$ 11.30 kN/m$^2$

Imposed: $=$ 5.00 kN/m$^2$

Design load $= 11.3 \times 1.4 + 5 \times 1.6 = 23.8$ kN/m$^2$

$M$ at $Fl/8 = 23.8 \times 6.35^2/8 = 120$ kN m

$M/bd^2 = (120 \times 10^6)/(10^3 \times 272^2) = 1.62$,

and from tables

$100A_s/bd = 0.43$.

So $A_s = (0.43/100) \times 1000 \times 272 = 1170\ \text{mm}^2/\text{m}$.
Use 16$\phi$ at 150 centres (1340 mm$^2$).

For deflection, $f_s = 288 \times 1170/1340 = 252\ \text{N/mm}^2$.

Modification factor for tension reinforcement = 1.29.
Allowable span/effective depth ratio = 20 × 1.29 = 25.8.
Actual ratio = 23.3.

It will have been noticed that the loading across the whole span has been taken as if the stair loading is the same as the landing. This is obviously not the case and if one wishes to carry out the design as accurately as possible the following method will be adopted.

## Landing

300 mm thick, as waist = 7.20 kN/m$^2$
Grano 0.015 × 24.0 = 0.36 kN/m$^2$

Total = 7.56 kN/m$^2$.

Design load = 7.56 × 1.4 + 5.0 × 1.6 = 18.6 kN/m$^2$

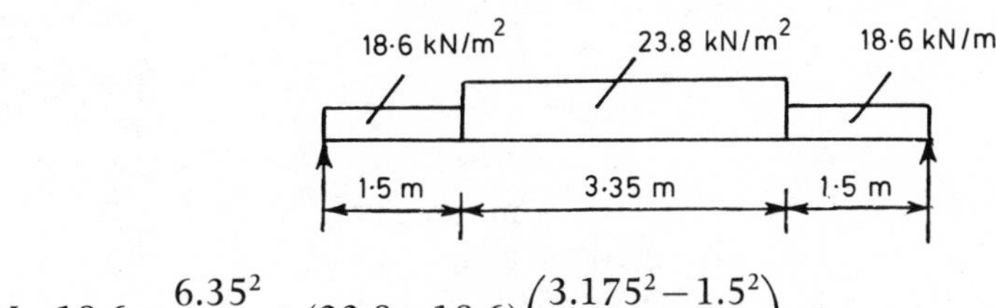


$$M = 18.6 \times \frac{6.35^2}{8} + (23.8 - 18.6)\left(\frac{3.175^2 - 1.5^2}{2}\right)$$

$$= 114\ \text{kN m}.$$

This is slightly less than before.

---

Up to now we have been considering the landings spanning in the same direction as the stairs, but there are many cases where the landings span at right angles to the stairs, classed as case (c) earlier on. These landings now become the supporting members within the definition in the Code and the effective span should be taken as the clear horizontal distance between the supporting members plus half the breadths of the supporting members subject to maximum additions of 900 mm at both ends. So if we have a staircase of eight goings at 250 mm with a landing at one end of 1.5 m and a landing at the other end of 2.2 m, the span of the stair is 8 × 0.25 + 1.50/2 + 0.9 (max) = 3.65 m. The following example will illustrate the procedure.

## EXAMPLE 15.4

Design the stairs shown where the risers are 175 mm, going is 250 mm, tread is 275 mm, 25 mm tile finish on treads, 15 mm plaster on underside and an imposed load of 3 kN/m$^2$.

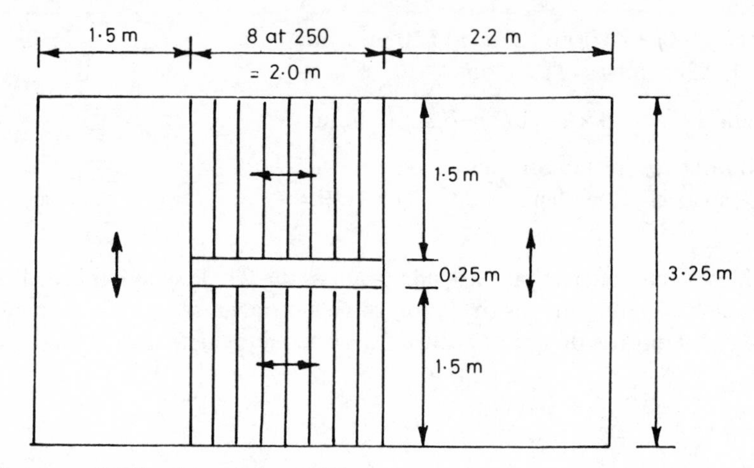


Span of stairs $=0.75+2.0+0.9=3.65$ m.
Using same grade of concrete as before, i.e. $f_{cu}=35, f_y=460$, and assuming continuity, take span/effective depth ratio as 30, $d=122$ mm. Allowing 26 mm for cover etc, $h=148$ mm. For landing if we assume a ratio of 24 (simply supported) we shall get $d=135$, which means $h=161$ mm.

Try a waist thickness of 150 mm and landing thickness of 160 mm.
Ratio for increase in weight due to slope $=1.22$.

## Stairs

Loading Waist $=0.150\times1.22\times24.0 \quad =4.40$ kN/m$^2$ on plan
Steps $=(0.175/2)\times24.0 \quad =2.10$ kN/m$^2$
Finish $=0.5\times275/250 \quad =0.55$ kN/m$^2$
Plaster $=(15/25)\times0.43\times1.36=0.35$ kN/m$^2$

$=$ Total dead $=7.40$ kN/m$^2$

Design load $=7.4\times1.4+3\times1.6=15.2$ kN/m$^2$.

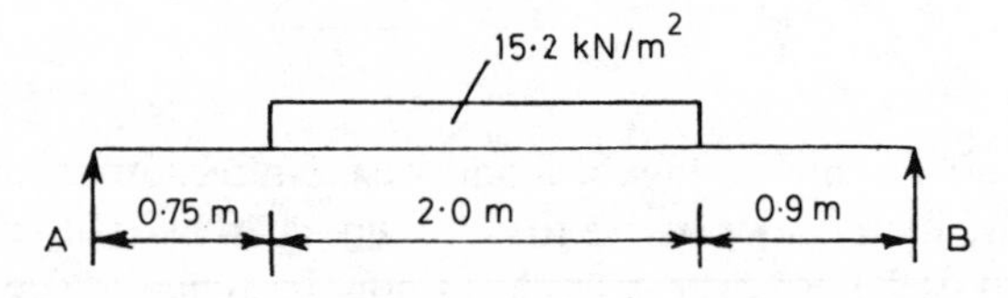


$R_A=15.2\times2.0\times1.9/3.65=15.8$ kN.
$R_B=30.4-15.8=14.6$ kN.
Maximum $M$ at $0.9+14.6/15.2=1.86$ m from $B$.
$M=14.6\times1.86-15.2\times0.96^2/2=20.2$ kN m.

$$\frac{M}{bd^2}=\frac{20.2\times10^6}{10^3\times124^2}=1.31,$$

and from tables

$100A_s/bd=0.35$.

$A_s=(0.35/100)\times1000\times124=434$ mm$^2$/m. Use 12$\phi$ at 250 centres (452 mm$^2$).

Factor for tension reinforcement $=1.3$.

Allowable $l/d = 26 \times 1.3 = 33.8$.

Actual ratio = 29.4, which is satisfactory.

Distribution reinforcement = $(0.13 \times 1000 \times 150)/100 = 195\ \text{mm}^2/\text{m}$.

Use $8\phi$ at 250 centres.

**Landing**

For small landing we shall assume load from stairs is carried uniformly on whole width.

**Loading**

Slab $0.16 \times 24.0 = 3.84\ \text{kN/m}^2$
Finish $= 0.50\ \text{kN/m}^2$
Plaster $= 0.26\ \text{kN/m}^2$
Total $= 4.60\ \text{kN/m}^2$

Design load $= 4.6 \times 1.4 + 3 \times 1.6 = 11.2\ \text{kN/m}^2$.

Total load per metre run of landing is $11.2 \times 1.5 + 15.8 = 32.6$ kN.

$M$ at $Fl/8 = 32.6 \times 3.25^2/8 = 43.1$ kN m.

$b = 1500$ mm, $d = 134$ mm.

$M/bd^2 = 1.60$, and from tables

$100A_s/bd = 0.42$.

$A_s = 844\ \text{mm}^2$.

This is total so use $8/12\phi$ bars ($905\ \text{mm}^2$).

$f_s = 288 \times 844/905 = 269\ \text{N/mm}^2$ so factor for tension reinforcement = 1.24.

Allowable $l/d = 20 \times 1.24 = 24.8$.

Actual ratio = 24.2, which is just satisfactory.

For the wider landing we shall assume the effective width carrying the load from the stairs is $0.9 \times 2 = 1.8$ m. The calculations would therefore be done as for the smaller landing. For the reinforcement in the remaining 400 mm it is suggested that bars of the same size and spacing as for the main part of the landing slab are put in.

## 15.5 Stairs with quarter landings

One further type of stair is that surrounding a lift well, an example being shown in Fig. 15.5.

At each floor level there is a beam and large landing, and between two successive floors occur two small landings. Each small landing slab receives steel from two flights and is supported by two walls at right angles to one another.

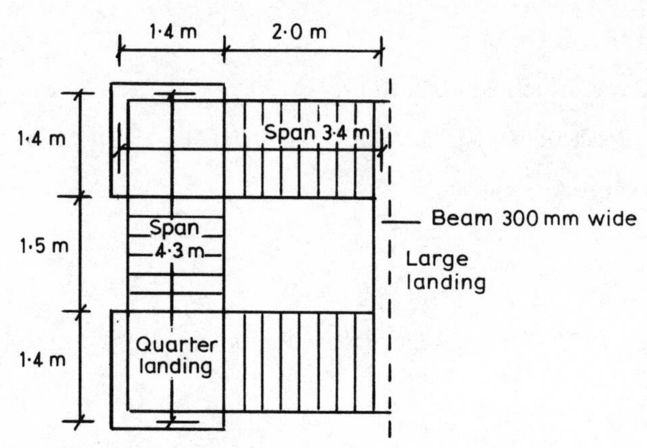


FIG. 15.5 Stairs surrounding a lift well.

The Code says that the loads on the small landings may be assumed to be divided equally between the two spans. So if $n_1$ is the load per unit area of the landings, and $n_2$ is the load per unit area of the stairs, the loading from the flight from the large landing to small landing is:

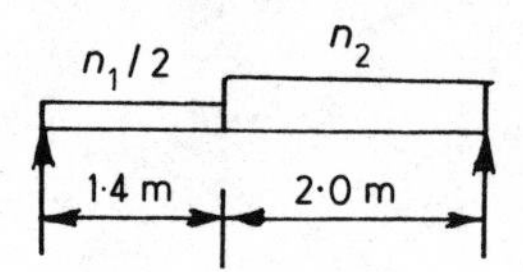


and for the flight between the small landings is:

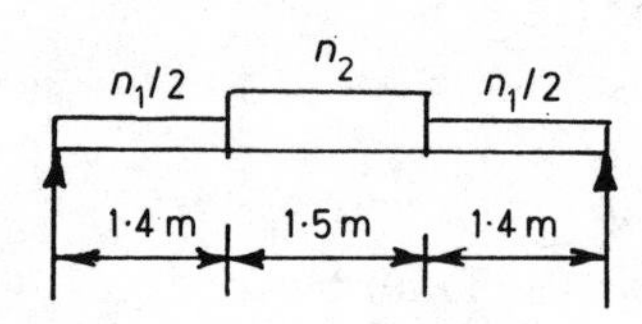

# BASES 16

The design of bases is dealt with in section 3.11 of the Code under pad footings and pile caps. Pad footings considered here will be strip footings to walls, and isolated bases carrying one or two columns.

In determining the size of a base it must be remembered that BS8004: 1986, the Code of Practice for Foundations, gives presumed bearing values for preliminary design, allowable bearing pressures being generally determined by permissible settlements. Ultimate bearing capacities will have to be considered, but we shall be considering serviceability conditions as the most critical criteria. This means that the column loads and moments on which the column has been designed, are not the loads and moments required in determining the base size, either pad footing or pile cap. The loads and moments required are those obtained from 1.0 $G_k$, 1.0 $Q_k$ and 1.0 $W_k$, where applicable. Where the main consideration is the maximum direct load on a column this will simply be a direct proportion using the ratio of the serviceability loads to ultimate loads at each floor to obtain the reactions due to dead and imposed loads. The reduction factor for the imposed load depending on the number of floors carried can then be applied. In other cases it will require an additional line of print-out results from the computer. Designers will quickly learn which is the better system, but loads and moments will be required at serviceability and ultimate limit states.

When the size of the base has been determined from serviceability loadings, the base will then be designed using ultimate loads.

For definitions of terms reference should be made to BS8004: 1986, but in the calculations that follow it is assumed that in values given for allowable bearing pressure allowance has been made in the net loading intensity so that the loads to be considered in determining the base size are those from the column and base only. In other words, the allowable bearing pressure given is the net loading intensity due to column and base.

The thickness of the base must be sufficient to resist the bending moments and shear forces at ultimate limit state. It must also be relatively thick, however, so that it will act as a rigid member and not a flexible member so that the assumption made of linear distribution of ground pressure (clause 3.11.2.1) holds good. The actual base will not be rigid, nor will the pressure be uniform beneath it, but solutions using this concept are usually satisfactory.

In calculations for bending moments and shear forces the weight of the base can be ignored, and the ground pressure is calculated from the column loads only. The uniform pressure caused by the uniform weight of the base will cancel one another. The load to be used for these calculations is therefore the column load only, and this will be the ultimate load $N$, as these calculations are at ultimate limit state.

## 16.1 Pad footings

### 16.1.1 Moments

In an isolated base, i.e. a single column or wall, the critical section for bending is taken at the face of the column or wall (see clause 3.11.2.2).

The moment at any vertical section passing completely across a base should be taken as that due to all external ultimate loads and reactions on one side of that section.

Consider a square column of side $c$ metres and a square base of side $l$ metres with a column load of $N$ kN, as shown in Fig. 16.1. The load on the shaded portion is

$$\frac{N}{l^2} \times \frac{l(l-c)}{2} = \frac{N}{l}\left(\frac{l-c}{2}\right) \text{ kN.}$$

The distance of the centre of gravity of the loaded area from the face of the column $= (l-c)/4$ so the moment at the column face is $(N/8l)\,(l-c)^2$ kN m. As the base and column are square the moment across the column face at right angles will be the same.

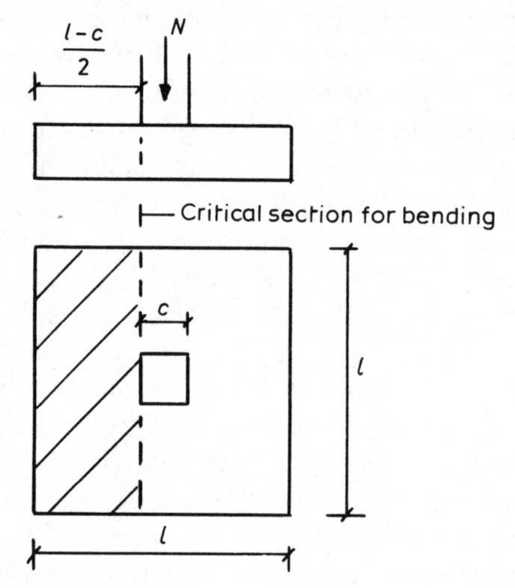


FIG. 16.1 Moment in square base.

The more general case of a rectangular column on a rectangular base is shown in Fig. 16.2. Note that $l_x$ is the longer side of the base, as given in the notation in clause 3.11.1, although it is not referred to subsequently.

It can be readily found that

$$M_{xx} = (N/8l_x)\,(l_x - c_x)^2$$

$$M_{yy} = (N/8l_y)\,(l_y - c_y)^2$$

By assuming an effective depth the area of reinforcement required can be found either from design charts or tables. The assessment of an effective depth is usually one of judgement or design office procedure, but a ratio of width of base to overall thickness of 5 is a fairly good starting point. In the case of a rectangular base the width is the smaller dimension.

Once one has found the area of reinforcement required, the distribution of the bars will generally be uniform across the section considered, but will also depend on the

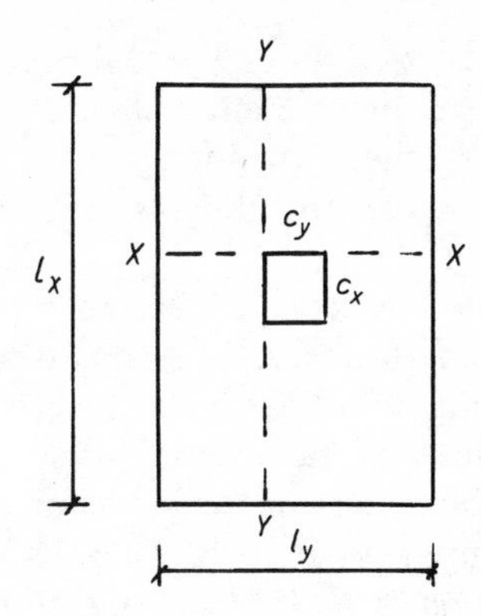


FIG. 16.2 Rectangular base.

relative size of the column and pad footing. This is described in 3.11.3.2 but using a different expression from that given below.

If $l_x$ is greater than $1.5(c_x + 3d)$, then two-thirds of the reinforcement spanning in the $l_y$ direction will be banded within a width of $(c_x + 3d)$. The remainder will be spread evenly over the outer parts of the section.

If $l_y$ is greater than $1.5(c_y + 3d)$, then two-thirds of the reinforcement spanning in the $l_x$ direction will be banded within a width of $(c_y + 3d)$. The remainder will be spread evenly over the outer parts of the section.

For example, if we have a 400 × 300 column on a pad footing with an effective depth of 450 mm, the maximum size of the base for uniform distribution of reinforcement in both directions is

$$l_y = 1.5(300 + 3 \times 450) = 2475 \text{ mm}$$
$$l_x = 1.5(400 + 3 \times 450) = 2625 \text{ mm}$$

For a combined footing, the moments at sections along the line joining the centre line of columns will be calculated as for beams, but will be taken at the column faces in the regions adjacent to the columns.

In determining the size and shape of the base it will usually be based on the service loads. If the ratio of the ultimate loads to service loads is not constant then the centre of gravity of the loads will not remain the same and the pressure distribution at ultimate will not be *pro rata* with the service pressures. This cannot be avoided, but it is felt that the pressure distribution at service loads is the most important.

There will usually be a moment in the top of the base between the columns, but the moments in the region of the columns may be in the top or bottom depending on how near the columns are to the edges of the base. For example, the moments in the longitudinal direction of the arrangement of columns shown will be as indicated by the tension lines in Fig. 16.3.

If the base projects beyond the column at the left-hand end then tension stresses would develop in the bottom of the base, similar to the right-hand end.

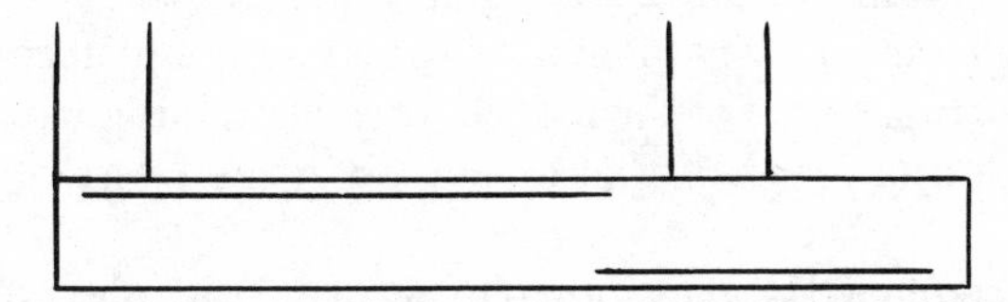

FIG. 16.3 Lines of tension stress in a particular combined base.

In the transverse direction, i.e. at right angles to the section shown in Fig. 16.3, the tension stresses and hence the reinforcement will always be in the bottom.

The Code does not explain how to calculate the moments and hence the reinforcement in the transverse direction, other than to say that the moments should be taken at the face of the column.

The recommendations for the layout of reinforcement that were given in the draft for comment are no longer included. However, the standard method of detailing structural concrete as prepared by the Joint Committee of the Concrete Society and the Institution of Structural Engineers does include the original recommendations. In effect, for the transverse direction, the combined base is treated as two individual column bases.

Consider the layout of columns shown in Fig. 16.4. This is a rectangular base $l_x$ by $l_y$ with the centres of the columns as $l_c$. The columns are different sizes, as indicated, where $c_{x2}$ is greater than $c_{x1}$.

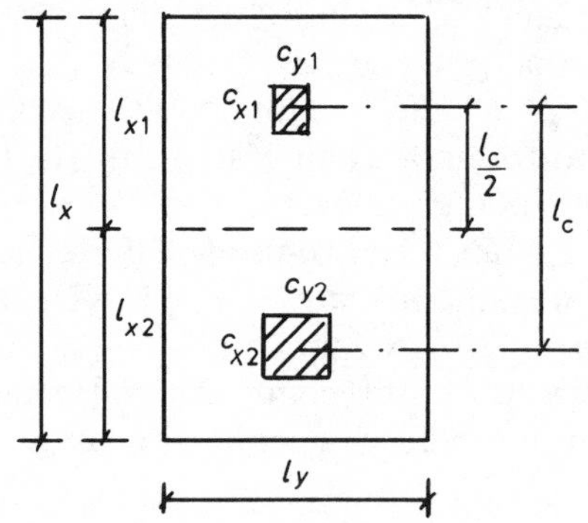


FIG. 16.4 Combined base.

A line is drawn across the base, midway between the column centres, giving us two bases $l_{x1}$ by $l_y$ and $l_{x2}$ by $l_y$. The moment at the face of both columns can be found from the pressure distribution as for an individual base. The banding of the reinforcement in the transverse direction is then carried out for each separate area as for an individual base, as previously described. For example, if $l_{x1}$ is greater than $1.5\,(c_{x1} + 3d)$, two-thirds of the reinforcement is concentrated in a band $(c_{x1} + 3d)$ centred on the column. A problem can arise, however, when the column face is closer than $1.5d$ to the edge of the base. The column may, in fact, be on the edge of the base. In cases where the column is closer to the edge of the base than $1.5d$, it is suggested that the band width should be taken as the distance from the edge of the base to a line $1.5d$ from the column face on the other side of the column. So for a column on the edge of the base the band width is $(c_{x1} + 1.5d)$.

If the reinforcement is concentrated in bands, the spacing of the reinforcement between these bands will have to be examined to ensure it complies with the bar spacing rules.

For the reinforcement in the longitudinal direction, i.e. the $l_x$ direction, we consider top and bottom reinforcement separately. For the bottom reinforcement we again use the rules for an individual base. For the top reinforcement, this is banded in a width the lesser of $(c_{y1} + 3d)$ and $(c_{y2} + 3d)$ if $l_y$ is greater than the lesser of $1.5(c_{y1} + 3d)$ and $1.5(C_{y2} + 3d)$.

For a trapezoidal combined base as shown in Fig. 16.5 the rules are the same as for a rectangular base except that for the top reinforcement between the columns, all the

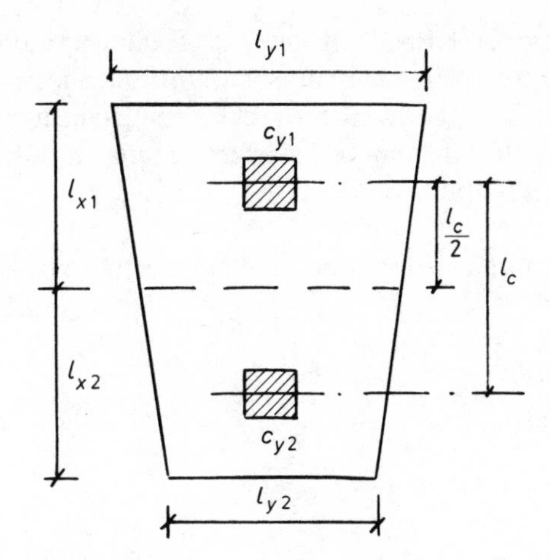


FIG. 16.5 Trapezoidal combined base.

reinforcement should lie within $l_{y2}$. If $l_{y2}$ is greater than $1.5(c_{y2} + 3d)$ two-thirds of the reinforcement should lie within a central band $(c_{y2} + 3d)$.

As will be seen later in a worked example, the traditional layout for the top reinforcement is rather different.

## 16.1.2 Shear

This is again at ultimate limit state and the design shear force is the algebraic sum of all ultimate vertical loads and reactions acting on one side of, or outside the periphery of, the critical section.

The critical sections are outlined below.

### (a) ALONG A VERTICAL SECTION EXTENDING ACROSS THE FULL WIDTH OF THE BASE

The rules as for slabs in clauses 3.5.5 and 3.5.6 will apply.

As discussed in Chapter 9, where we have uniformly distributed loads the critical section will be $2d$ from the face of the support if the shear span is $4d$ or more. In the case of an isolated column base it is possible that the shear span will be less than $4d$ and so the critical section will be half the shear span from the face of the column (see Fig. 16.6) and there will be an enhancement factor of $2d/a_v$.

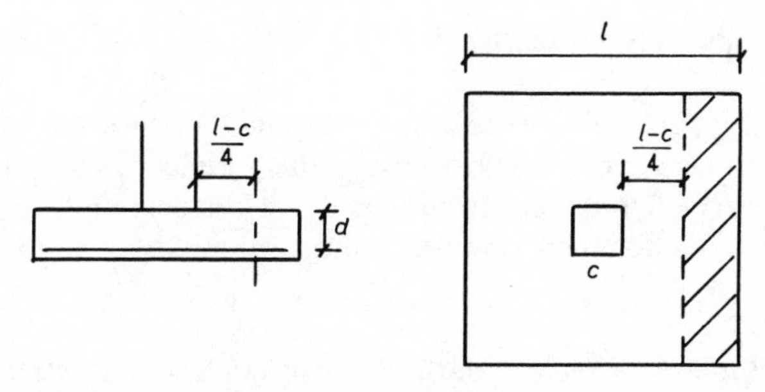


FIG. 16.6 Critical section for shear on single column base.

The allowable stress $v_c$, is based on the area of effective tension reinforcement, which means that any tension reinforcement must continue an effective depth, or be provided with an equivalent anchorage beyond the section being considered. Where the distance cannot be achieved so that the amount of effective tension reinforcement is zero, then the value in Table 3.9 of the Code for $100A_s/bd$ of less than or equal to 0.15 will be taken.

If the column is square of side $c$, and the base is square of side $l$, the shear force is

$$\frac{N}{l}\left(\frac{l-c}{4}\right).$$

The shear stress equals

$$\frac{N}{l}\left(\frac{l-c}{4}\right)\times\frac{10^{-3}}{ld}\ \text{N/mm}^2$$

if $N$ is in kN and $l$, $c$, $d$ are in metres. If we assume that $l=5.5d$ and $c=0.5d$, the expression can be simplified to $N/24000d^2$. $a_v=(l-c)/4=1.25d$, so $2d/a_v=1.6$. Assuming 0.25% of reinforcement giving a $v_c$ of 0.4 N/mm$^2$ we can arrive at an equation $d=8\sqrt{N}$ where $d$ is in millimetres and $N$ is in kilonewtons.

It is assumed that the reinforcement is effective at the critical section and if the percentage of reinforcement is more than 0.25 then the answer given by the above formula will be much greater than required.

For a combined base the shear force will be taken from the shear force diagram at the critical section $2d$, or half the shear span if less, from the face of the column.

### (b) PUNCHING SHEAR

We consider punching shear along a vertical section on the perimeter at a distance $1.5d$ from the column, where $d$ is the average effective depth of the reinforcement spanning in two directions (see Fig. 16.7). The clause referred to here is 3.7.7, the clause dealing with concentrated loads on slabs.

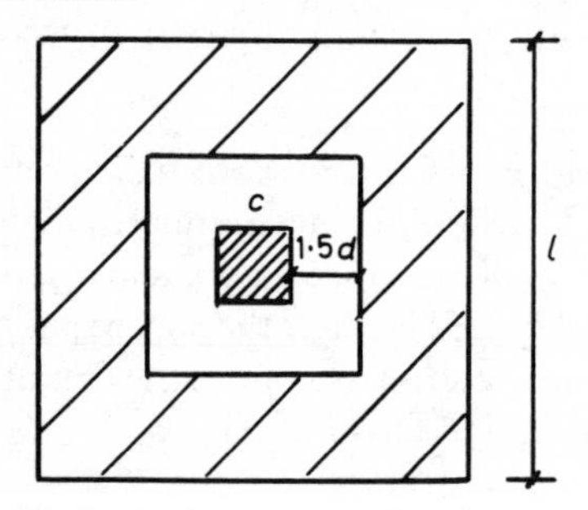


FIG. 16.7 First critical shear perimeter.

Assuming the base and column are square, the length of the first critical perimeter is $4(c+3d)$ and the shear force to be considered on this perimeter $=N-p(c+3d)^2$ where $p$ is the pressure distribution under the base, and $N$ is the column load.

The shear stress $v$ is obtained from the formula

$$v=V/ud$$

where $V$ is the ultimate shear force, $u$ is the effective length of the perimeter, and $d$ is the average effective depth.

In obtaining $v_c$, the shear resistance stress, it should be remembered that the enhancement factor of $2d/a_v$ cannot be used for perimeters of $1.5d$ or more from the face of the column. Where it is desired to check on perimeters closer than $1.5d$, $v_c$ may be increased by $1.5d/a_v$, but as the maximum shear stress on the perimeter of the column itself is to be checked this should be satisfactory in itself. The maximum shear stress on the column perimeter is the lesser of $0.8\sqrt{f_{cu}}$ and 5 N/mm$^2$.

The effective steel area to determine $v_c$ is given in clause 1.2.3.5 and is the total area of tension reinforcement that passes through the perimeter being considered and extends at least an effective depth or 12 times the bar size beyond the perimeter.

Where the column is close to the edge of a base, as often occurs in a combined base, the requirements of 3.7.7.8 should be considered. The effective perimeter is the lesser of the two perimeters indicated in Fig. 16.8.

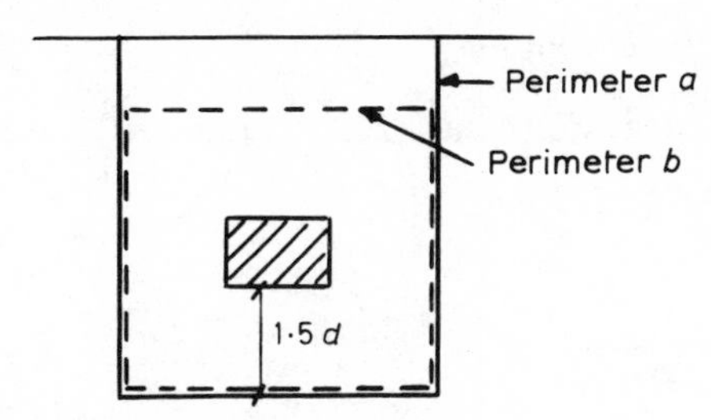


FIG. 16.8 Shear perimeters with load close to free edge.

If the base is subjected to a bending moment, the pressure under the base will not be uniform. To check for shear across the full width of the base is a relatively simple operation, but to check for punching shear is not so straightforward. We are no longer interested in the average punching shear stress as before; we require the maximum shear stress on one face of the shear perimeter. A conservative approach is to assume that the maximum average pressure due to the load and moment outside the critical perimeter acts uniformly over the whole base. With this assumption it can be shown that the shear stress obtained from a uniform pressure considering the vertical load only is increased by a factor

$$1+\frac{3M}{Vl^2}\left(l+a\right),$$

where $M$ is the moment in kN m, $V$ is the column load in kN, $l$ is the length of base in metres in the direction of bending, and $a$ is the length of the side of the critical perimeter, in metres, in the direction of bending.

An alternative method is to use the enhancement factor in clause 3.7.6.2 which states that $V_{eff}=1.15V_t$ for internal columns in braced structures with approximately equal spans.

## 16.1.3 Bond and anchorage

Anchorage bond lengths will be determined from the stress in the bars at the face of the column, and the allowable bond stresses can be obtained from clause 3.12.8.4.

As footings are to be considered as beams and slabs the minimum percentage of reinforcement as given in clause 3.12.5.3 will apply and for crack control the spacing of bars is given in clasue 3.12.11. Deflection is not considered in bases.

## EXAMPLE 16.1

A column 400 × 400 carries a service load of 1100 kN and an ultimate load of 1650 kN. The allowable bearing pressure on the soil is 200 kN/m$^2$. Design a square base using concrete Grade 35 and high-yield reinforcement, type 2.

Neglecting weight of base in first calculation, area required = 1100/200 = 5.5 m$^2$.

Try base 2.4 m square × 0.5 deep.

Weight of base = 2.4 × 2.4 × 0.5 × 24 = 69 kN.

Total load = 1169 kN.

Area required = 1169/200 = 5.84 m$^2$.

This is more than suggested, so try base 2.5 m square × 0.5 m deep.

Pressure = 1175/6.25 = 188 kN/m$^2$.

### Preliminary check for shear

$N = 1650$ kN
$d = 8\sqrt{1650} = 325$ mm.

Overall depth of 500 mm would appear satisfactory.

### Bending moment

Critical bending moment $= \frac{1650}{8 \times 2.5}(2.5 - 0.4)^2 = 364$ kN m.

Assuming adequate blinding layer, minimum cover is 40 mm.
Using 20$\phi$ bars, average effective depth = 500 − 40 − 20 = 440 mm.

$\frac{M}{bd^2} = \frac{364 \times 10^6}{2500 \times 440^2} = 0.75$, and from tables $\frac{100A_s}{bd} = 0.195$.

$A_s = (0.195/100) \times 2500 \times 440 = 2145$ mm$^2$.

We could use 7/20$\phi$ bars, but the spacing would then exceed 300 mm, so suggest 8/20$\phi$ bars (2510 mm$^2$) at 300 centres.

Width of base for banding = 1.5(400 + 3 × 440) = 2580 mm.

This is greater than actual width so reinforcement will be distributed uniformly across base.

Anchorage bond length = (2145/2510) × 34$\phi$ (from Table 3.29 of the Code)

= 29$\phi$ = 580 mm.

Distance from column face to edge of base less end cover is $1250-200-40=1010$ mm. So we can use straight bars without bends unless we need the reinforcement effective for shear.

## Check for shear

(1) Across width of base

Shear span = 1050 mm which is less than $4d$, so check shear at 525 m.

$$\text{Force} = \frac{1650}{2.5} \times 0.525 = 346.5 \text{ kN.}$$

$$\text{Stress} = \frac{346.5 \times 10^3}{2500 \times 440} = 0.32 \text{ N/mm}^2.$$

Reinforcement continues more than an effective depth beyond the section, so

$$\frac{100A_s}{bd} = \frac{100 \times 2510}{2500 \times 440} = 0.23.$$

$v_c = 0.43$ N/mm$^2$ and $2d/a_v = 2 \times 440/525 = 1.67$.

Allowable shear stress $= 0.43 \times 1.67 = 0.72$ N/mm$^2$.

Note: From Table 3.2 it is obvious that the shear stress is satisfactory, but the calculation has been included to show the procedure.

Without bends on the ends of the bars the tension reinforcement becomes ineffective for shear resistance at a distance of $440+40=480$ mm from the edge of the base.

Shear stress here is

$$\frac{1650}{2.5} \times \frac{0.480 \times 10^3}{2500 \times 440} = 0.29 \text{ N/mm}^2.$$

This is less than the minimum value in Table 3.9 of the Code so satisfactory as straight bar.

(2) Punching shear

Perimeter $= 4(400 + 3 \times 440) = 6880$ mm.

Pressure at ultimate $= 264$ kN/m$^2$.

$$\text{Shear Force} = 1650 - 264(0.4 + 3 \times 0.44)^2$$
$$= 869 \text{ kN}$$

$$\text{Shear stress} = \frac{869 \times 10^3}{6880 \times 440} = 0.29 \text{ N/mm}^2.$$

Perimeter is at a distance of $1.5 \times 440 = 660$ mm from face of column and so, to be effective, tension reinforcement must extend 435 mm beyond this to a distance of 1095 mm from face of column. This is not necessary as shear stress is less than minimum value.

A detail of the base is shown in Fig. 16.9.

In this example, if we consider an ultimate bending moment from the column of 80 kN m, the enhancement factor is

$$1+\frac{3\times 80}{1650\times 2.5^2}(2.5+1.72)=1.10$$

and so the shear stress would become 0.32 N/mm$^2$, which is still less than the minimum.

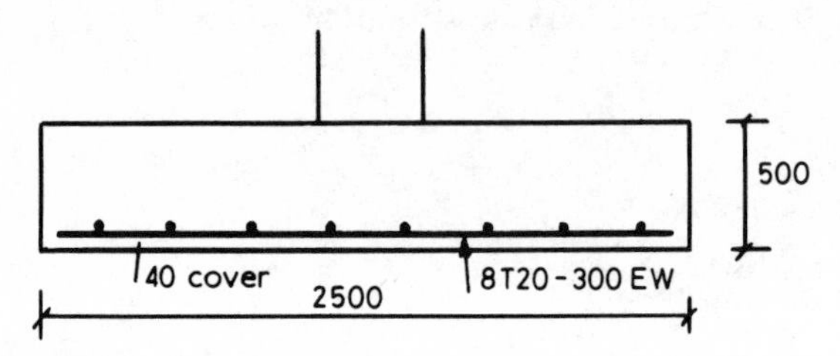


FIG. 16.9 Detail of base.

## EXAMPLE 16.2. Combined footing supporting one exterior and one interior column

An exterior column, 600 × 450, with service loads of 760 kN (dead) and 580 kN (imposed) and an interior column, 600 × 600, with service loads of 1110 kN (dead) and 890 kN (imposed) are to be supported on a rectangular footing whose outside end cannot protrude beyond the outer face of the exterior column. The distance centre to centre of columns is 5.5 m with the exterior column having the 450 dimension at right angles to the line joining the centre line of columns. The allowable bearing pressure is 200 kN/m$^2$. Design the base using concrete Grade 35 and high-yield reinforcement, type 2.

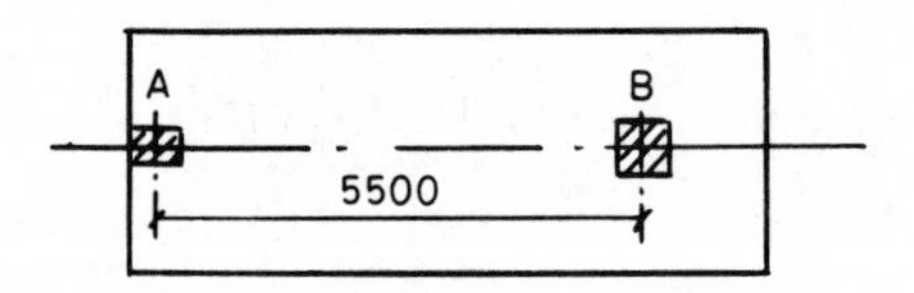


### Service loads

Column $A$: 1340 kN, column $B$: 2000 kN.

Centre of gravity of loads = (2000 × 5.5)/3340 = 3.293 m from centre of column $A$.

Length of base required = 2(0.3 + 3.293) = 7.186, say 7.2 m.

Assuming base is 500 mm thick, allowable pressure for column loads is 188 kN/m$^2$, so width of base

$= 3340/(7.2 \times 188)$

$= 2.47$ m, say 2.5.

## Ultimate loads

Using factors of 1.4 m on dead loads and 1.6 on imposed loads, the ultimate loads are column $A$: 1992 kN, column $B$: 2978 kN.

Centre of gravity of loads $= 2978 \times 5.5/4970 = 3.295$ m from centre of column $A$. This is virtually the same as for service loads and at the centre of the base, so a uniform pressure will be assumed.

Net pressure $= 4970/(7.2 \times 2.5) = 276$ kN/m$^2$ $= 690$ kN/m run in longitudinal direction.

The maximum negative moment between the column occurs at the section of zero shear. Let $x$ be the distance from the inside face of column $A$; then

$$x = \frac{1992 - 690 \times 0.6}{690} = 2.29 \text{ m},$$

i.e. 2.89 m from the edge of the base.

At this section, moment
$= (690 \times 2.89^2)/2 - 1992 \times 2.59$
$= -2277$ kN m.

$M/bd^2 = 4.8$ assuming 40 mm cover and 25$\phi$ bars.

This is within the limits of tension reinforcement only, but would require approximately 16 000 mm$^2$ of reinforcement.

The overall thickness originally chosen would, therefore, appear to be unsatisfactory. Try a base 1000 mm thick.

This will reduce pressure available for column loads to 176 kN/m$^2$, so breadth of base will be increased to $3340/(7.2 \times 176) = 2.63$ m, say 2.7 m.

At ultimate net pressure $= 4970/(7.2 \times 2.7)$
$= 256$ kN/m$^2$
$= 690$ kN/m run (as before).

Base is 7.2 m long by 2.7 m wide by 1.0 m thick.

Distribution of forces and shear force diagram are shown in Fig. 16.10.

If one treats the column loads as uniformly distributed loads the bending moment diagram is as shown in Fig. 16.11.

## Flexure

Longitudinal direction

Maximum moment $= 2277$ kN m.
Assume $d = 1000 - 40 - 25/2 = 947.5$, say 945.

$$\frac{M}{bd^2} = \frac{2277 \times 10^6}{2700 \times 945^2} = 0.94,$$

and from tables $\dfrac{100A_s}{bd} = 0.25.$

$A_s = 6379$ mm$^2$.

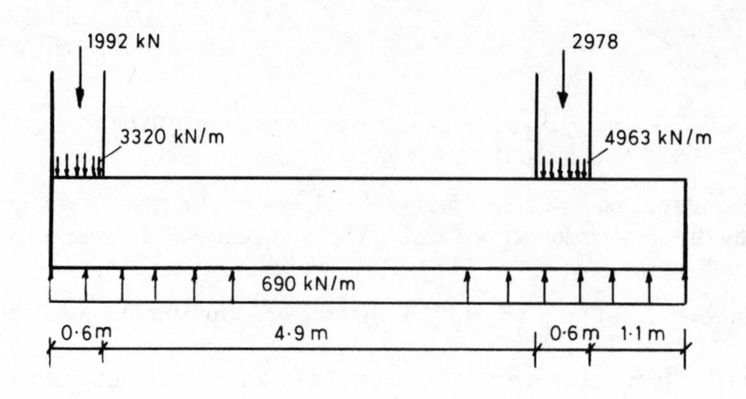


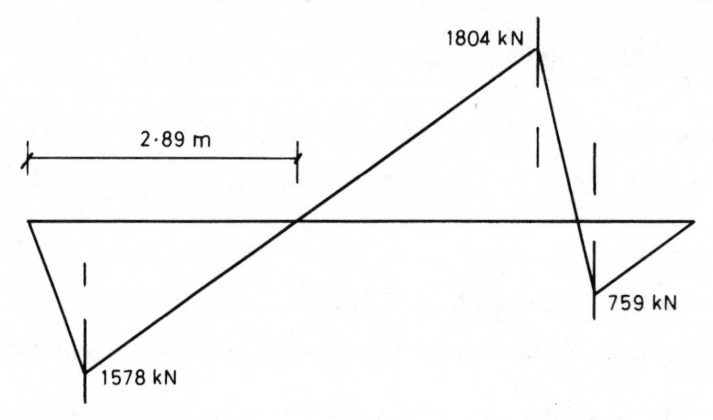


FIG. 16.10 Shear force diagram.

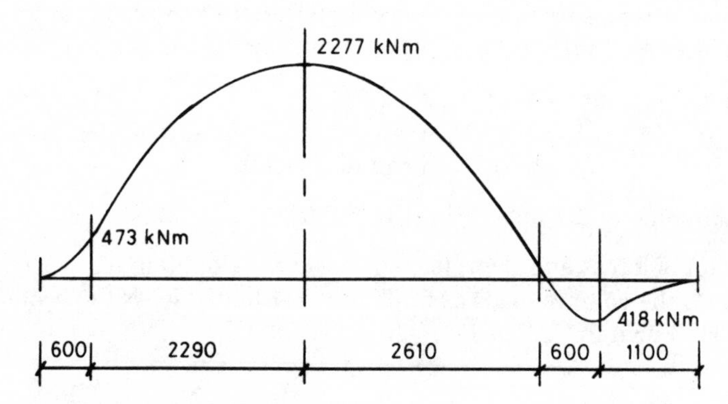


FIG. 16.11 Bending moment diagram.

Suggest $13/25\phi$ (6380 mm$^2$).

Width of base = 2700, which is less than 1.5 $(450 + 3 \times 945)$, so reinforcement distributed uniformly across base in top.

Moment at interior face of column $A = 473$ kN m.

$M/bd^2 = 0.2$, so minimum reinforcement must be provided.
Minimum area of reinforcement $= 0.13 \times 2700 \times 1000/100$
$= 3510$ mm$^2$ (or 1300 mm$^2$/m).

This requires $8/25\phi$ ($3930\ \text{mm}^2$), which can be obtained by curtailing some of the $14/25\phi$ bars in the top.

The moment in the bottom at the exterior face of column $B$ is less than above, so again minimum area of reinforcement will control. The area of reinforcement that would be required is $1163\ \text{mm}^2$, so the anchorage bond length can be reduced in the ratio of the reinforcement required to that supplied.

Transverse direction

At column $A$, the 'effective' base width to be considered is $0.3 + 5.5/2 = 3.05$ m.

This again is less than $1.5(0.6 + 3 \times 0.92)$, the effective depth in this direction being taken as 920 mm, so banding of reinforcement will not be necessary.

$$\text{Moment at face of column} = 256 \times 3.05 \times (1.35 - 0.225)^2/2$$
$$= 494\ \text{kN m}.$$

From previous calculations it can be seen that minimum area of reinforcement will control.

At column $B$, the situation will be very similar.

So, for reinforcement in transverse direction use $20\phi$ at 225 centres ($1396\ \text{mm}^2/\text{m}$) distributed along the whole length of the base.

This reinforcement will also be required for the top bars as secondary reinforcement.

## Shear

Before doing the shear calculations it is necessary to determine the effective area of tension reinforcement. From the previous calculations for flexure, it can be seen that the top bars could be curtailed, particularly at the column faces. Bearing in mind the fact that they must continue at least an effective depth beyond the theoretical cut-off point it is suggested that this is not done. As the point of contraflexure is virtually at the inside face of Column $B$ the top reinforcement will continue an effective depth beyond this point.

(a) Across the full width of the base

In the longitudinal direction check the shear at half the shear span from the inside face of the interior column, i.e. at a distance of 1.305 m from the face of the column.

Shear force is 902 kN.

$$\text{Shear stress} = \frac{902 \times 10^3}{2700 \times 945} = 0.35\ \text{N/mm}^2.$$

$$\frac{100A_s}{bd} = \frac{100 \times 6380}{2700 \times 945} = 0.25.$$

$v_c = 0.25\ \text{N/mm}^2$.

This is satisfactory without using the enhancement factor for $2d/a_v$.

(b) Punching shear

The first critical perimeter is outside the base, so all we can check is the perimeter of the column.

Column $A$: Load $=1992$ kN; perimeter $=2\times600+450=1650$ mm.

Average effective depth, $d=1000-40-25=935$.

$$v=\frac{1992\times1000}{1650\times935}=1.29\ \text{N/mm}^2,$$

which is less than $0.8\sqrt{35}$.

Column $B$: Load $=2978$ kN; perimeter $=4\times600=2400$.

$v=1.32\ \text{N/mm}^2$.

## EXAMPLE 16.3

In Example 16.2 the base could not protrude beyond the outer face of the exterior column. If we now assume that the criterion is reversed and that the base cannot protrude beyond the inner face of the interior column it can quickly be established that a rectangular base will not be applicable.

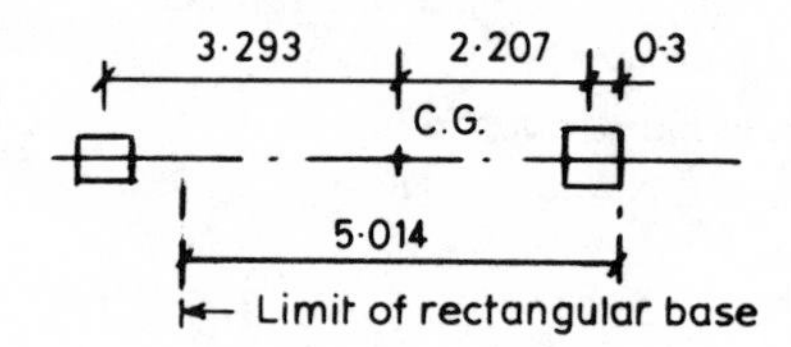


We shall therefore need a trapezoidal base as shown:

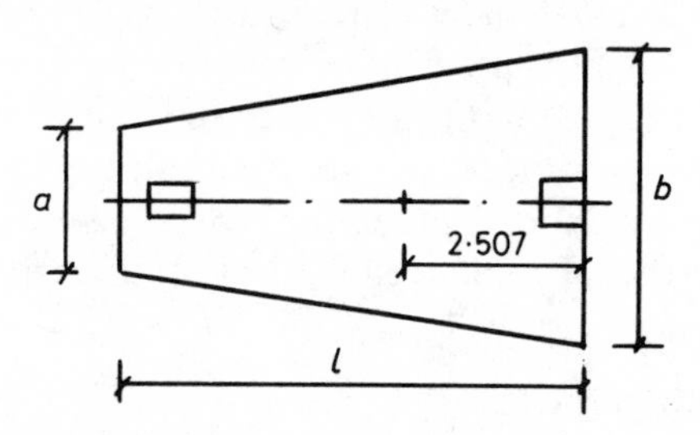


Assuming a base 1 m thick then

$\frac{1}{2}(a+b)\times l=3340/176=19$

and

$[(2a+b)/3(a+b)]\times l=2.507$.

There are several solutions to these equations, but $a$ cannot be less than 0.45 m and $l$ cannot be less than 6.1 m.
Taking $l=6.1$ m, it can be found that $a=1.45$ m and $b=4.8$ m gives a satisfactory solution, and the base will not protrude beyond the column faces at both ends.

$$\text{Pressure under the base}=\frac{4970\times2}{6.1\times6.25}=260.7\ \text{kN/m}^2\ \text{at ultimate.}$$

The loads on the base, the shear force diagram and BM diagram in the longitudinal direction are as shown in Figure 16.12.

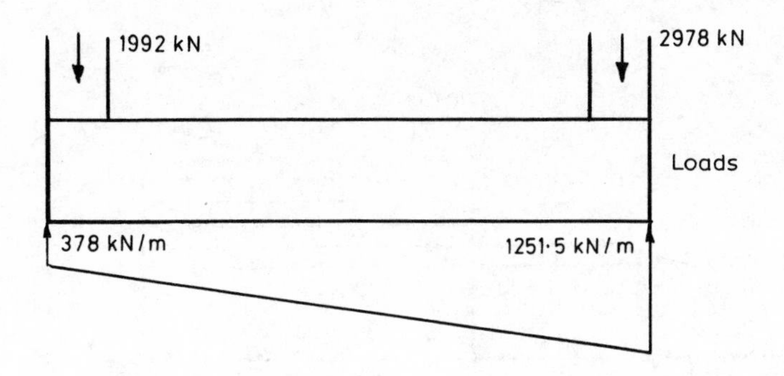


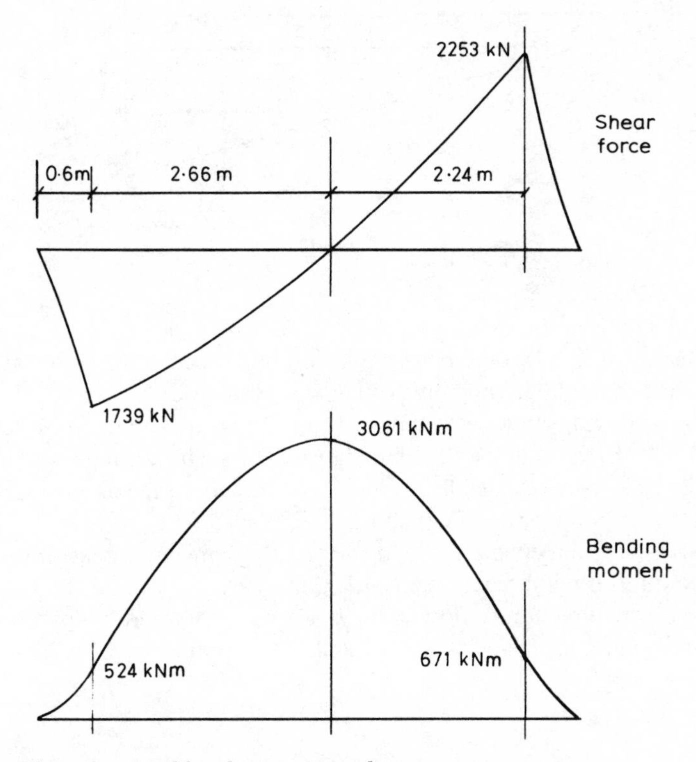


FIG. 16.12 Shear force and bending moment diagram.

The maximum negative moment is 3061 kN m.
Assume $d = 1000 - 40 - 15 = 945$.

At point of maximum bending,
width of base $= 1.45 + (3.26/6.10)/3.35 = 3.24$ m.

$$\frac{M}{bd^2} = \frac{3061 \times 10^6}{3240 \times 945^2} = 1.06 \text{ and from tables } \frac{100A_s}{bd} = 0.28.$$

$A_s = 8573$ mm$^2$.

This could be $7/40\phi$ (8800 mm$^2$) or $11/32\phi$ (8850 mm$^2$).

If the recommendations of the detailing committee are followed, then this reinforcement will be concentrated in a band 1.45 m along the length of the base. The regions of

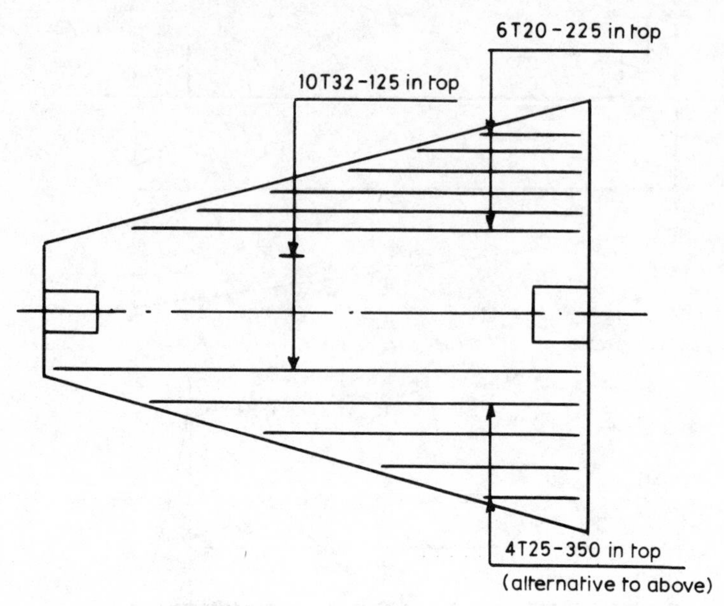


FIG. 16.13 Longitudinal reinforcement.

the base outside this band would need minimum reinforcement and the layout of the top steel in the longitudinal direction would be as shown in Fig. 16.13.

An alternative arrangement, which would appear to comply with one interpretation of clause 3.11.3.2 in the Code, is shown in Figure 16.14. This arrangement is shown in books on reinforced concrete detailing and is certainly more economical. The decision must be left to the designer as to which method is adopted.

The transverse reinforcement in the top will be the same in both arrangements and will be the minimum reinforcement, i.e. 1300 mm$^2$/m.

In the transverse direction we now have to divide the base into separate areas and calculate the bending moments at the faces of the columns.

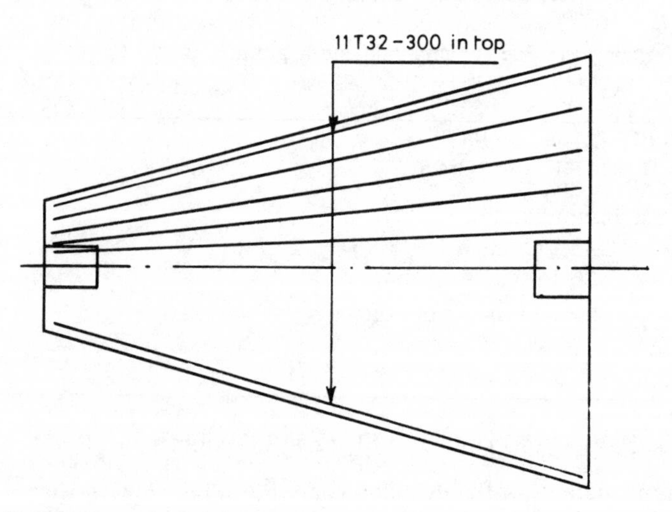


FIG. 16.14 Alternative to Fig. 16.13.

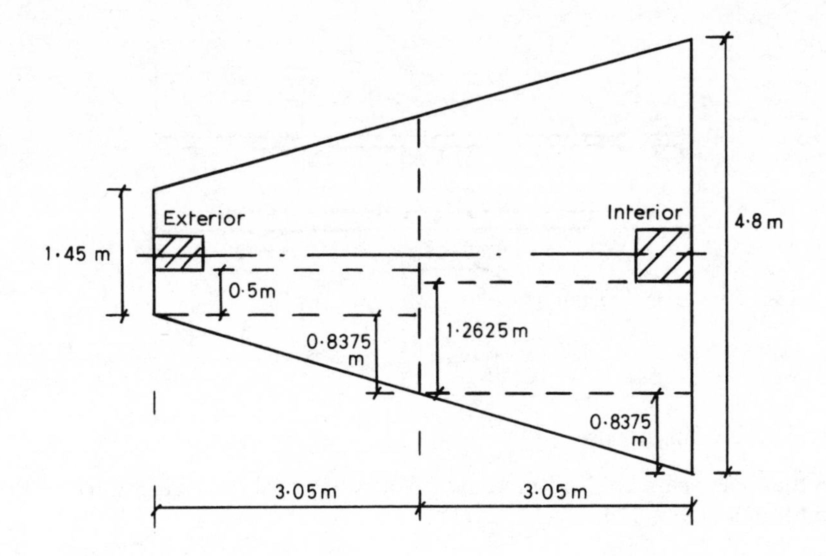


Moment at face of interior column

$$= [3.05 \times (1.2625^2/2) + (3.05/2) \times 0.8375 \times 1.542] \times 260.7$$
$$= 1147 \text{ kN m}.$$

Assuming $d = 1000 - 40 - 15 = 945$,

$$\frac{M}{bd^2} = \frac{1147 \times 10^6}{3050 \times 945^2} = 0.42.$$

This requires less than the minimum area of tension reinforcement, but as we may need the area for banding it will be calculated:

$z/d = 0.95$.

So

$$A_s = \frac{1147 \times 10^6}{0.87 \times 460 \times 0.95 \times 945} = 3193 \text{ mm}^2 = 1047 \text{ mm}^2/\text{m}.$$

As the column is on the edge of the base, the criteria for banding will be taken as comparing the width of the base, 3050 mm, with $1.5(c_x + 1.5d) = 1.5(600 + 1.5 \times 945) = 3026$ mm. These values are almost the same, and even if the actual area of reinforcement is calculated and then concentrated in a band approximately 2 m wide it will be found that the minimum reinforcement still covers this. So use the minimum reinforcement of 25$\phi$ at 350 centres.

The same will apply to the exterior column so we end up with the minimum reinforcement in both directions in the bottom of the base.

## Shear

For shear calculations we need a layout of the reinforcement, and for the reinforcement in the top in the longitudinal direction we will assume the alternative arrangement as shown in Fig. 16.14. A longitudinal section will be as shown in Fig. 16.15.

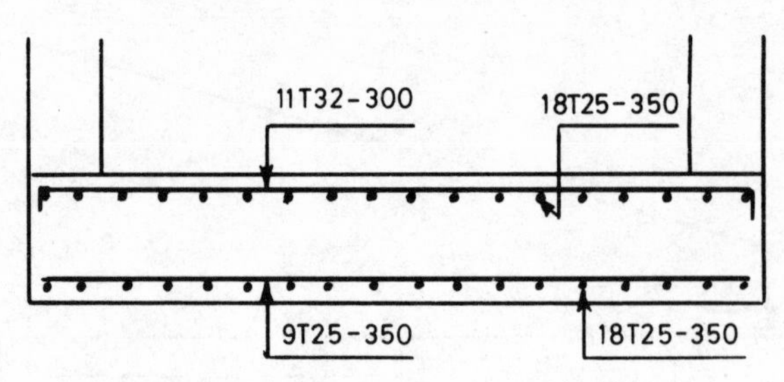


FIG. 16.15 Longitudinal section.

(a) Across full width of base

At half the shear span, i.e. 1.33 m from face of exterior column, shear force = 996 kN and width of base = 2.51 m.

$$v = \frac{996 \times 10^3}{2510 \times 940} = 0.42 \text{ N/mm}^2.$$

$$\frac{100A_s}{bd} = \frac{100 \times 8850}{2510 \times 940} = 0.375.$$

$v_c = 0.50$ N/mm$^2$.

This is satisfactory without using enhancement factor of $2d/a_v$.

At half the shear span, i.e. 1.12 m from face of interior column, shear force = 1037 kN and width of base = 3.85 m.

$v = 0.29$ N/mm$^2$. This is less than 0.34 N/mm$^2$, so is satisfactory.

(b) Punching shear

$d$ for top steel is 940 mm, and for bottom steel is 945 mm, so average = 942.5 mm; but use 940 mm.

Exterior column does not have a critical perimeter inside the base.

Interior column would appear to have a three-sided critical perimeter at $1.5 \times 940 = 1410$ from face of column, but from the diagram a more critical perimeter would be a line completely across the base. From the calculations above this would be satisfactory.

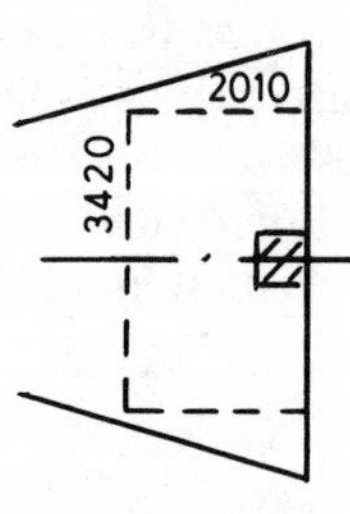


Shear stress on column perimeter only

$$= \frac{2978 \times 10^3}{1800 \times 940} = 1.76 \text{ N/mm}^2, \text{ i.e. } < 0.8 \times \sqrt{35}.$$

## 16.2 Pile caps



Pile caps are designed either on the beam theory or truss theory to determine the main tension reinforcement in the bottom of the cap.

The truss theory only applies to pile caps with up to five piles, and in this method the load from the column is transmitted to the piles by inclined thrust and the tie necessary to maintain equilibrium is provided by reinforcement.

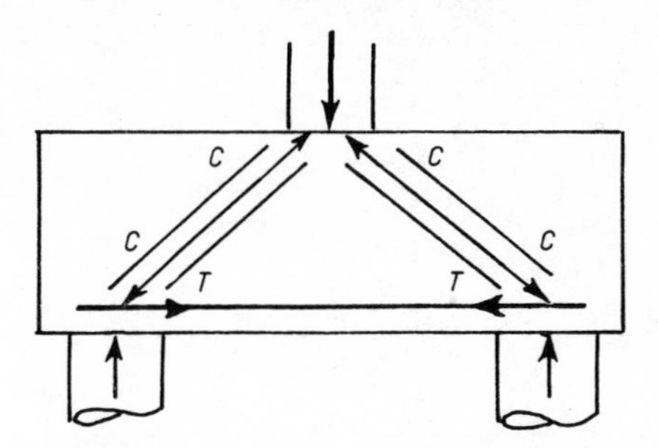


If the ultimate load on the column is N and we have two piles the load on each pile is N/2. The forces can be represented diagrammatically thus:

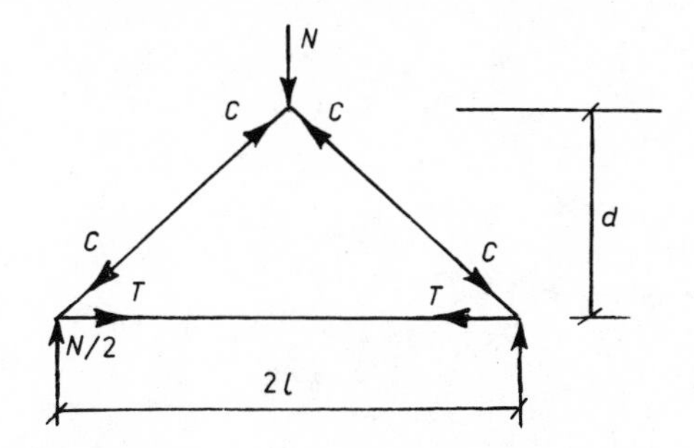


From the diagram of forces,

$$\frac{T}{(N/2)} = \frac{l}{d}, \text{ i.e. } T = \frac{Nl}{2d}.$$

Area of reinforcement required is

$$\frac{Nl}{2d \times 0.87 f_y}.$$

In the simple frame described above, the dimensions of the column have been ignored. In the May and June 1954 issues of *Civil Engineering and Public Works Review*, Mr H. T. Yan has analysed the effects of taking the column size into consideration. If the column is square of side $2a$, it will be found that

$$T = \frac{N}{6ld}(3l^2 - a^2).$$

(If $a$ is put equal to zero the answer becomes the same as above.) Mr Yan has extended the analysis to three and four-pile groups and a summary of the results is given in Fig. 16.16. It should be noted that the four-pile group has been extended to cover a five-pile group where one pile is immediately beneath the column. It will be seen from the diagrams in Fig. 16.16 that the tie force is given between a pair of piles, as this was the

| *Pile group* | *Tensile forces across pile cap* — *Column size taken into account* | *Column size ignored* |
|---|---|---|
| 2 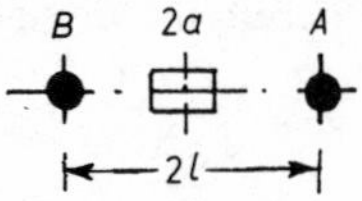 | $T_{AB} = \frac{N}{6ld}(3l^2 - a^2)$ | $T_{AB} = \frac{Nl}{2d}$ |
| 3 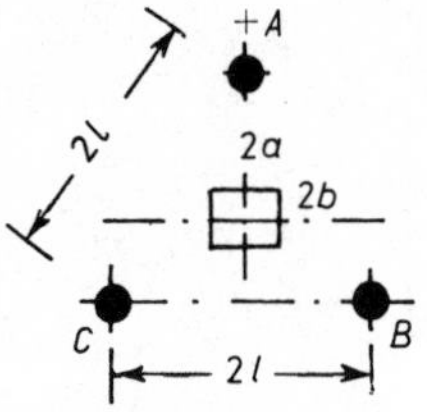 | $T_{BC} = \frac{N}{18ld}(4l^2 + b^2 - 3a^2)$ | $T_{AB} = T_{BC} = T_{CA}$ |
| | $T_{AB} = T_{AC} = \frac{N}{9ld}(2l^2 - b^2)$ | $= \frac{2Nl}{9d}$ |
| 4 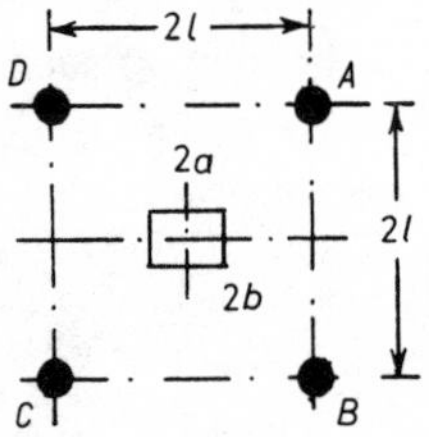 | $T_{AD} = T_{BC} = \frac{N}{12ld}(3l^2 - a^2)$ | $T_{AB} = T_{BC} = T_{CD} = T_{DA}$ |
| | $T_{AB} = T_{CD} = \frac{N}{12ld}(3l^2 - b^2)$ | $= \frac{Nl}{4d}$ |
| 5 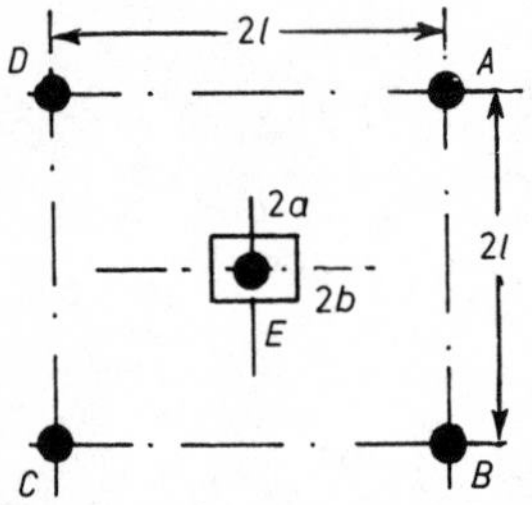 | $T_{AD} = T_{BC} = \frac{0.8N}{12ld}(3l^2 - a^2)$ | $T_{AB} = T_{BC} = T_{CD} = T_{DA}$ |
| | $T_{AB} = T_{CD} = \frac{0.8N}{12ld}(3l^2 - b^2)$ | $= \frac{0.8Nl}{4d}$ |

FIG. 16.16 Pile caps – truss theory. Notation: Distance between centre of piles $= 2l$; ultimate load on pile cap $= N$; with a square base plate $a = b$.

way the cap was analysed. In using the truss theory, therefore, it has usually been the practice to band the reinforcement along the lines joining the piles. The Code now suggests that this method of banding is only necessary if the piles are spaced at more than three times the pile diameter. For the more normal spacing of three times the pile diameter the total reinforcement forming the tie force in one direction can be distributed uniformly across the cap.

With a three-pile cap designed on the truss theory, it is difficult to see how this can be done and it is suggested that the reinforcement is banded along the centre lines joining the piles.

In all cases when the truss theory is used it must be remembered that the reinforcement has a constant force in it between the centres of piles. It must therefore have an anchorage for this force beyond the centre of the piles, i.e. the lower nodes of the truss.

This can sometimes present problems as large radius bands are required. Where the reinforcement has been banded, the high compression from the pile in this area will mean that a shorter anchorage length would be satisfactory. Tests carried out at the Cement and Concrete Association have shown this to be the case if flexure governs. If the reinforcement is uniformly distributed then this 'pinching' action does not occur and a proper anchorage is required.

As with pad footings, pile caps must also be relatively thick so that they can be classed as rigid. In the case of pile caps designed using the truss theory it is suggested that the effective depth is approximately half the distance between the centre of piles. This means the truss has an angle of approximately 45°.

With pile groups of more than five the beam theory should be used. For this reason many designers use the beam theory for all pile caps. The bending moment is now found at the face of the column and the reinforcement requires a satisfactory anchorage length from this point. The pile cap is considered as having simply supported ends so there must be a $12\phi$ anchorage beyond the centre line of the piles. With the beam theory the reinforcement is usually distributed uniformly across the cap.

As with pad footings we have shear across the width of the cap and also punching shear.

For shear across the full width of the cap, the Code refers the reader to the slab clauses.

For the enhancement factor of $2d/a_v$, the dimension $a_v$ is the distance between the face of the column and a line 20% of the pile diameter inside the pile as illustrated.

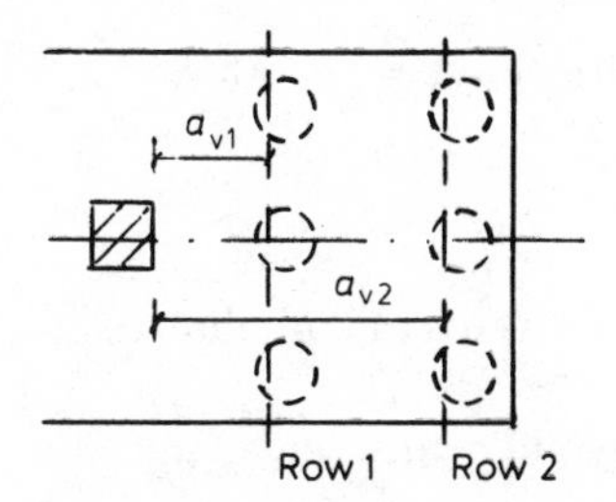


If the spacing of the piles across the section is less than three times the pile diameter the enhancement factor may be applied over the whole of the section. Where the spacing is greater, the enhancement is only applied to strips equal to three times the pile diameter centred on each pile.

As with the banding of the reinforcement using the truss theory where the pile centres are more than three times the pile diameter, a problem can arise if the pile cap does not extend a pile diameter beyond the pile. It is suggested for both cases that the band width should be the distance from the edge of the cap to the pile face plus two pile diameters or three pile diameters, whichever is the lesser.

For punching shear the main requirement is around the perimeter of the column itself and this must not exceed $0.8\sqrt{f_{cu}}$.

However, if the spacing of the piles is greater than three times the pile diameter, punching shear should be checked on a perimeter as given in Fig. 16.7 for a pad footing.

## EXAMPLE 16.4

A four-pile group supports a 500 mm square column carrying an ultimate load of 2800 kN. The piles are 450 mm diameter and are spaced at 1350 mm centres. Using concrete Grade 35 and high-yield reinforcement type 2, design the cap.

Try an effective depth of cap as 650 mm, which is approximately half the pile spacing.

Assume depth is 750 mm, with cover of 75 mm, so average $d = 650$ mm. Extend the cap 150 mm beyond the piles, so the cap will be 2100 mm square.

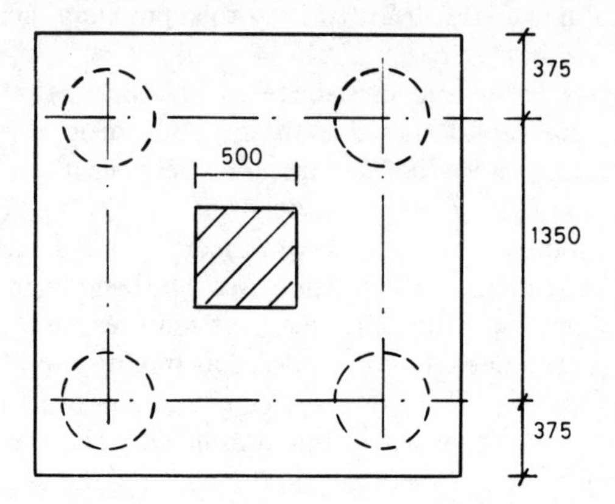


Using truss theory and taking column size into account,

$$\text{total force in each direction} = 2T = \frac{2 \times 2800(3 \times 675^2 - 250^2)}{12 \times 675 \times 650}$$

$$= 1387 \text{ kN.}$$

Note that, ignoring column size, force $= 1454$ kN.

$$A_s = \frac{1387 \times 10^3}{0.87 \times 460} = 3466 \text{ mm}^2.$$

Try 8T25 (3930 mm$^2$) at 275 centres in both directions as pile centres do not exceed three times pile diameter.

Anchorage bond length beyond centre line of pile

$$= 34 \times (3466/3930) \times \phi = 30\phi = 750 \text{ mm}$$

and stress in bars

$$= 0.87 \times 460 \times 3466/3930 = 353 \text{ N/mm}^2.$$

Allowing for 75 mm end cover, the distance available is 300 mm, so the bar will need to turn up the cap. As the bar will be stressed beyond the end of the bend, the bearing stress inside the bend must be checked.

The distance $a_b$ is 275 mm and from Appendix 3 it can be seen that an internal radius of $5\phi$ will be required. Using BS4466, the length of the bar from the centre line of the pile and turning up to a distance of 100 mm from the top is $300 + 550 - (\text{radius}/2) - \phi = 762$ mm.

So this will be satisfactory.

## Shear

(a) Punching shear around column perimeter

$$= \frac{2800 \times 10^3}{4 \times 500 \times 650} = 2.15 \text{ N/mm}^2\text{, i.e. less than } 0.8\sqrt{35} \text{ (4.7 N/mm}^2\text{).}$$

(b) Across the full width of the cap

$V=1400$ kN, $v=\dfrac{1400\times10^3}{2100\times650}=1.03$ N/mm².

$a_v=675-250-225+0.2\times450=290.$

So $2d/a_v=1300/290=4.48.$

$$\frac{100A_s}{bd}=\frac{100\times3930}{2100\times650}=0.29.$$

$v_c=0.47$ N/mm².

Allowable shear stress $=4.48\times0.47=2.09$ N/mm².

Shear is satisfactory as reinforcement projects more than 650 mm beyond the section.

For horizontal binders it is suggested that 25% of the main tension steel be used.

A detail of the reinforcement will be as shown.

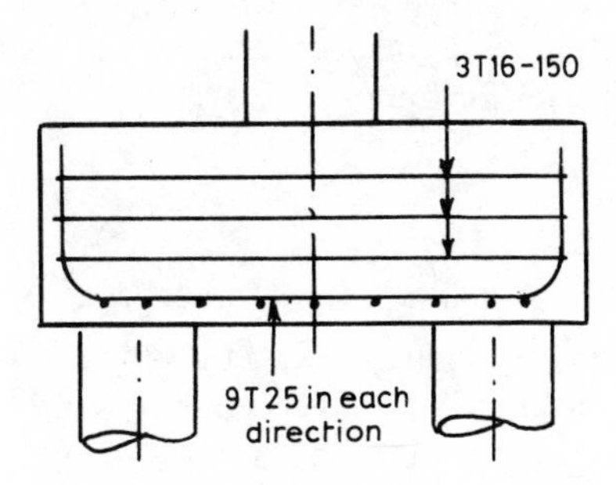


## Alternative design using beam theory

Bending moment at face of column $=1400(0.675-0.25)$
$=595$ kN m.

$$\frac{M}{bd^2}=\frac{595\times10^6}{2100\times650^2}=0.67.$$

From tables $\dfrac{100A_s}{bd}=0.175$, $A_s=2389$ mm².

$8/20\phi$ bars (2510 mm²) will do this.

This is far less than for the truss theory and the anchorage length will be from the face of the column so there will be ample room for this and a standard $3\phi$ radius bend will be sufficient, if required.

Shear. $\dfrac{100A_s}{\text{bd}}=\dfrac{100\times2510}{2100\times650}=0.18$, so $v_c=0.4$ N/mm².

Enhancement factor, as before, $=4.48$ so allowable shear stress $=1.8$ N/mm², which is satisfactory.

As this reinforcement must project an effective depth beyond the critical section for shear the bars will have to be turned up the side of the cap.

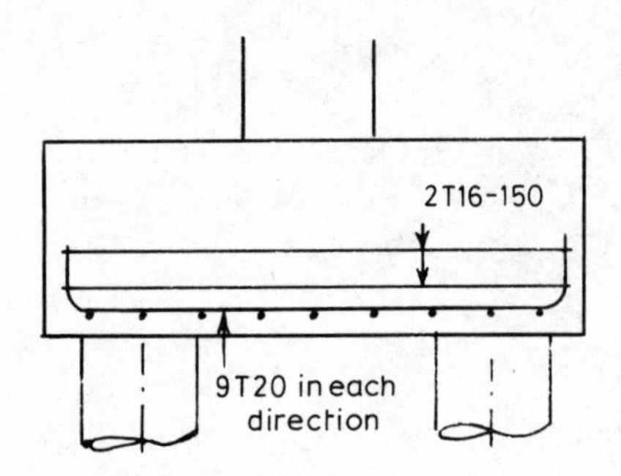


From this example it can be seen that the beam theory requires far less reinforcement than the truss theory. It is felt, however, that the cap will behave more as a truss than a beam.

When the pile cap is subjected to a bending moment in addition to a direct force, the loads on the piles will not be equal. Using the truss theory for small pile groups it is suggested that to find the maximum tensile force $T$, the load $N$ is made equal to the maximum pile reaction (excluding the weight of the cap), multiplied by the number of piles, and then substituted in the formula.

For larger pile groups when using the beam theory, the loads on individual piles can be found using the well known formula

$$Q_n = \frac{N}{n_{\text{tot}}} \pm \frac{Ne_x x_n}{\sum(x_1^2 + x_2^2 \ldots)} \pm \frac{Ne_y y_n}{\sum(y_1^2 + y_2^2 \ldots)}$$

where $Q_n$ is the load on a pile with appropriate suffixes (kN), $N$ is the column load (kN), $n_{\text{tot}}$ is the total number of piles, $e_x$, $e_y$ are the eccentricities of column load in relation to group axes (m), $x_n$, $y_n$ are the coordinates of the pile $Q_n$ in relation to the group of axes (m), and $x_1$, $y_1$ are the coordinates of the piles (m).

Moments and shears in particular directions can then be found using the appropriate pile loads.

---

# TORSION 17

CP110 dealt with torsion for the first time in a structural code, and although BS8110 has continued with this, it has been put into Part 2 of the Code. Presumably this is because – to quote 2.4.1 in Part 2 – 'in normal slab and beam construction specific calculations are not usually necessary, torsional cracking being adequately controlled by shear reinforcement.'

The clause goes on to say that when the design relies on torsional resistance of a member the recommendations that are given should be taken into account. So when is torsion to be taken into account?

Quite obviously, if the main effect is due to torsion, then it must be considered. But in many cases it is not always so clear. Take a beam and slab construction as shown in Figure 17.1.

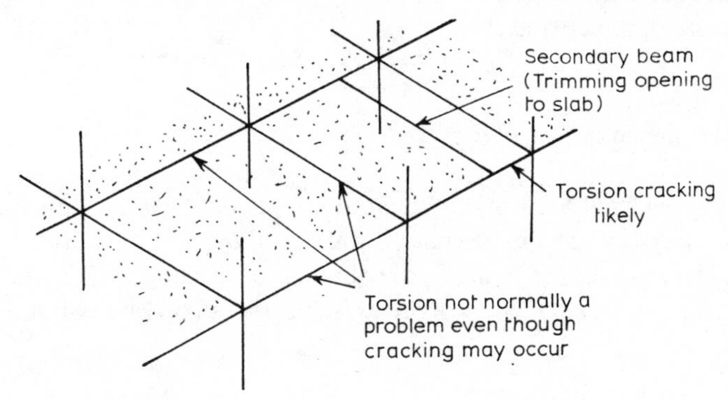


FIG. 17.1 Torsion due to rotation.

Owing to asymmetric loading of the slabs torsion is induced in the beams supporting the slabs, but this is generally relieved by the beam twisting slightly and shedding this torsion into the supported slabs in the form of bending moment. The beam, in fact, undergoes twist which does not significantly affect its flexural strength, and it is not relied upon to resist torsion. Broadly speaking experience has shown that torsion arising from statical indeterminancy or continuity may be ignored because lack of torsional resistance does not cause collapse, whereas that arising due to statical determinancy or non-continuity may not be ignored because lack of torsional resistance would cause collapse.

There are obvious exceptions to this rule. First, experience has shown that small amounts of torsion in statically determinate cases causes no significant reduction in flexural capacity and can be resisted provided nominal stirrup resistance is provided. So the Code suggests that torsion may be ignored in conventional beam and slab design. One should, however consider cases that may arise during construction, e.g. unbalanced loading.

Secondly, some statically indeterminate members may have to undergo exceptionally larger amounts of twist to enable them to shed the torque back into the members causing it. This may cause severe cracking which could be detrimental to the shear and flexural strength. In this case the torque should be assessed from analysis and taken into account. In Fig. 17.1 the short lengths of beams are subject to a high rate of twist due to the deflection of the secondary beam. The torque in these beam lengths would have to be assessed from their torsional stiffness and the end rotation of the trimming beam.

The flexural rigidity of a member is $EI$ where $E$ is obtained from Section Seven of Part 2 and $I$ is calculated from the concrete section. The torsional rigidity is $GC$ where $G$ shall be taken as $0.42E$ and $C$, the torsional constant, is half the St Venant value associated with the plain concrete section. The justification for taking the value of $C$ as half the theoretical stiffness of the plain section is the result of a series of tests.

These values are only used when a frame is analysed and the following examples should clarify what is meant by $C$.

If we call $J$ the torsion constant for the plain concrete section then $C=\frac{1}{2}J$.

(a) RECTANGULAR SECTION

Minimum dimension $b$
Maximum dimension $h$

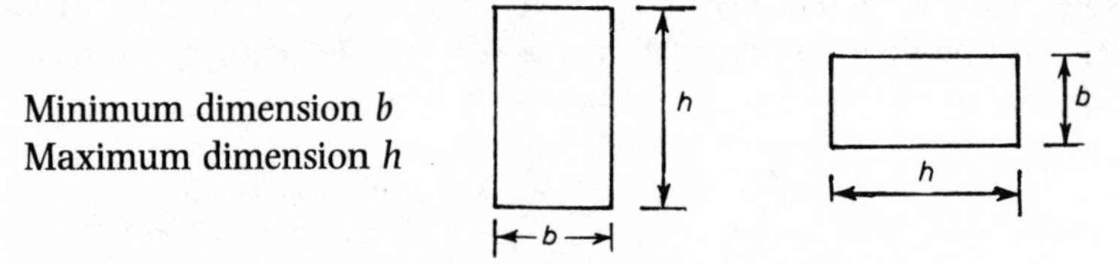


$J=\gamma b^3h$ where $\gamma$ depends on the ratio of $h/b$.

This was discovered by St Venant and in various text books can be found tables giving values for $\gamma$ (it may not always be called $\gamma$ but it will usually be a Greek symbol).

The Code uses the terms $h_{max}$ and $h_{min}$ in equation (1) of clause 2.4.3, and the symbol should be $J$, not $C$ as printed. In view of the example that follows we will stay with $h$ and $b$.

Some values of $\gamma$ are as follows:

| $h/b$ | 1 | 1.5 | 2 | 3 | 5 | >5 |
|---|---|---|---|---|---|---|
| $\gamma$ | 0.14 | 0.20 | 0.23 | 0.26 | 0.29 | 0.33 |

A formula which will give values within 4% is

$$\gamma=0.33-0.21\frac{b}{h}\left(1-\frac{b^4}{12h^4}\right).$$

(b) TEE, I OR ELL SECTIONS

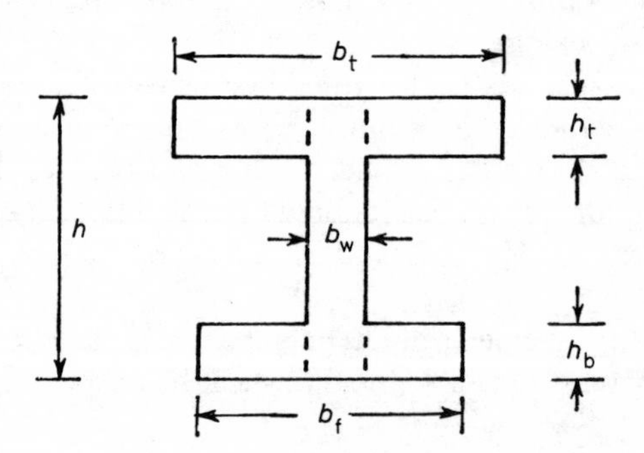

$$J=\gamma_1 b_w^3 h+\gamma_2 h_t^3(b_t-b_w)+\gamma_3 h_b^3(b_f-b_w).$$

The $\gamma$'s depend on the individual $h/b$ ratios for each rectangle, where it should be remembered that $h$ and $b$ are the maximum and minimum dimensions respectively, e.g. for the top flange $h$ is $(b_t-b_w)$ and $b$ is $h_t$.

A different value will be found if the section is divided up as follows:

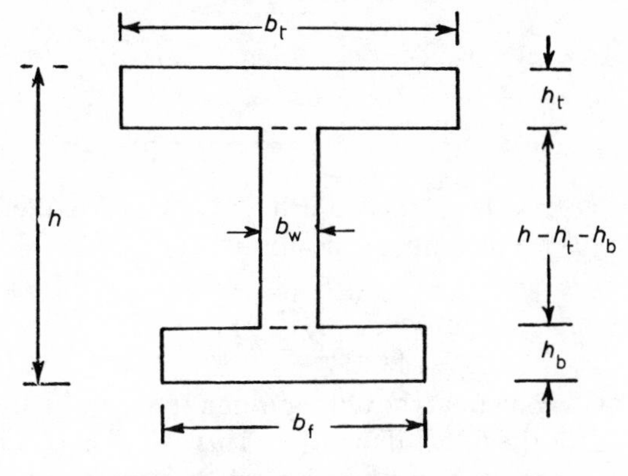


$$J=\gamma_1 h_t^3 b_t+\gamma_2 h_b^3 b_t+\gamma_3 b_w^3(h-h_t-h_b).$$

In codes used by other countries where torsion clauses are included and are based on the work of Professor Cowan, it is the latter method of dividing up the section which is suggested. BS8110 suggests that the division of the section should be arranged so as to maximize the calculated stiffness. This means generally taking the widest rectangle as long as possible. So if $b_w$ is greater than $h_t$ and $h_b$ use the first method of dividing up the section. If not, then use the second method.

A comparison of numerical values obtained for the same section is as follows:

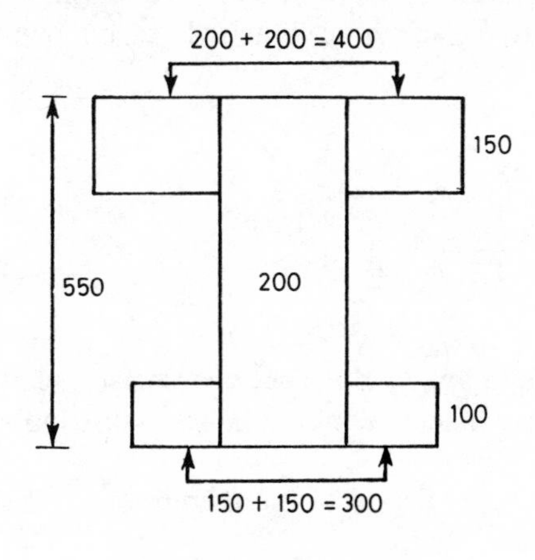


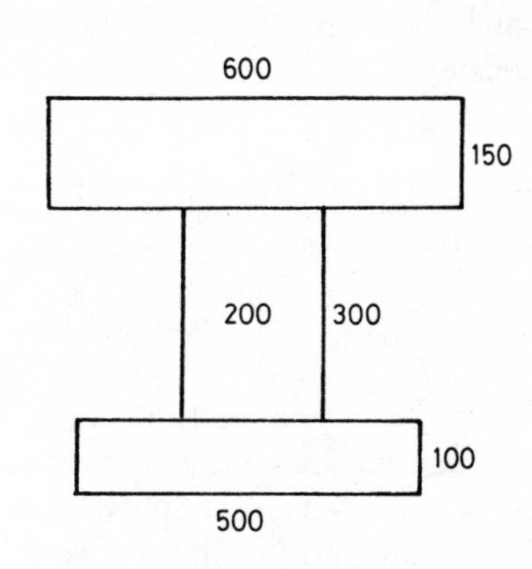


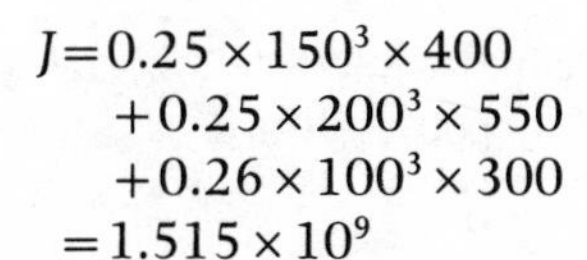

$$\begin{aligned}J&=0.25\times 150^3\times 400\\&\quad+0.25\times 200^3\times 550\\&\quad+0.26\times 100^3\times 300\\&=1.515\times 10^9\end{aligned}$$

$$\begin{aligned}J&=0.28\times 150^3\times 600\\&\quad+0.20\times 200^3\times 300\\&\quad+0.29\times 100^3\times 500\\&=1.192\times 10^9\end{aligned}$$

The method suggested in the Code gives the greater value.

So if we have to find the torsional moment in a statically indeterminate frame, we can do this by distributing the moment at the joint in the ratio of the stiffness. A simple example of this will be done later.

A more normal case will be a statically determinate one, i.e. where the torsional moment is readily calculated. For example:

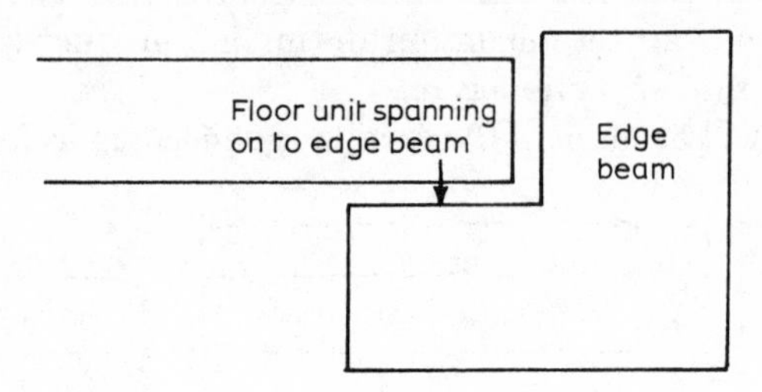


The torsional shear stress $v_t$ is calculated using plastic distribution formulae and the Code gives equation (2) for rectangular beams as

$$v_t = \frac{2T}{h_{min}^2(h_{max} - h_{min}/3)},$$

where $T$ is the torsional moment at the ultimate limit state, $h_{min}$ is the smaller dimension of the section, and $h_{max}$ is the larger dimension. This equation is derived from the sand heap analogy and is in several text books but will be summarized briefly.

The sand heap analogy is developed from the membrane analogy, so we start with that.

## 17.1 Membrane analogy

The membrane analogy establishes certain relations between the deflected surface of a uniformly loaded membrane and the distribution of stresses in a twisted member. Imagine a soap film on the end of a circular hollow box. Blow into the box; the film will expand outwards. The film is subjected to a uniform pressure (it may be a child with a clay pipe).

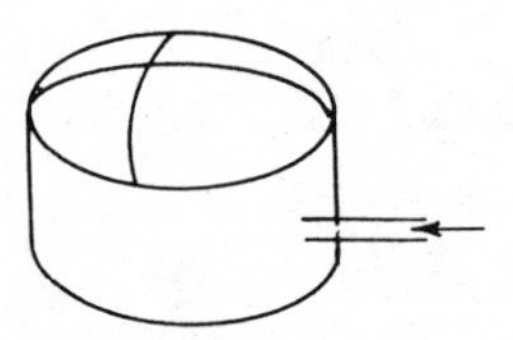

It can be shown that the differential equation of the deflected surface of this membrane has the same form as the equation which determines the stress distribution over the cross-section of the twisted member.

The following relationships exist between the surface of the membrane and the distribution of shearing stresses in a twisted member.

1. The tangent to a contour line at any point of the deflected membrane gives the direction of the shearing stress at the corresponding point in the cross section of the twisted member;

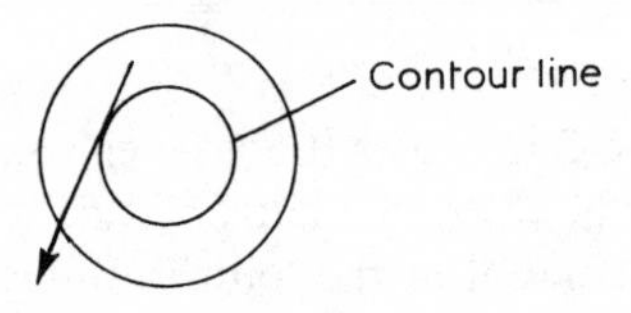

2. The slope of the membrane at any point is equal to the magnitude of the torsional shearing stress at the corresponding point in the twisted member;

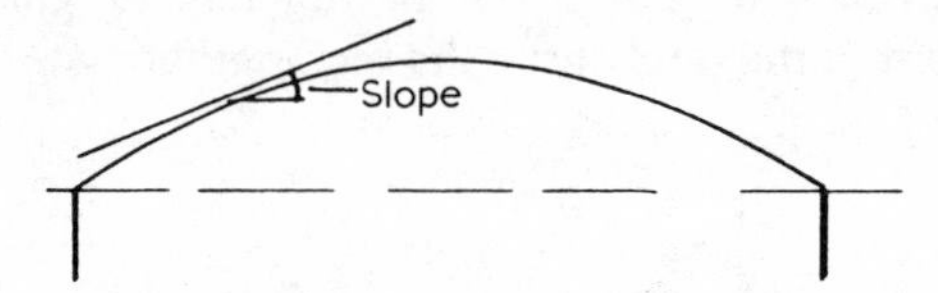


3. Twice the volume included between the surface of the deflected membrane and the plane of its outline is equal to the torque in the twisted member.

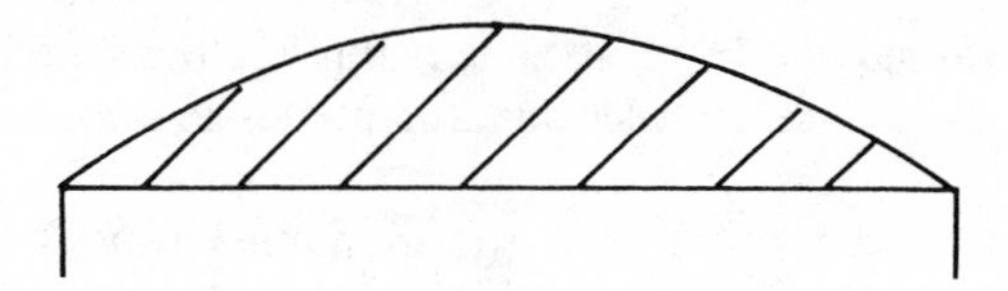

With a square or rectangular section the surface of the deflected membrane may be represented by contour lines.

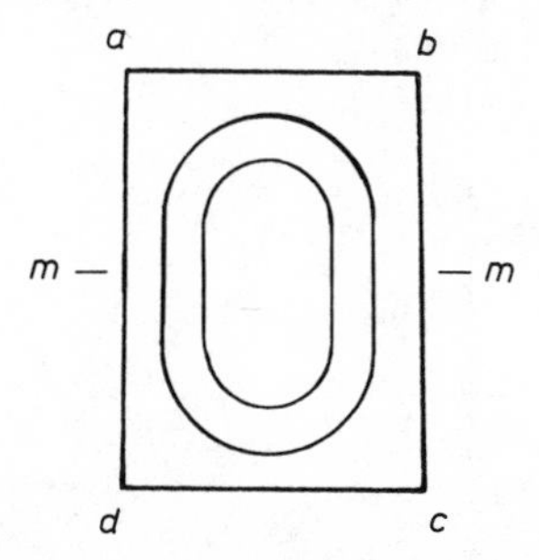


The stress is inversely proportional to the distance between these lines: hence it is larger where the lines are closer together, i.e. the slope is greater. The maximum stress occurs at points *m*, where the slope is largest. At the corners *a*, *b*, *c*, *d* where the surface of the membrane coincides with the plane of the contour *abcd*, the slope of this surface is zero. Hence the torsional shearing stress at these points is zero, and is also zero at the centroid.

As the pressure increases the height of the bubble increases and one comes to the point where the material of the member yields. This obviously starts at the outside, as the stresses are greater. So we have the outer portion yielding while the inner portion is still behaving elastically. To extend the membrane analogy it is now necessary to use a rigid cone together with the membrane.

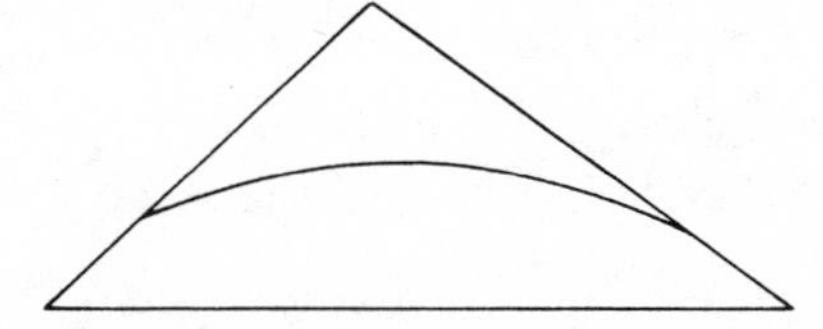

The slope of the cone represents the yield stress $v_{tu}$ to the proper scale. So once the material yields the membrane coincides with the slope of the cone in the areas where yielding has taken place, while the inner portion still has the curved membrane. In the limit the membrane has the same shape as the cone.

When we say a cone, it is really only a cone for a circular member. For a square section it is a pyramid and for a rectangular section it is like a hipped roof surface or the lines showing the loads carried by beams supporting a rectangular two-way spanning slab. Alternatively, if one pours sand onto a flat rectangular plate this shape is obtained; that is where the 'sand heap' analogy comes in.

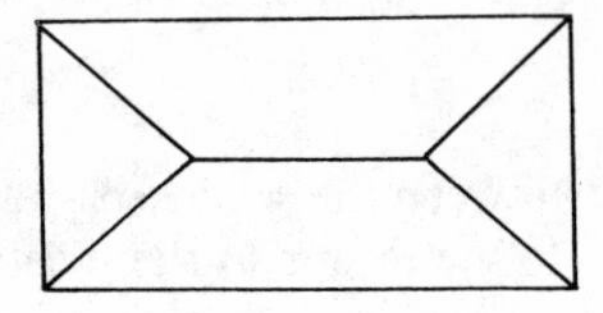

So the membrane analogy is for elastic behaviour and the sand heap is for plastic or ultimate behaviour. As we are considering ultimate this straight-line shape is what we shall consider.

Twice the volume under this roof is the ultimate torque. So if you consider a rectangular section we can derive the Code equations as follows.

## 17.2 Derivation of Code equations for rectangular sections

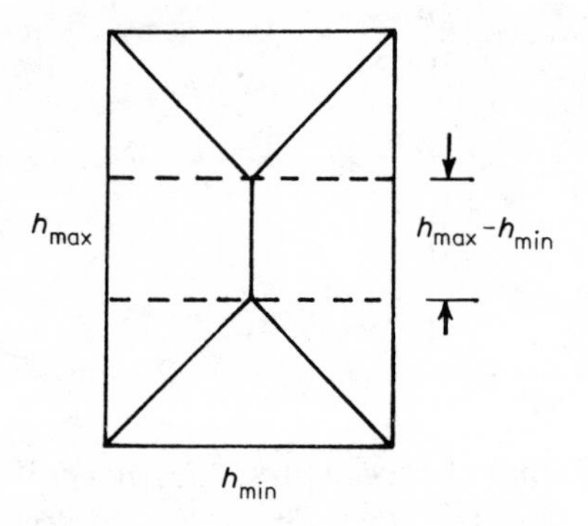


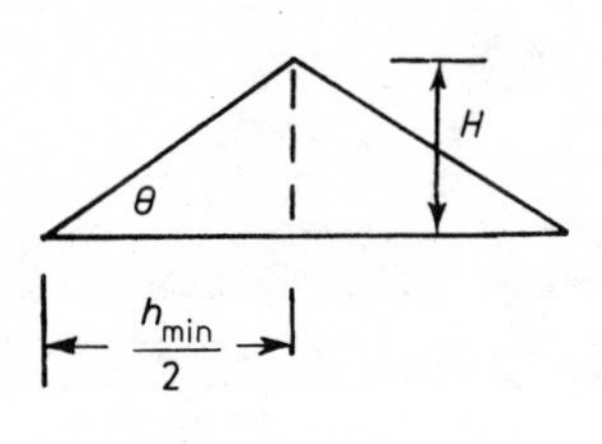


$$\text{Volume} = \tfrac{1}{3}h_{min}^2 H + \tfrac{1}{2}h_{min}(h_{max} - h_{min})H$$

$$= h_{min}H\left(\frac{h_{min}}{3} + \frac{h_{max} - h_{min}}{2}\right)$$

$$= \frac{h_{min}H}{2}(h_{max} - h_{min}/3).$$

$$\text{Slope } \theta = \frac{H}{h_{min}/2} = v_t,$$

i.e. $H = v_t h_{min}/2$.

$$\text{So volume} = \frac{v_t h_{min}^2}{4}(h_{max} - h_{min}/3).$$

$T =$ twice volume.

$$\text{So } v_t = \frac{2T}{h_{min}^2(h_{max} - h_{min}/3)},$$

which is equation (2) of the Code.

Table 2.3 of the Code gives values for $v_{t,min}$ and $v_{tu}$ with Table 2.4 giving recommendations for reinforcement for combinations of shear and torsion.

If the torsional shear stress, $v_t$, as calculated from equation (2) is less than the appropriate value of $v_{t,min}$, calculations for torsion reinforcement are not necessary. Shear reinforcement, either minimum or designed, will be provided depending on whether $v$ is less than or greater than $(v_c + 0.4)$. If $v_t$ is greater than $v_{t,min}$ then calculations for torsion reinforcement are required. If the shear stress is greater than $(v_c + 0.4)$ then design shear reinforcement is required and will be added to that required for torsion.

On the other hand, if the shear stress is less than $(v_c + 0.4)$, the Code says that torsion reinforcement only is required, but this must not be less than minimum shear reinforcement. This may be reasonable if the shear stress is very low, but if it is approaching $(v_c + 0.4)$ it is difficult to see how minimum links would cope with shear and torsion. A better approach would seem to be to say that if $v$ is greater than $v_c$, but less than $v_c + 0.4$, the torsion reinforcement should be added to the minimum shear reinforcement.

In no case, however, can the total shear stress arising from shear force and torsion exceed the appropriate value of $v_{tu}$ in Table 2.3.

Where $v_t$ exceeds $v_{t,min}$ the torsional reinforcement should consist of rectangular closed links together with longitudinal reinforcement and is additional to reinforcement required for bending and shear.

As will be realized the cracks developing from torsion are inclined diagonally on all four faces of a beam.

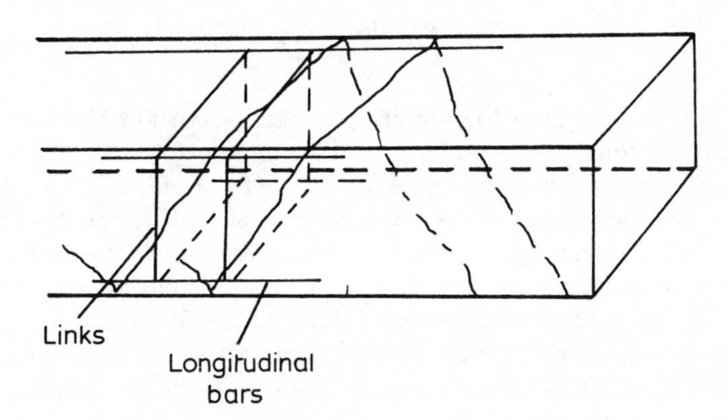


The most effective way to reinforce would therefore be two diagonal spirals which would have to go in opposite directions. This is not very easy to make, so vertical stirrups are used to cope with the vertical component and horizontal bars to take the horizontal component. The corner bars are very important as not only do they prevent bearing failure inside the bends, but they also help to control the resisting forces where they turn from one plane to the next plane at right angles.

For links we have

$$\frac{A_{sv}}{s_v} = \frac{T}{0.8x_1y_1(0.87f_{yv})},$$

where $A_{sv}$ is the area of two legs of closed links at a section, $s_v$ is the spacing of the links, $x_1$ is the smaller centre-to-centre dimension of the links, $y_1$ is the larger centre-to-centre dimension of the links, and $f_{yv}$ is the characteristic strength of the links. It should be noted that for $A_{sv}$, if the section is reinforced with multiple links, only the area of the legs closest to the outside of the section should be used.

The 0.8 in the denominator is an efficiency factor on the reinforcement stress as the full design stress cannot be achieved without spalling the concrete.

For sections where $y_1$ is less than 550 mm, the Code says that the torsional shear stress alone shall not exceed $v_{tu}y_1/550$. This again has been derived from tests where, in order to keep the 0.8 efficiency factor, a limit had to be set on the torsional stress in cases where $y_1$ is less than 550. There is no magic about the figure 550. It is just that this value came out of test results and is the best value to be taken when the spalling at corners had a serious effect.

The spacing of links should not exceed the least of $x_1$, $y_1/2$ or 200 mm. So where there is a square section such that $x_1$ equals $y_1$, then the spacing is $x_1/2$ or 200. The links themselves are to be of the closed type similar to shape code 74 in BS4466, i.e.

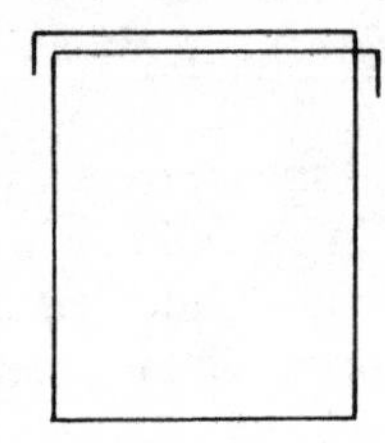

Longitudinal reinforcement is given by the formula

$$A_{sl} = \frac{A_{sv}}{s_v}\frac{f_{yv}}{f_y}(x_1 + y_1).$$

This formula is derived from the requirement that tests show the best results are obtained by using equal volumes of hoop and longitudinal steel.

This longitudinal reinforcement should be distributed evenly around the perimeter, with at least one bar in each corner. Where flexural reinforcement is required the additional area for torsion can be added to this so that larger bars than required for flexure can be used or, of course, additional bars can be put in. The clear distance between longitudinal reinforcing bars should not exceed 300 mm.

All torsion reinforcement should extend a distance at least equal to the largest dimension of the section beyond where it ceases to be required.

## 17.3 Tee, Ell and I beams

For apportioning the torsional moment in individual rectangles the Code gives the following formula:

$$T' = T\frac{(h_{min}^3 h_{max})}{\sum(h_{min}^3 h_{max})},$$

where $T'$ is for an individual rectangle and $\sum h_{min}^3 h_{max}$ is the sum of the component rectangles.

As mentioned previously the section is divided up into component rectangles so as to maximize $\sum h_{min}^3 h_{max}$ and this is done by making wide rectangles as long as possible.

For example:

Tee beam

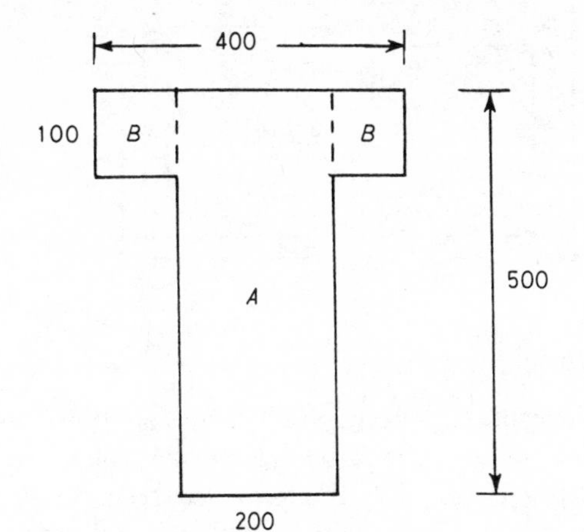


$$\sum h_{min}^3 h_{max} = (200)^3 \times 500 + (100)^3 \times 200$$
$$= 4 \times 10^9 + 0.2 \times 10^9$$
$$= 4.2 \times 10^9 \text{ mm}^4.$$

So

$$T'_A = T \times 4/4.2 = 0.95T.$$
$$T'_B = T \times 0.2/4.2 = 0.05\ T.$$

Having found the individual torsional moments we can now put these into the equation and find the individual torsional shear stress, and reinforcement provided accordingly. In the final detailing stirrups should be extended so that the component rectangles are properly tied together, thus:

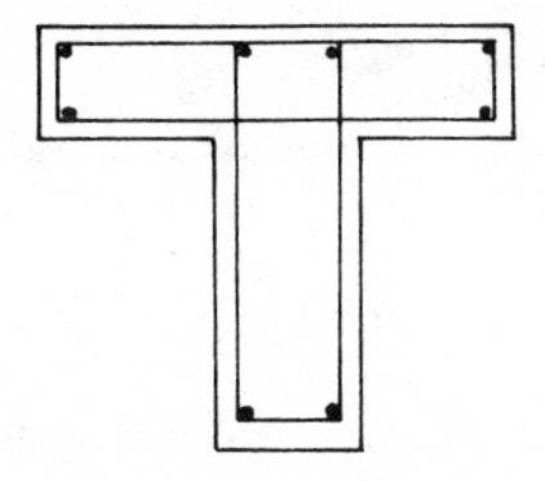

## 17.4 Box sections

Where the wall thicknesses are greater than one-quarter of the overall thickness in the direction being measured, they can be considered as solid sections. For other sections, specialist literature should be consulted, but for a thin-walled section as shown, the Bredt equation is often used, which gives:

$$\text{torsional stress in wall} = \frac{T}{2t_1 A_0}$$

$$\text{torsional stress in top or bottom} = \frac{T}{2t_2 A_0}$$

where $A_0$ = area within median line, i.e. $(B - t_1)(H - t_2)$.

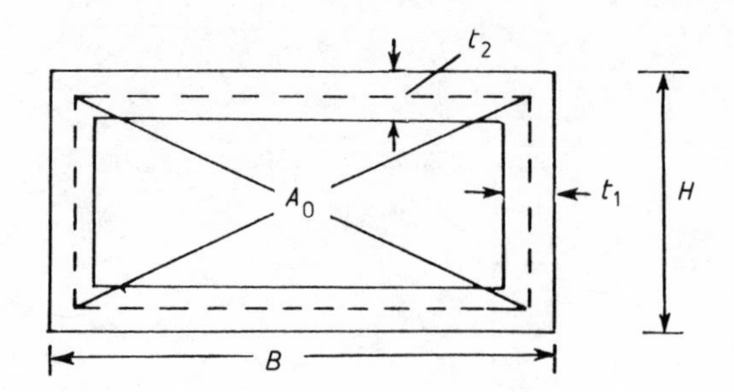


## 17.5 Shear centre

An important consideration in designing for torsion is in determining what the torsional moments are. In other words, what is the point about which moments should be taken? This point is generally referred to as the shear centre or in some cases the flexural centre and is the point about which the section will rotate. For no rotation, the centre of force of applied and self loads must go through the shear centre.

For a rectangular section this point is the geometric centre of the section. For other sections the point is not so easy to find and reference should be made to text books or books like *Formulae for Stress and Strain* by Roark. A section being used in buildings is a vertical channel section positioned thus [ . Services can be hidden inside the recess. If the section is of uniform thickness and has equal flanges as indicated,

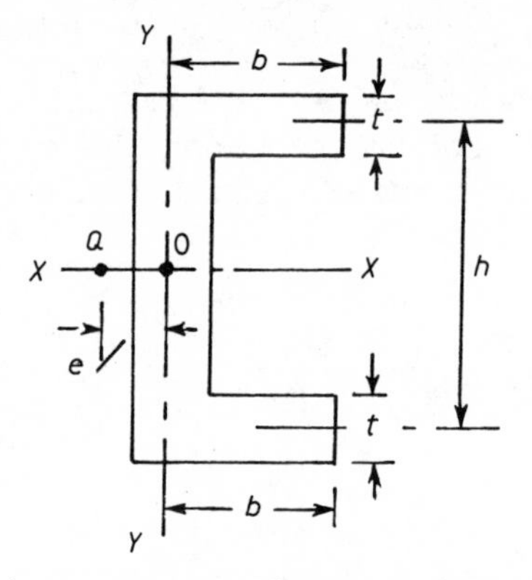


the distance from $Q$, the shear centre, to the centre line of the web is given by the equation

$$e = b^2h^2t/4I_x$$

where $b$, $h$, $t$ are as indicated, and $I_x$ is the second moment of area of the whole section with respect to $X$–$X$.

If the web is a different thickness from the flange it will be necessary to calculate the position more accurately, but for a small difference of up to 25% it will be sufficiently accurate to substitute the flange thickness for $t$. As example using the shear centre is given in Example 17.1.

## EXAMPLE 17.1

The edge beam shown below has a span of 14 m and is fully restrained at the ends. The ends of simply supported floor slabs of 8.0 m span rest on the lower flange with the centre of bearing at a distance of 145 mm from the inside face of the vertical leg. The self load of the floor slabs is 3.5 $kN/m^2$ and the imposed load on the floor is 2.5 $kN/m^2$.

Assume $d = 1445$ mm and 0.25% tensile reinforcement is required for flexure. Cover to links is 30 mm. Concrete is Grade 40 and reinforcement is Grade 460. Calculate the maximum values for the reinforcement required to resist torsion and shear.

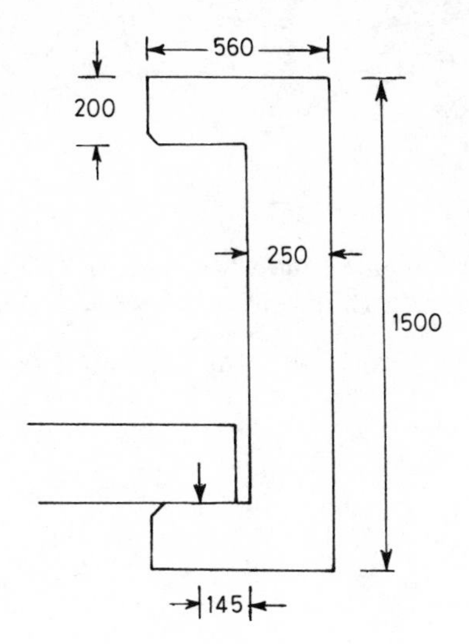


## Properties of section

Area $= 2 \times 310 \times 200 + 1500 \times 250 = 499 \times 10^3$ $mm^2$.
Self load $= 0.499 \times 24 = 12$ kN/m.
Distance of C.G. from external vertical face

$$= \frac{2 \times 310 \times 200 \times 405 + 1500 \times 250 \times 125}{499 \times 10^3} = 195 \text{ mm}$$

$$I_{xx} = \frac{560 \times 1500^3}{12} - \frac{310 \times 1100^3}{12} = 123 \times 10^9 \text{ mm}^4.$$

Distance of shear centre from external vertical face

$$= \frac{435^2 \times 1300^2 \times 200}{4 \times 123 \times 10^9} - 125 = 130 - 125$$

$$= 5 \text{ mm (outside unit).}$$

Design load on flange from floor slabs $= (8/2)(3.5 \times 1.4 + 2.5 \times 1.6)$
$= 35.6$ kN/m.

Design self load $= 12.0 \times 1.4 = 16.8$ kN/m.

## Shear

This will be taken on the web only.

Design shear $= 7(35.6 + 16.8) = 367$ kN.

$$v = \frac{367 \times 10^3}{250 \times 1445} = 1.02 \text{ N/mm}^2.$$

$100A_s/bd = 0.25$, so $v_c = 0.47$ N/mm$^2$.

$v_c + 0.4 < v < 5.0$ so design shear reinforcement required:

$$\frac{A_{sv}}{s_v} = \frac{250(1.02 - 0.47)}{0.87 \times 460} = 0.34.$$

## Torsion

Distance from shear centre to load from slabs $= 145 + 250 + 5 = 400$ mm.
Distance from shear centre to C.G. of unit $= 195 + 5 = 200$ mm.

Taking moments about the shear centre, the maximum design torsion

$T = (14/2)(35.6 \times 0.400 + 16.8 \times 0.2) = 123$ kN m.

$$\begin{aligned}\sum h^3_{min} h_{max} &= 2 \times 200^3 \times 310 + 250^3 \times 1500 \\ &= 2 \times 2.5 \times 10^9 + 23.4 \times 10^9 \\ &= 28.4 \times 10^9 \text{ mm}^4.\end{aligned}$$

## Large rectangle, i.e. web of unit

$$T' = \frac{23.4}{28.4} \times 123 = 101.2 \text{ kN m}.$$

$$v_t = \frac{2 \times 101.3 \times 10^6}{250^2(1500 - 250/3)} = 2.29 \text{ N/mm}^2.$$

This is greater than $v_{t,min}$ so design for torsion:

$v + v_t = 1.02 + 2.29 = 3.30$, i.e. less than $v_{tu}$ (5.0 N/mm$^2$).

Assuming 10 mm links, $x_1 = 1250 - 35 \times 2 = 180$
$y_1 = 1500 - 35 \times 2 = 1430$.

$$\frac{A_{sv}}{s_v}\text{(torsion)} = \frac{101.3 \times 10^6}{0.8 \times 180 \times 1430 \times 0.87 \times 460} = 1.23.$$

$$\text{Total } \frac{a_{sv}}{s_v} = 0.34 + 1.23 = 1.57.$$

Using $10\phi$ links $s_v = 157/1.57 = 100$ mm.

Using $12\phi$ links $s_v = 226/1.57 = 144$ mm.

The second arrangement would appear to be better, as we shall also need reinforcement to transfer the load from the flange to the top of the main beam.

$$\text{Area required} = \frac{35.6^3 \times 10^3}{0.87 \times 460} = 89 \text{ mm}^2\text{/m}.$$

So $12\phi$ at 130 centres will do.

$A_{sl} = 1.23 \times (460/460)(180 + 1430) = 1980$ mm$^2$.

As maximum distance between these bars is 300 mm we shall require six levels of bars with two bars in each level – but see notes at end of calculation.

## Small rectangles, i.e. top and bottom flanges

$$T' = \frac{2.5}{28.4} \times 123 = 10.8 \text{ kN m}.$$

$$v_t = \frac{2 \times 10.8 \times 10^6}{200^2(310 - 200/3)} = 2.22 \text{ N/mm}^2.$$

$y_1$ for small rectangles $= 310 - 35 = 275$ mm.

So allowable $v_t = 5.0 \times 275/550 = 2.5$ N/mm$^2$.

This is satisfactory, but if concrete had been Grade 30 it would not have been and we should have had to change section size.

$v_t$ is greater than $v_{t,min}$ so design for torsion.

$x_i = 130$, $y_1 = 275$.

$$\frac{A_{sv}}{s_v} = \frac{10.8 \times 10^6}{0.8 \times 130 \times 275 \times 0.87 \times 460} = 0.94.$$

For the top rectangle we provide for these links only and although 10$\phi$ at 167 centres would be satisfactory, they will have to be tied into the web reinforcement and so use the same centres as in the web.

For the bottom rectangle we shall also need reinforcement for this to act as a continuous concrete nib. It is suggested that vertical loops are provided by combining the requirements for torsion and flexure of the nib and using the same centres (if possible) as the links in the web. It is not proposed to carry out these calculations in this section.

The longitudinal reinforcement for torsion will be the same in both rectangles and is given by

$A_{sl} = 0.94(130 + 275) = 381$, say 4/12$\phi$.

The suggested beam detail is as shown, but it must be remembered that the reinforcement for flexure is to be added in either as separate bars, or larger bars can be used, which would cater for flexure and torsion. It should be also remembered that as the beam is greater than 750 mm deep, bars will be required in the side faces to prevent cracking and the maximum spacing is 250 mm. As the width of the beam is 250 the maximum spacing using 12$\phi$ bar is 236 m. The detail shown complies with this.

A quite usual example of torsion is when a cantilever slab is used as a canopy over an

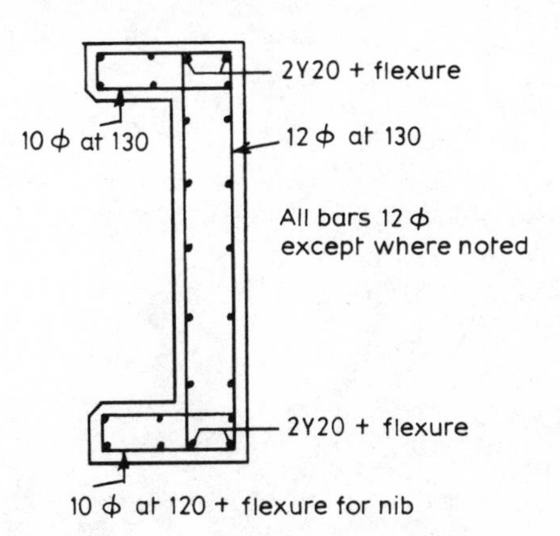

entrance and as this occurs generally between floors there will not be a floor slab behind the canopy. The beam supporting the canopy will usually span between two columns and will therefore be restrained from rotating by the columns, thus setting up torsional moments. An example is given in Example 17.2.

## EXAMPLE 17.2

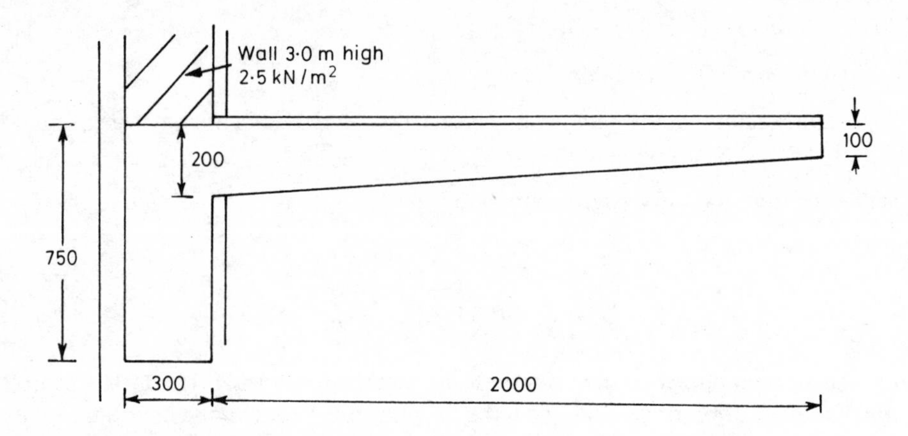


Span of beam = 6.0 m.
Finishes on cantilever = 0.4 kN/m$^2$.
Imposed load = 1.0 kN/m$^2$.
$f_{cu}$ = 40 N/mm$^2$.
$f_y$ = 460 N/mm$^2$.
Condition of exposure – moderate.

### Slabs

Dead loads:

Self (average) = 0.150 × 24 = 3.60 kN/m$^2$
Finishes = 0.40 kN/m$^2$
4.00 kN/m$^2$

Imposed load: 1.0 kN/m$^2$.

### Beam

Dead load:

Self = 0.75 × 0.3 × 24 = 5.4 kN/m
Slab = 4.0 × 2 = 8.0 kN/m
Wall = 2.5 × 3 = 7.5 kN/m
20.9 kN/m

Imposed:

Slab 1.0 × 2 = 2.0 kN/m.

Maximum design load on beam $=20.9 \times 1.4+2.0 \times 1.6=32.5$ kN/m.
Assuming factors of $\frac{1}{12}$ at support and $\frac{1}{24}$ at midspan, and $d=700$ mm,

$$M_{support}=\tfrac{1}{12} \times 32.5 \times 6^2=97.5 \text{ kN m}$$
$$M_{span}=\tfrac{1}{24} \times 32.5 \times 6^2=48.7 \text{ kN m.}$$

## Support

$M/bd^2=(97.5 \times 10^6)/(300 \times 700^2)=0.66$.

From design chart or tables $100A_s/bd=0.17$.

$A_s=357$ mm$^2$, say 2/16 (402 mm$^2$).

## Midspan (assuming rectangular section)

$M/bd^2=(48.7 \times 10^6)/(300 \times 700^2)=0.33$,

which is less than in tables.

$z/d=0.95$, so $A_s=183$ mm$^2$.

Note: The minimum percentage is 0.13 of gross section so the minimum $A_s=293$ mm$^2$ will have to be provided in the final arrangement.

## Shear

Design shear $=32.5 \times 3=97.5$ kN.

$v=(97.5 \times 10^3)/(300 \times 700)=0.46$ N/mm$^2$.

At support we shall probably need more than the 2/16 provided for flexure, but use this amount at this stage.

$100A_s/bd=(402 \times 100)/(300 \times 700)=0.19$.

From Table 3.9 of the Code,

$v_c=0.42$ N/mm$^2$.

So shear reinforcement required.

$$\frac{A_{sv}}{s_v}=\frac{300(0.46-0.42)}{0.87 \times 460}=0.03.$$

## Torsion

Assuming shear centre (i.e. centre of rotation) is at centre of gravity of rectangular section,

$$\begin{aligned}\text{torsion moment} &= 3.60 \times 2 \times 1.4 \times (0.90+0.15)\\ &\quad +0.4 \times 2 \times 1.4 \times (1.0+0.15)+1.0 \times 2 \times 1.6 \times 1.15\\ &=10.58+1.29+3.68=15.55 \text{ kN m/m.}\end{aligned}$$

Total torsional moment at column $=15.55\times 3=46.65$ kN m.

$$v_t=\frac{2\times 46.65\times 10^6}{300^2(750-100)}=1.59\ \text{N/mm}^2.$$

This is greater than $v_{t,min}$ so torsion reinforcement required.

$v+v_t=0.46+1.59=2.05$, i.e. $<v_{tu}$.

For 'moderate' condition, cover to stirrups $=30$ mm, and assuming 10 mm stirrups we have 35 mm to centre line of stirrups, so $x_1=230$, $y_1=680$.

$$A_{sv}=\frac{46.65\times 10^6}{0.8\times 230\times 680\times 0.87\times 460}=0.93.$$

Total $A_{sv}/s_v=0.03+0.93=0.96$.

Using 10 mm stirrups $s_v=157/0.96=163$.

$A_{s1}=0.96\times(460/460)\times(230+680)=874\ \text{mm}^2$.

Torsional reinforcement not required when $v_t=0.40\ \text{N/mm}^2$, i.e. when

$$T=\frac{0.40\times 300^2\times 650}{2\times 10^6}=11.7\ \text{kN m},$$

which occurs at a distance of $11.7/15.55=0.75$ m from centre of span. Torsion reinforcement has to be continued a distance equal to the largest dimension of the section so this will take us up to the centre line. The torsion reinforcement is therefore required along the complete span, but as the torsion moment reduces towards the centre of the span so will the reinforcement required to resist it.

### At 1.0 m from support

Torsion $T=31.1$ kN m, $v_t=1.06\ \text{N/mm}^2$.

$A_{sv}/s_v=0.62$.

Shear $v=\frac{2}{3}\times 0.46=0.31\ \text{N/mm}^2$, i.e. $<v_c$, so torsion reinforcement only.
Using $10\phi$ bars $s_v=157/0.62=253$ mm.

The maximum spacing for torsion must not exceed the least of 230, 680/2 or 200, so use 10 mm at 200.

$A_{s1}=0.62\times 1\times 910=564\ \text{mm}^2$.

### At 2.0 m from support

Torsion $A_{sv}/s_v=0.31$. This will also be the total and maximum spacing will be 200 mm as before.

$A_{s1}=282\ \text{mm}^2$.

## Minimum stirrups

The minimum area of stirrups required is obtained from

$A_{sv}/s_v = (0.4 \times 300)/(0.87 \times 460) = 0.3,$

i.e. 10 mm at $157/0.3 = 523$ mm.
The minimum we are providing is 10 at 200 so we are covered.

## Arrangement of reinforcement

For the longitudinal bars we need one in each corner and as the spacing between the bars should not exceed 300 mm we shall need two intermediate bars in each side. So dividing the area by 4 we can find the area required at each level for any section.

From the assumed bending moment diagram the point of contraflexure will be 1.30 m from the support, and the minimum distance the top bars have to be continued into the span is 700 mm (i.e. effective depth *d*). Similarly for the bottom bars the requirements for flexure are as shown.

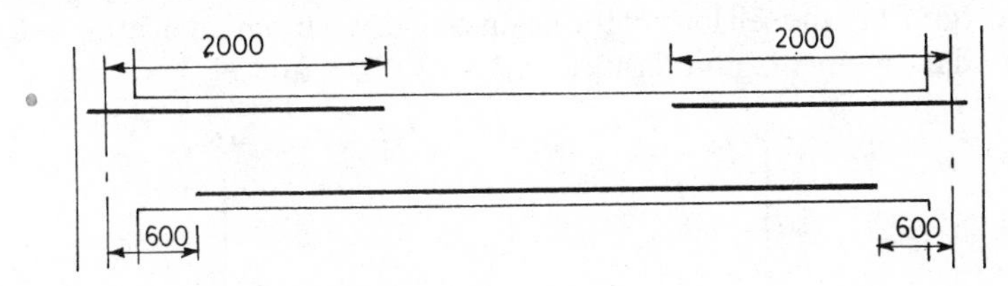


We can tabulate requirements as follows:

## Area of longitudinal bars required

| Position | Torsion (per level) mm² | Flexure Top mm² | Flexure Bottom mm² | Required Top mm² | Required Bottom mm² |
|---|---|---|---|---|---|
| Support | $\frac{874}{4} = 219$ | 357 | — | 576 | 219 |
| 1.0 m from support | $\frac{564}{4} = 141$ | 357 | 183 | 498 | 324 |
| 2.0 m from support | $\frac{282}{4} = 71$ | — | 183 | 71 | 254 |
| Midspan | $\frac{282}{4} = 71$ | — | 183 | 71 | 254 |

For intermediate bars the maximum is 219 mm², so provide 2/12 (226).

At the support provide 2/20 (628) in the top.

By providing 2/16 (402) in the bottom throughout the span we comply with all requirements. As the bars for torsion are only stressed to 0.8 times the design stress take the bottom bars anchorage length of 0.8 times a full anchorage bond length. The top bars would need a full anchorage length.

A section through the beam would be:

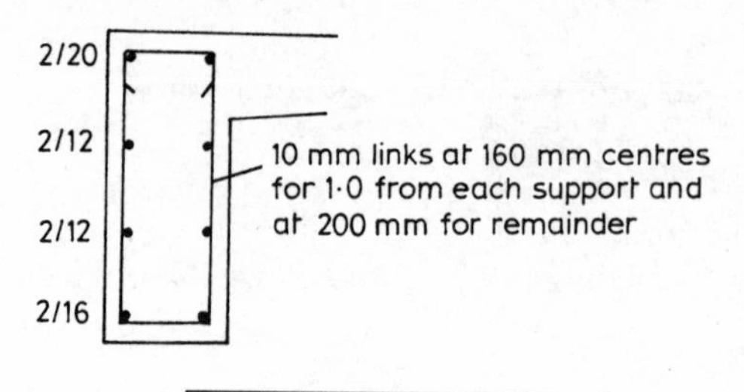


A simple example to illustrate the use of the torsional rigidity constant, $C$, is given below.

## EXAMPLE 17.3

A single vertical point load of 180 kN at ultimate limit state acts on the centre of a beam $BE$. This beam is supported by two beams $AC$ and $DF$ as shown. All beams are 300 wide by 600 deep. Ignoring the self load of the beams, by using moment distribution find the resulting bending moments and torques at $A$, $B$, $C$, $D$, $E$ and $F$.

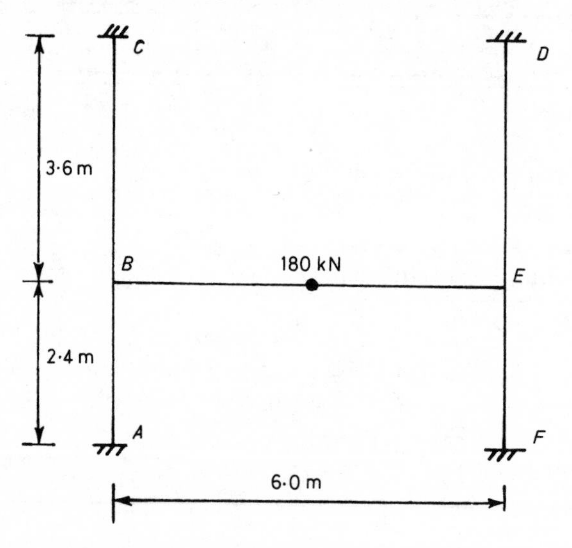


$I = 300 \times 600^3/12 = 5.4 \times 10^9$ mm$^4$.
$J_{gross} = \gamma b^3 h = 0.23 \times 300^3 \times 600 = 3.72 \times 10^9$ mm$^4$.
$C = \frac{1}{2}J = 1.86 \times 10^9$ mm$^4$.

As the frame is symmetrical

$$K_{BE} = \tfrac{1}{2}\left(\frac{4EI}{L}\right) = \frac{2 \times 5.4 \times 10^9}{6 \times 10^3} E = 1.8 \times 10^6 E$$

$$K_{TBA} = \frac{CG}{L_{BA}} = \frac{1.86 \times 10^9 \times 0.4E}{2.4 \times 10^3} = 0.31 \times 10^6 E$$

$$K_{TBC}=\frac{CG}{L_{BC}} \quad =\frac{1.86\times 10^9\times 0.4E}{3.6\times 10^3}=0.21\times 10^6E$$

$$\sum K=2.32\times 10^6E.$$

The distribution factors are:

$D_{BE}=1.8/2.32 \;=0.776$

$D_{BA}=0.31/2.32=0.134$

$D_{BC}=0.21/2.32=0.090.$

Fixed end moment at $B$ due to load $=180\times 6.0/8=135$ kN m.
Releasing joints $B$ and $E$ simultaneously.

$M_{BE}=135-0.776\times 135=135-105=30$ kN m

$T_{BA}\;=135\times 0.134=18$ kN m

$T_{BC}\;=135\times 0.09=12$ kN m.

The carry-over factors for torque are $+1.0$ so

$T_{AB}=18$ kN m

$T_{CB}=12$ kN m.

The bending moments in $AC$ and $DF$ can be calculated as a beam carrying a point load of 90 kN at $B$ and $E$ respectively.

If we wish to find the rotation at $B$ we can use the formula $\theta=TL/CG$.

Using the length $AB$, $T=18$ kN m, $L=2.4$ m, $C=1.86\times 10^{-3}$ m$^4$,

$G=0.42\times 28\times 10^6$ kN/m$^2$ (assume Grade 40 concrete)
$=11.76\times 10^6$.

$$\text{So } \theta=\frac{18\times 2.4}{1.86\times 10^{-3}\times 11.76\times 10^6}=1.97\times 10^{-3} \text{ radians.}$$

# Appendix 1

# CALCULATION FOR DEFLECTION

This is dealt with in Part 2 of the Code. In clause 3.4 it states that in general it will be sufficiently accurate to assess the moments and forces at serviceability limit states by using an elastic analysis. It should be pointed out here that a completely new analysis will have to be carried out, using the appropriate load combinations. It is not possible to use an analysis carried out for ultimate limit state. The Code also points out that although the calculations may show the members to be cracked, a more accurate picture of the moment and force fields is more likely to be provided by the use of the concrete section for stiffness rather than the transformed section.

In the calculations that follow any analysis will be based on the concrete section without any modifications. In clause 3.7 the Code lists four factors which are difficult to allow for in calculations, but which can have a considerable effect on the reliability of the calculations. The methods of calculation given in the Code are not the only acceptable methods, but should give reasonable results for short-term and long-term loading. In general, however, it will be found that there is little advantage over the 'deemed-to-satisfy rules' unless the factors in this clause are taken into account.

The object of this Appendix is to illustrate the procedure only and as the factors listed in clause 3.7 are ignored, the answers obtained could be reached much more quickly from the 'deemed-to-satisfy' rules.

The approach used is to assess the curvature of sections under the appropriate moments and then calculate the deflections from the curvature. So the procedure is as follows:

1. Calculate the moments;
2. Calculate the curvature or curvatures;
3. Calculate the deflection.

### (a) MOMENTS

As stated earlier we shall use an elastic analysis using the concrete section (as at ultimate limit state) taking loads of $1.0G_k$ and $1.0Q_k$ with appropriate loading patterns. No redistribution of moments is allowed.

### (b) CURVATURE

For calculating the curvature due to loading, clause 3.6 in Part 2 gives a procedure which employs an appropriate set of assumptions depending on whether the section is cracked ($A$), or uncracked ($B$), whichever gives the larger value. This will generally be ($A$).

The assumptions for $(A)$ are straightforward and can be illustrated diagrammatically as shown in Fig. A1.1.

Values for $E_c$ for short-term loading may be obtained from clause 3.5 which refers to Table 7.2 of the Code. For long-term loading the effective $E$ is modified by a value $1/(1+\phi)$, where $\phi$ is the appropriate creep coefficient obtained from clause 7.3.

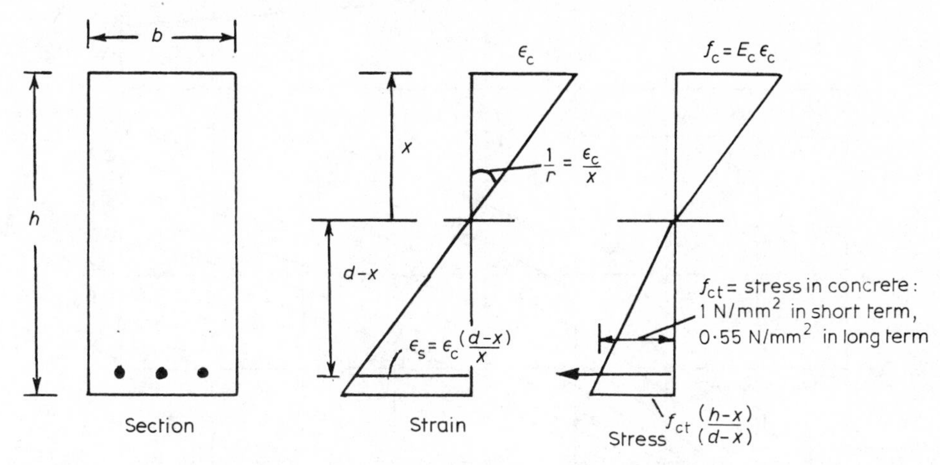


FIG. A1.1 Assumptions for calculating curvatures.

As can be seen from Fig. A1.1, equation (7) of the Code gives

$$1/r_b = f_c/xE_c = f_s/[(d-x)E_s].$$

Assessment of the stresses requires a trial-and-error approach. Calculation by means of a computer or programmable calculator is straightforward, but by hand is rather tedious.

In the previous book on CP110 the author proposed an alternative method, the details of which had been devised by Tony Threlfall. In view of the fact that the Code, rather than the Handbook, contains the equations given the author has compared the results of the two methods and has found them to give virtually the same answer.

The alternative method calculates the neutral-axis depth on the basis of zero stress in the concrete in the tension zone. Moments are then taken about the neutral axis taking into account the tensile stress in the concrete. So what it does in effect is to ignore the concrete in tension in deriving the neutral-axis depth, but takes it into consideration when calculating the resistance moment. The properties of transformed sections can be obtained from any well known text book, but Fig. A1.2 is a design chart for $x/d$ and $z/d$ for $d'/d = 0.1$ and Fig. A1.3 is a design chart for $I_e/bd^3$ for $d'/d = 0.1$.

The design charts in Figs A1.2 and A1.3 are typical charts and if compression reinforcement is ignored are all that is necessary for rectangular sections or Tee sections, where the neutral axis is within the flange.

The general force diagram for the assumptions in $(A)$ can be drawn as shown in Fig. A1.4.

Using the force diagram, and noting from Fig. A1.1 that the curvature $1/r$ is $\varepsilon_c/x$ we can find a relationship between $M$ and $1/r$ as follows.

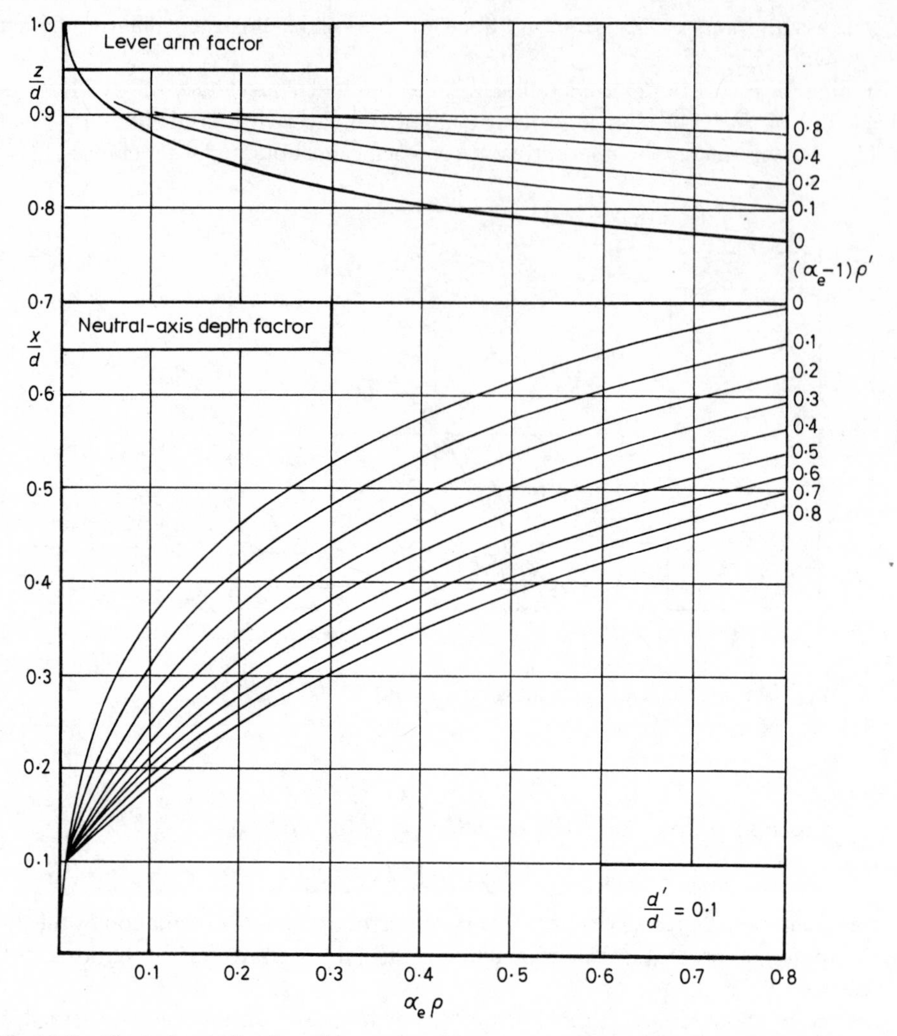


FIG. A1.2 Neutral-axis depth and lever arm factors for transformed rectangular sections.

## *Rectangular section*

Taking moments about the neutral axis for all the forces on the section,

$$M=\frac{1}{3}bx^2E_e\varepsilon_c+(\alpha_e-1)\rho'bd\frac{(x-d')^2}{x}E_e\varepsilon_c+\alpha_e\rho bd\frac{(d-x)^2}{x}E_e\varepsilon_c$$

$$+\frac{1}{3}b\frac{(h-x)^3}{(d-x)}f_{ct}$$

$$=\frac{E_e\varepsilon_c}{x}\left\{\frac{1}{3}bx^3+(\alpha_e-1)\rho'bd(x-d')^2+\alpha_e\rho bd(d-x)^2\right\}+\frac{1}{3}b\frac{(h-x)^3}{(d-x)}f_{ct}$$

$$=\left(\frac{1}{r}\right)E_eI_e+\frac{(h-x)^3}{(d-x)}\frac{bf_{ct}}{3},$$

where $I_e$ is the second moment of area of the equivalent transformed section.

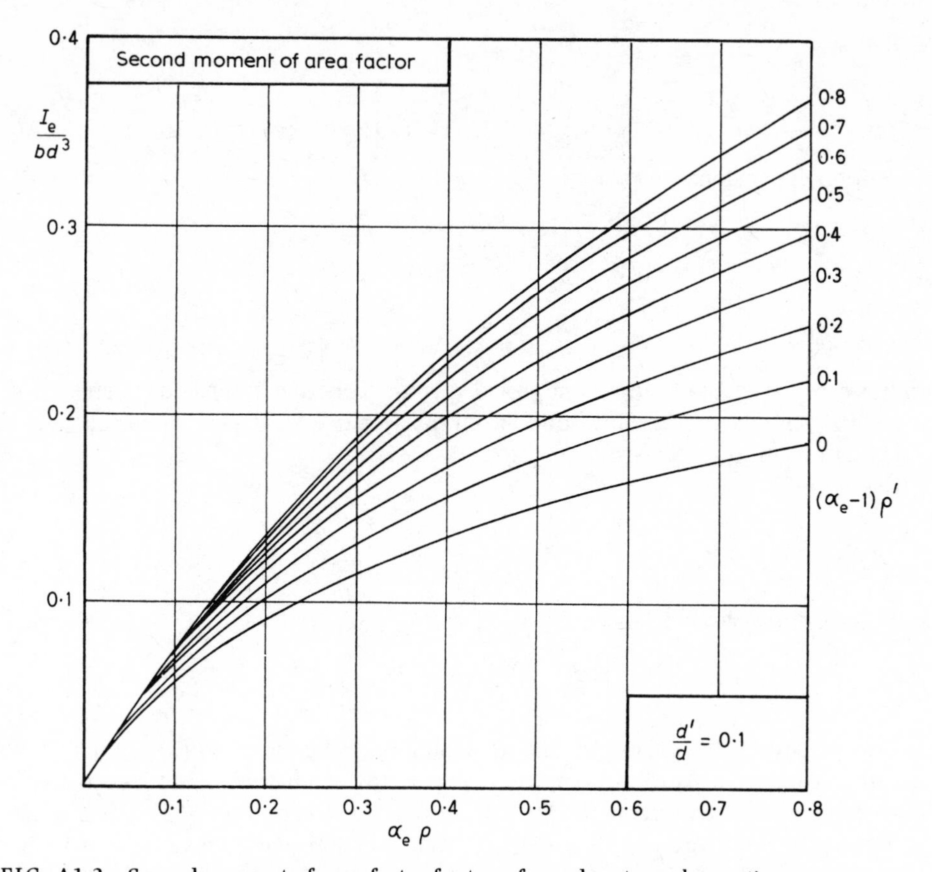


FIG. A1.3 Second moment of area factor for transformed rectangular sections.

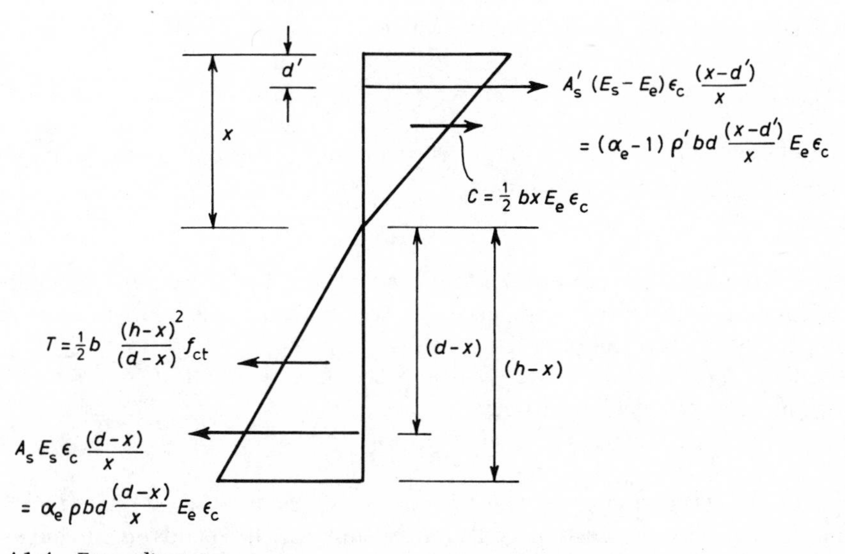


FIG. A1.4 Force diagram.

Hence,

$$\frac{1}{r}=\left[M-\frac{(h-x)^3}{(d-x)}\frac{bf_{ct}}{d}\right]\Big/E_eI_e,$$

or in 'non-dimensional' terms,

$$\frac{d}{r}=\left[\frac{M}{bd^2}-\frac{(h/d-x/d)^3}{(1-x/d)}\frac{f_{ct}}{3}\right]\Big/\frac{E_eI_e}{bd^3}.$$

*Tee section*

In this case the resistance moment provided by the concrete in tension depends on $b_w$ instead of $b$, ignoring the contribution from the area $(b-b_w)(h_f-x)$ when the neutral axis occurs within the flange.

Hence

$$\frac{1}{r}=\left[M-\frac{(h-x)^3}{(d-x)}\frac{b_wf_{ct}}{3}\right]\Big/E_eI_e,$$

or in 'non-dimensional terms,

$$\frac{d}{r}=\left[\frac{M}{bd^2}-\frac{(h/d-x/d)^3}{(1-x/d)}\frac{b_w}{b}\frac{f_{ct}}{3}\right]\Big/\frac{E_eI_e}{bd^3}.$$

Note: in assessing the curvature due to loading by using equation (7) of the Code, a useful starting point is to estimate the neutral axis depth from what has just been described as the alternative method. We know that this will be too small but it does give a starting point for the iterative method. A comparison will be given in the design example in this Appendix.

In assessing the long-term curvature the procedure is outlined in clause 3.6 and is set out in four stages. This is shown diagrammatically in Fig. A1.5.

For long-term effects the designer will need to modify the modulus of elasticity so $E_e=E_c/(1+\phi)$, and also determine the shrinkage curvature $1/r_{cs}$. The equation for $1/r_{cs}$ is equation (9) of clause 3.6, and as for determining the creep coefficient, $\phi$, Section 7 of Part 2 gives charts for determining this value.

## (c) DEFLECTION

The final step is the calculation of the deflection from the curvature. Equation (10) of clause 3.7.2 gives the relationship as

$$1/r_x=d^2y/dx^2,$$

where $r_x$ is the curvature at $x$ and $y$ is the deflection at $x$. The deflection can be obtained by calculating the curvatures at successive sections along the member and using a numerical integration technique such as that proposed by Newmark.

The Code does, however, suggest a simplified approach and in equation (11) the deflection $a$ can be calculated from

$$a=Kl^2(1/r_b)$$

where $l$ is the effective span of the member, $1/r_b$ is the curvature at midspan or, for a cantilever, at the support section, and $K$ is a constant which depends on the shape of the bending moment diagram.

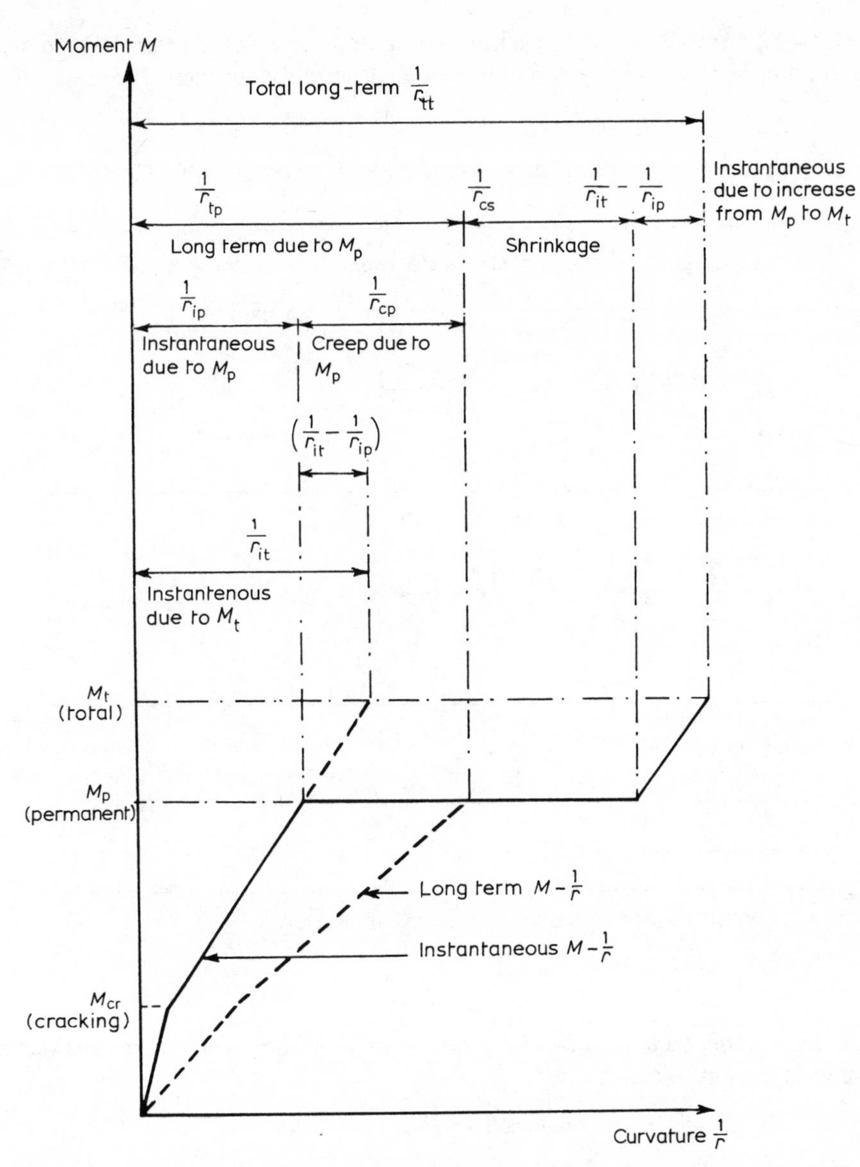


FIG. A1.5 Loading history for serviceability limit state – deflection.

The Code, in Table 3.1, gives values for $K$ for various loadings and support conditions. For combinations of loads the principles of superposition can be used.

As the curvature is calculated at midspan for members other than cantilevers, the deflection will be measured at midspan. If the loads are unsymmetrical this may not be the worst case, but the difference is small.

For a cantilever, it should be noted that the coefficients for $K$ assume that it is rigidly fixed at the support, i.e. it is horizontal. In practice this may not be so, because the loading on adjoining members may cause the root of the cantilever to rotate. If this rotation is $\theta_b$, the cantilever deflection may be increased or decreased by $l\theta_b$, depending on the direction of rotation at the root.

If $M_1$ ($=M_b$ normally) and $M_2$ are the restraint moments at the ends of the adjacent span, $L_1$, and $M_s$ is the simply supported moment at midspan, then

$$\theta_b = K_1 L_1 (M_s/E_e I_1) - \tfrac{1}{6} L_1 (2M_1 + M_2)/E_e I_1,$$

where $I_1$ is the second moment of area of span $L$ and $K_1$ can be obtained from Fig. A1.6.

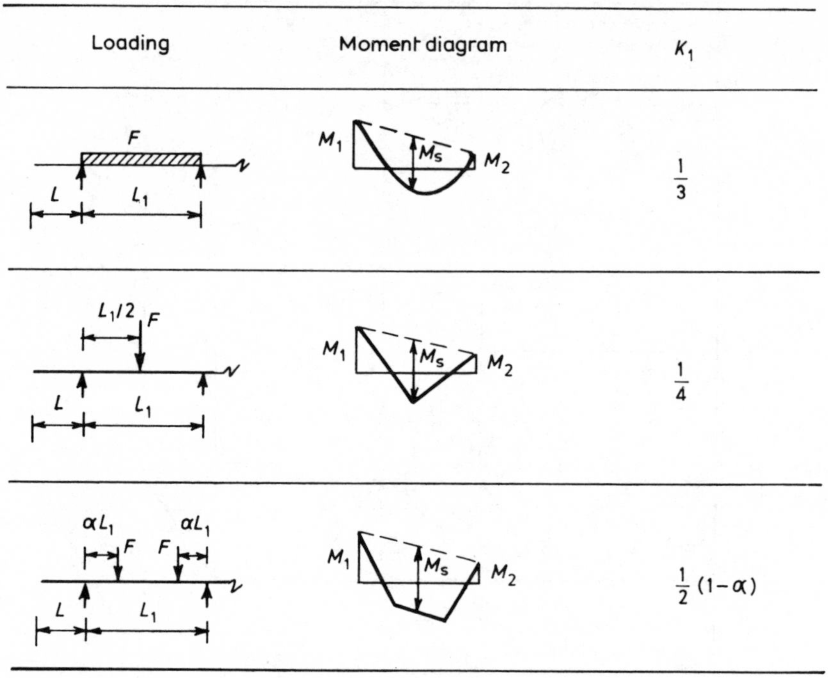

| Loading | Moment diagram | $K_1$ |
| --- | --- | --- |
| $F$; $L$; $L_1$ | $M_1$; $M_s$; $M_2$ | $\frac{1}{3}$ |
| $L_1/2$; $F$; $L$; $L_1$ | $M_1$; $M_s$; $M_2$ | $\frac{1}{4}$ |
| $\alpha L_1$; $\alpha L_1$; $F$; $F$; $L$; $L_1$ | $M_1$; $M_s$; $M_2$ | $\frac{1}{2}(1-\alpha)$ |

FIG. A1.6 Coefficients for rotation at supports of cantilever.

As with the other factors when using the values for $K$ from Table 3.1 in the Code we can rewrite the formula as

$$\theta_b = K_1 L_1 (1/r_s) - \tfrac{1}{6} L_1 (2/r_1 + 1/r_2).$$

In order to calculate the long-term deflection it is necessary to assess how much load is permanent. The proportion of imposed load which may be regarded as permanent depends on the occupancy of the building. The Handbook suggests that for domestic or office occupancy 25% of the imposed load should be considered as permanent, and for storage purposes 75% should be considered permanent.

## EXAMPLE A.1

A roof slab and canopy is to be constructed as indicated below. Calculate the following deflections at the end of the canopy:

1. instantaneous deflection due to dead load
2. long-term deflection due to dead load and shrinkage
3. instantaneous deflections due to imposed load.

$f_{cu}=40\ \text{N/mm}^2$, $\phi=1.8$, $\varepsilon_{cs}=0.0003$,
$g_k=5\ \text{kN/m}^2$ (canopy), $6.2\ \text{kN/m}^2$ (roof slab),
$q_k=0.75\ \text{kN/m}^2$.

Consider $d=(h-26)$ for both top and bottom reinforcement.

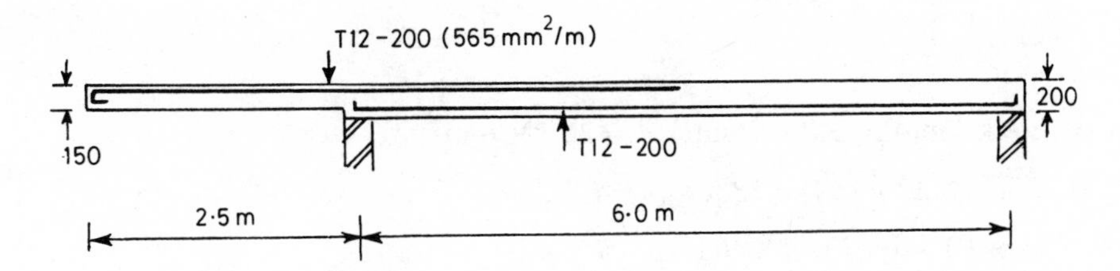


(1) Calculate the cantilever moment $M_b$ and the midspan simply supported moment $M_s$ due to dead and imposed loads.

Due to dead load,

$M_b=5\times2.5^2/2=15.6$ kN m, $M_s=6.2\times6^2/8=27.9$ kN m.

Due to imposed load,

$M_b=0.75\times2.5^2/2=2.35$ kN m, $M_s=0.75\times6^2/8=3.35$ kN m.

(2) Calculate the properties of the equivalent transformed sections.

For the canopy, $d=124$,

$$\frac{100A_s}{bd}=\frac{100\times565}{1000\times124}=0.46.$$

For the roof slab, $d=174$,

$$\frac{100A_s}{bd}=\frac{100\times565}{1000\times174}=0.32.$$

(a) For short-term load, $E_c=28\ \text{kN/mm}^2$, $\alpha_e=200/28=7$.

For the canopy, $\alpha_e\rho=7\times0.46/100=0.032$, $(\alpha_e-1)\rho'=0$.

From Fig. A1.2 $x/d=0.223$, from Fig. A1.3 $I_e/bd^3=0.023$.

For the roof slab, $\alpha_e\rho=7\times0.32/100=0.022$, $(\alpha_e-1)\rho'=0$.

From Fig. A1.2 $x/d=0.19$, from Fig. A1.3 $I_e/bd^3=0.017$.

(b) For long-term loading, $E_e=28/(1+1.8)=10\ \text{kN/mm}^2$, $\alpha_e=200/10=20$.

For the canopy, $\alpha_e\rho=20\times0.46/100=0.092$, $(\alpha_e-1)\rho'=0$.

From Fig. A1.2 $x/d=0.347$, from Fig. A1.3 $I_e/bd^3=0.053$.

For the roof slab, $\alpha_e\rho=20\times0.32/100=0.064$, $(\alpha_e-1)\rho'=0$.

From Fig. A1.2 $x/d=0.3$, from Fig. A1.3 $I_e/bd^3=0.04$.

(1) Calculate instantaneous curvature at support section of canopy due to dead load only.

From relationship for rectangular section

$$\frac{1}{r_b}=\left[M_b-\frac{(h-x)^3}{(d-x)}\frac{bf_{ct}}{3}\right]\Big/E_eI_e$$

where

$M_b=15.6$ kN m, $f_{ct}=1.00$ N/mm², $E_e=28$ kN/mm²

$x=0.223d=0.223(124)=28$ mm, $I_e=0.023bd^3$.

$$\frac{1}{r_b}=\frac{15.6\times10^6-[(150-28)^3/(124-28)]\times10^3\times1.00/3}{28\times10^3\times0.023\times10^3\times124^3}=7.6\times10^{-6}.$$

Note: If the iterative procedure as suggested by the Code is carried out it will be found that the neutral axis depth is 37 mm. The stress in the tension steel is 131 N/mm² and the maximum compressive stress in the concrete is 8 N/mm². Hence $1/r_b=7.7\times10^{-6}$, which is virtually the same as calculated above.

Calculate instantaneous rotation, due to dead load, at support section of canopy.

$\theta_b=K_1L_1(1/r_s)-\frac{1}{6}L_1\{2(1/r_1)+1/r_2\}$

where $K_1=\frac{1}{3}$, $1/r_1=1/r_b$ and $1/r_2=0$ so that

$\theta_b=\frac{1}{3}L_1\{(1/r_s)-(1/r_b)\}$.

Since the reinforcement in the roof slab is the same at midspan and support sections, the section properties and hence the term involving $f_{ct}$ in the moment curvature relationship equation is also the same, in which case the term is cancelled out by subtraction, so that

$$\theta_b=\tfrac{1}{3}L_1\frac{(M_s-M_b)}{E_eI_1}$$

where $M_s=27.9$, $M_b=15.6$ kN m,

$L_1=6$ m, $E_e=28$ kN/mm², $I_1=0.017\ bd^3$.

$$\theta_b=\frac{6000(27.9-15.6)10^6}{3\times28\times10^3\times0.017\times10^3\times174^3}=9.8\times10^{-3}.$$

Instantaneous deflection, due to dead load, at edge of canopy

$a=KL^2(1/r_b)-L\theta_b$

where $K=\frac{1}{4}$ and $L=2500$.

$a=\frac{1}{4}\times2500^2\times7.6\times10^{-6}-2500\times9.8\times10^{-3}$
$=11.9-24.5=-12.6$ mm (upwards).

(2) Calculate long-term curvature, due to dead load, at support section of canopy.

$M_b=15.6$ kN m, $f_{ct}=0.55$ N/mm², $E_e=10$ kN/mm²,
$x=0.347d=0.347(124)=43$ mm, $I_e=0.053\ bd^3$.

$$\frac{1}{r_b} = \frac{15.6\times 10^6 - [(150-43)^3/(124-43)]\times 10^3 \times 0.55/3}{10\times 10^3 \times 0.053\times 10^3 \times 124^3} = 12.7\times 10^{-6}.$$

Calculate shrinkage curvature at support section of canopy:

$S_s = 565\times 81 = 45.7\times 10^3$, $\varepsilon_{cs} = 0.0003$.

$$\frac{1}{r_{cs}} = \frac{0.0003\times 20\times 45.7\times 10^3}{0.053\times 10^3\times 124^3} = 2.7\times 10^{-6}.$$

Total long-term curvature due to dead load and shrinkage,

$$\left(\frac{1}{r_b}\right)_t = \frac{1}{r_b} + \frac{1}{r_{cs}} = (12.7+2.7)\times 10^{-6} = 15.4\times 10^{-6}.$$

Calculate long-term rotation due to dead load and shrinkage at support section of canopy.

Since the reinforcement in the roof slab is the same at midspan and support sections, the shrinkage curvature, $\varepsilon_{cs}$, is also the same, in which case it is cancelled out by subtraction, so that

$$\theta_{bt} = \tfrac{1}{3}L_1 \frac{(M_s - M_b)}{E_e I_1}$$

where $M_s = 27.9$, $M_b = 15.6$ kN m,

$L_1 = 6$ m, $E_e = 10$ kN/mm$^2$, $I_1 = 0.04\ bd^3$.

$$\theta_{bt} = \frac{6000(27.9-15.6)10^6}{3\times 10\times 10^3\times 0.04\times 10^3\times 174^3} = 11.7\times 10^{-3}.$$

Long-term deflection due to dead load and shrinkage at edge of canopy,

$a = KL^2(1/r_{bt}) - L\theta_{bt}$,

where $K = \tfrac{1}{4}$ and $L = 2500$.

$$a = \tfrac{1}{4}\times 2500^2\times 15.4\times 10^{-6} - 2500\times 11.7\times 10^{-3}$$

$$= 24.1 - 29.3 = -5.2 \text{ mm (upwards).}$$

(3) (a) Imposed load on canopy only – $M_b = 2.35$ kN m, $M_s = 0$.

Additional instantaneous curvature and rotation due to imposed load at support section of canopy.

$$\left(\frac{1}{r_b}\right)_a = \frac{M_b}{E_c I_e} = \frac{2.35\times 10^6}{28\times 10^3\times 0.023\times 10^3\times 124^3} = 1.9\times 10^{-6}$$

$$(\theta_b)_a = \tfrac{1}{3}L_1 \frac{(-M_b)}{E_c I_1} = \frac{-6000\times 2.35\times 10^6}{3\times 28\times 10^3\times 0.017\times 10^3\times 174^3} = -1.9\times 10^{-3}.$$

Additional instantaneous deflection due to imposed load, at edge of canopy,

$$a = \tfrac{1}{4}\times 2500^2\times 1.9\times 10^{-6} - 2500\times(-1.9\times 10^3)$$
$$= 3.0 + 4.8$$
$$= 7.8 \text{ mm.}$$

(b) Imposed load on roof slab only – $M_b = 0$, $M_s = 3.35$ kN m.

Additional instantaneous rotation due to imposed load, at support section of canopy,

$$(\theta_b)_a = \tfrac{1}{3}L_1 \frac{M_s}{E_c I_1} = \frac{6000 \times 3.35 \times 10^6}{3 \times 28 \times 10^3 \times 0.017 \times 10^3 \times 174^3} = 2.7 \times 10^{-3}.$$

Additional instantaneous deflection due to imposed load, at edge of canopy,

$a = -2500 \times 2.7 \times 10^3 = -6.8$ mm (upwards).

## Note

Maximum total long-term *downwards* deflection at edge of canopy $= -5.2 + 7.8 = 2.6$ mm.

Limiting deflection = span/250 = 2500/250 = 10 mm.

Actual span/effective depth ratio = 2500/124 = 20.2.

From the Code, to find the limiting value:

$M$ (ultimate) $= 15.6 \times 1.4 + 2.35 \times 1.6 = 25.6$ kN m/m.

$M/bd^2 = (25.6 \times 10^6)/(1000 \times 124^2) = 1.66.$

From Tables $100A_s/bd = 0.44$.

$A_s$ required = 546 mm²/m. $A_s$ provided = 565 mm²/m.

$f_s = \tfrac{5}{8} \times 460 \times 546/565 = 278$ N/mm².

From Table 3.11 of the Code, modification factor for tension reinforcement = 1.2.

Allowable ratio $= 7 \times 1.2 = 8.4$.

In this example account has been taken of the rotation at the support adjacent to the cantilever. This rotation can only take place completely where the support can be classed as 'simple'. If there was restraint, such as a monolithic beam and slab construction or a wall above the support, then the rotation would be reduced. If we assume a rigid support then we should get the following deflections:

## Long-term deflection due to permanent load and shrinkage

From item (2), curvature $= 15.4 \times 10^{-6}$, and deflection = 24 mm.

## Instantaneous deflection due to difference between total and permanent loads

This will be due to the imposed load on the canopy only, and from item (3), curvature $= 1.9 \times 10^{-6}$, and deflection = 3.0 mm.

## Total deflection

This will be 27 mm.

Using the simplified approach, the estimated deflection

$$= \frac{\text{Actual span/effective depth ratio}}{\text{Allowable span/effective depth ratio}} \times \frac{\text{span}}{250}$$

$$= \frac{20.2}{8.4} \times \frac{2500}{250}$$

$= 24$ mm.

# Appendix 2

# CALCULATION FOR CRACK WIDTHS

Clause 3.8 of Part 2 of the Code gives equations for assessing crack widths for flexure and direct tension. The maximum surface width of crack will generally be 0.3 mm, but if this would impair the efficiency of the structure a smaller value may be more appropriate. For example, for watertightness values of 0.2 mm or even 0.1 mm may be required. An analysis will be required at service loads to determine the moments and stresses. We shall deal with flexure and direct tension separately.

## (a) FLEXURE

The design surface crack width at any point on the surface of the tension zone may be calculated from the following equation, which is equation (12) in Part 2.

$$w_{cr} = \frac{3a_{cr}\varepsilon_m}{1+2(a_{cr}-c_{min})/(h-x)}$$

where $a_{cr}$ is the distance from the point considered to the surface of the nearest longitudinal bar, $c_{min}$ is the minimum cover to tension steel, $h$ is the overall depth of member, $x$ is the depth of the neutral axis assuming a cracked section and determined by normal elastic theory, and $\varepsilon_m$ is the average strain at the level where the cracking is being considered.

$$\varepsilon_m = \varepsilon_1 - \frac{b_t(h-x)(a'-x)}{3E_sA_s(d-x)}$$

where $\varepsilon_1$ is the strain at the level considered, ignoring the stiffening effect of the concrete and determined from the same analysis as calculating $x$, $b_t$ is the width of section at the centroid of the tension steel, $a'$ is the distance from the compression face to the point at which the crack width is being calculated, $E_s$ is the modulus of elasticity of the reinforcement, and $A_s$ is the area of tension reinforcement.

The significance of the above terms may be illustrated diagrammatically as shown in Fig. A2.1.

The minimum cover, $c_{min}$, will not be the same if there is different side and bottom cover, so the bottom cover applies to points along the bottom edge of the beam, and the side cover to points along the side.

The Code also states that the formula only applies if the strain in the tension reinforcement is limited to $0.8f_y/E_s$. On occasions when 30% redistribution has been carried out at ultimate limit state, it will be found that the serviceability stress exceeds $0.8f_y$. If the increase is only small the calculation will be satisfactory.

It should also be noted that in calculating strains the modulus of elasticity of the

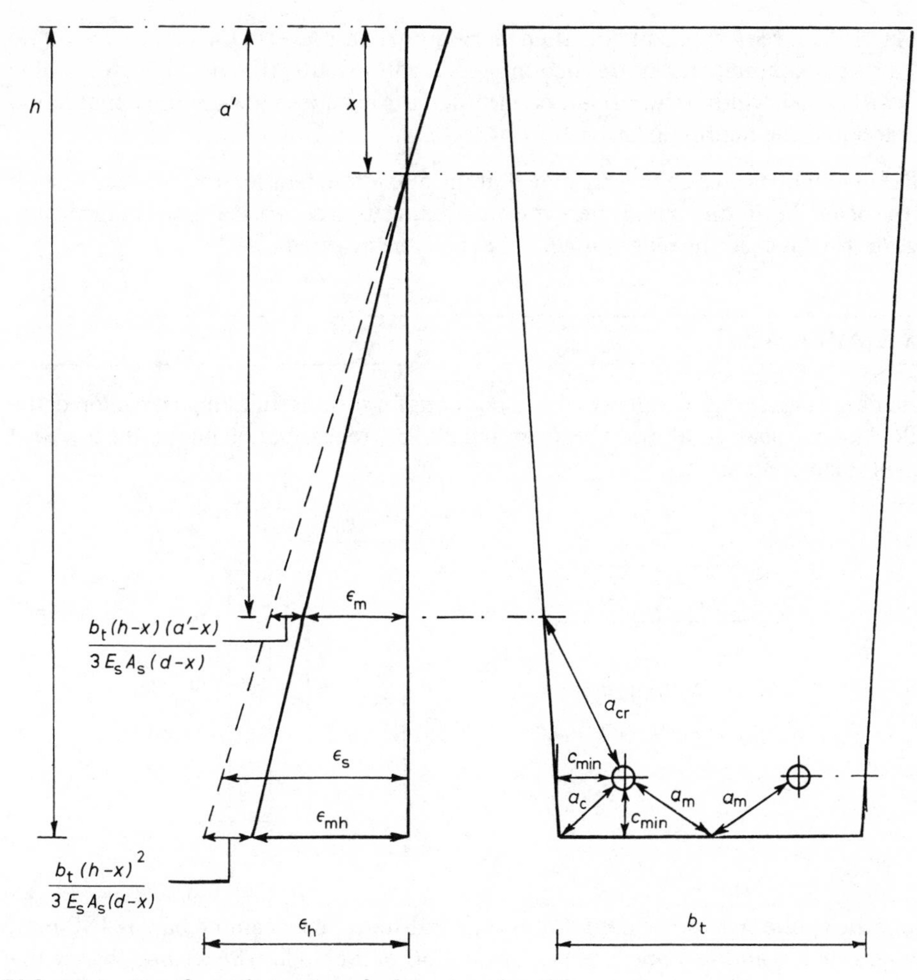


FIG. A2.1 Significant factors in calculating crack width.

concrete should be taken as half the instantaneous value obtained from Table 7.2 in Part 2.

Considering the formula at different positions in the beam as shown in Fig. A2.1, it can be seen that:

1. Along the bottom edge, in regions of maximum tension, $a' = h$, so $\varepsilon_m$ becomes

$$\varepsilon_1 - \frac{b_t(h-x)^2}{3E_sA_s(d-x)};$$

   a constant value. So in the crack width formula, $a_{cr}$ is the only variable, and as the maximum value of $a_{cr}$ is midway between the bars, the maximum crack width occurs at this point.
2. Immediately below the reinforcing bar, $a_{cr}$ is a minimum and equals $c_{min}$, so crack width $= 3c_{min}\varepsilon_m$ and is the minimum value.
3. Moving towards the corner, $a_{cr}$ becomes $a_c$, which is greater than $c_{min}$ and so the crack is wider.
4. Moving up the side of the beam, $\varepsilon_m$ decreases linearly from a maximum at the corner of the beam to zero at the neutral axis. The value of $a_{cr}$ decreases to a minimum at the

level of the reinforcement, and then incrases up to the level of the neutral axis, if no further longitudinal bars are encountered. It will be found that the maximum value of the crack width occurs about one-third of the distance between the longitudinal steel and the neutral axis.

All the comments given above assume that the maximum tensile stress is in the bottom of the beam, i.e. in the span. Where the maximum stress occurs in the top of the beam, e.g. over a support, the diagram will, of course, be inverted.

## EXAMPLE A2.1

Calculate design crack widths at critical positions for the internal support section of the following two-span continuous beam for which 15% redistribution has been allowed at the ultimate limit state.

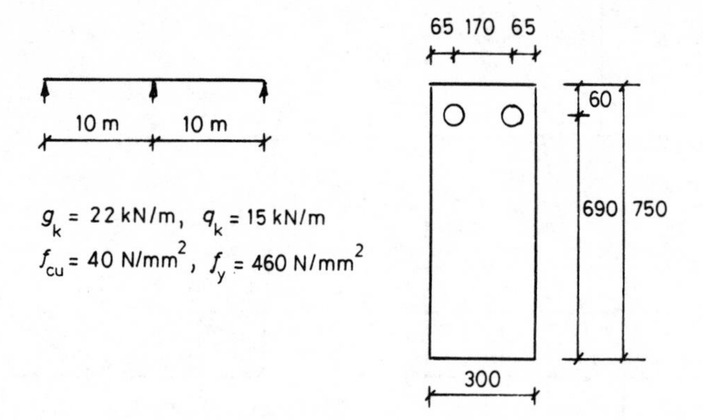


From the diagram it can be seen that the clear distance between the bars is 130 mm, which is less than the value given in Table 3.30 of the Code. The formula below the table gives 138.6 mm. The distance to the corner of the beam is 68.5 mm, which is also within the limit.

(1) Calculate internal support moment due to service load.

$M_s = (22 + 15) \times 10^2/8 = 463$ kN m.

(2) Calculate properties of equivalent transformed section with $E_e = E_c/2$.

$E_c = 28$ kN/mm$^2$ from Table 7.2 in Part 2, so $E_e = 14$ kN/mm$^2$.

$\alpha_e = E_s/E_c = 200/14 = 14$,

so $\alpha_e\rho = (14 \times 2510)/(300 \times 690) = 0.17$.

$x/d = \alpha_e\rho + \sqrt{[\alpha_e\rho(2 + \alpha_e\rho)]} = 0.437$.

So $x = 302$ mm and $z = 589$ mm.

(3) Calculate average surface strain at top of beam.

$$f_s = \frac{M_s}{A_s z} = \frac{463 \times 10^6}{2510 \times 589} = 313 \text{ N/mm}^2 (< 0.8 f_y = 368 \text{ N/mm}^2).$$

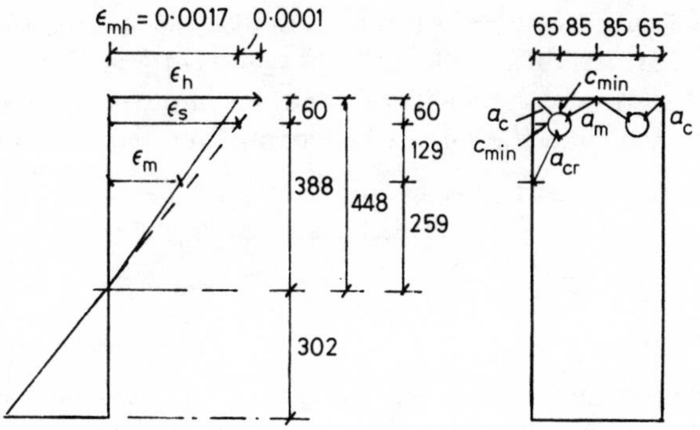

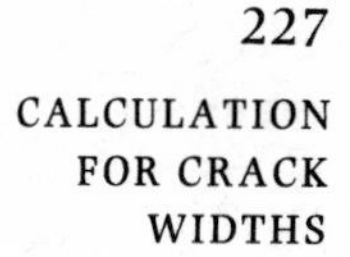



$$\varepsilon_s = \frac{f_s}{E_s} = \frac{313}{200 \times 10^3} = 0.001565.$$

$$\varepsilon_h = \frac{h-x}{d-x}\varepsilon_s = \frac{448}{388} \times 0.001565 = 0.0018.$$

$$\varepsilon_{mh} = \varepsilon_h - \frac{b_t(h-x)^2}{3E_sA_s(d-x)} = 0.0018 - \frac{300(448)^2}{3 \times 200 \times 10^3 \times 2510 \times 388}$$

$$= 0.0018 - 0.0001 = 0.0017.$$

(4) Calculate crack widths at critical positions

(a) Top of beam
Maximum crack width occurs midway between bars, where

$a_m = \sqrt{(60^2 + 85^2)} - 20 = 84$ mm

$c_{min} = 40$ mm (top face)

$$w_m = \frac{3 \times 84 \times 0.0017}{1 + 2(84 - 40)/448} = 0.36 \text{ mm } (>0.3 \text{ mm}).$$

(b) At corner of beam, $a_c = 68.5$ mm.

$$w_c = \frac{3 \times 68.5 \times 0.0017}{1 + 2(68.5 - 40)/448} = 0.31 \text{ mm } (>0.3 \text{ mm}).$$

(c) On side of beam
Critical position approximately $(d-x)/3$ from reinforcement, which is 259 mm from neutral axis.

$a' = 302 + 259 = 561$ mm.

$$\varepsilon_m = \frac{a'-x}{h-x}\varepsilon_{mh} = \frac{259}{448} \times 0.0017 = 0.00098.$$

$a_{cr} = \sqrt{[65^2 + (d-a')^2]} - 20 = \sqrt{(65^2 + 129^2)} - 20 = 124.5$ mm.

$c_{min} = 45$ mm (side face).

$$w_{cr} = \frac{3 \times 124.5 \times 0.00098}{1 + 2(124.5 - 45)/448} = 0.27 \text{ mm } (<0.3 \text{ mm}).$$

From these calculations it can be seen that although the spacing of the bars comply with the bar spacing rules, the calculated crack widths in 4(a) and (b) exceed 0.3 mm. This is one of the cases where the calculation does not improve on the rules, and in deep beams this does appear to be the case. A better condition would be achieved by using $3/32\phi$ bars.

---

### (b) DIRECT TENSION

Direct tension as a predominant force is unlikely to occur in normal building structures, but direct tension forces and bending moments may occur. In this case the neutral axis will be calculated taking into account the direct tension; this is more complicated than as carried out in the previous example. Having done this the calculations for flexural crack width will be as before.

The limit of flexure being predominant is when the neutral axis is at the top face of the member shown in Fig. A2.1, i.e. when $x=0$. The equation for the crack width now becomes

$$w_{cr}=\frac{3a_{cr}\varepsilon_m}{1+2(a_{cr}-c_{min})/h}$$

and the tension stiffening effect $=\dfrac{b_t h^2}{3E_s A_s d}$.

When the whole section is in tension it is suggested that the modification suggested in 3.8.3(b) for axial tension is used. The equation for the crack width now becomes

$$w_{cr}=3a_{cr}\varepsilon_m$$

and

$$\varepsilon_m=\varepsilon_1-\frac{2b_t h}{3E_s A_s},$$

where $A_s$ is the total area of steel in the section, equally divided between the two faces.

The most obvious case of axial tension is the hoop tension in a circular tank containing water. In the Code for liquid-retaining structures the maximum crack width is 0.2 mm. Designers of these structures are interested primarily in serviceability limit state of cracking. A calculation is required at ultimate limit state, but if the walls are cracked and water is leaking out, the tank is unserviceable and a factor of safety at ultimate is irrelevant.

Tables have been prepared whereby a designer can select an arrangement of bars to suit a particular wall thickness, tensile force, cover and crack width requirement. To illustrate the calculation procedure the following example uses values from these tables.

## EXAMPLE A2.2

The hoop tension force in the wall of a tank is 570 kN/m. The wall is 200 mm thick, cover to reinforcement is 40 mm, and the limiting crack width is 0.2 mm. Check that T16 at 150 centres each face is satisfactory.

Total $A_s = 2680$ mm$^2$/m.

$$\varepsilon_1 = \frac{570 \times 10^3}{2680 \times 200 \times 10^3} = 0.00106.$$

$$\text{Tensioning stiffening} = \frac{2 \times 1000 \times 200}{3 \times 200 \times 10^3 \times 2680} = 0.00025.$$

$\varepsilon_m = 0.00106 - 0.00025 = 0.00081.$

$a_{cr} = \sqrt{(48^2 + 75^2)} - 8 = 81.$

$w_{cr} = 3 \times 81 \times 0.00081 = 0.197$ mm.

# Appendix 3
# RADII OF BENDS TO LIMIT BEARING STRESSES

Concrete Grade 25
Radii of bends in bar sizes ($r = K\phi$).

| Bar size | (mm) | Stress $f$ (N/mm$^2$) 100 | 150 | 200 | 250 | 300 | 350 | 400 |
|---|---|---|---|---|---|---|---|---|
| 16 | 40 | 2.8 | 4.2 | 5.7 | 7.1 | 8.5 | 9.9 | 11.3 |
| | 50 | 2.6 | 3.9 | 5.2 | 6.4 | 7.7 | 9.0 | 10.3 |
| | 60 | 2.4 | 3.6 | 4.8 | 6.0 | 7.2 | 8.4 | 9.6 |
| | 70 | 2.3 | 3.4 | 4.6 | 5.7 | 6.9 | 8.0 | 9.2 |
| | 80 | 2.2 | 3.3 | 4.4 | 5.5 | 6.6 | 7.7 | 8.8 |
| | 90 | 2.1 | 3.2 | 4.3 | 5.3 | 6.4 | 7.5 | 8.5 |
| | 100 | 2.1 | 3.1 | 4.1 | 5.2 | 6.2 | 7.3 | 8.3 |
| | 150 | 1.9 | 2.9 | 3.8 | 4.8 | 5.7 | 6.7 | 7.6 |
| | 200 | 1.8 | 2.7 | 3.6 | 4.6 | 5.5 | 6.4 | 7.3 |
| | 250 | 1.8 | 2.7 | 3.5 | 4.4 | 5.3 | 6.2 | 7.1 |
| | 300 | 1.7 | 2.6 | 3.5 | 4.3 | 5.2 | 6.1 | 7.0 |
| 20 | 40 | 3.1 | 4.7 | 6.3 | 7.9 | 9.4 | 11.0 | 12.6 |
| | 50 | 2.8 | 4.2 | 5.7 | 7.1 | 8.5 | 9.9 | 11.3 |
| | 60 | 2.6 | 3.9 | 5.2 | 6.5 | 7.9 | 9.2 | 10.5 |
| | 70 | 2.5 | 3.7 | 4.9 | 6.2 | 7.4 | 8.6 | 9.9 |
| | 80 | 2.4 | 3.5 | 4.7 | 5.9 | 7.1 | 8.2 | 9.4 |
| | 90 | 2.3 | 3.4 | 4.5 | 5.7 | 6.8 | 7.9 | 9.1 |
| | 100 | 2.2 | 3.3 | 4.4 | 5.5 | 6.6 | 7.7 | 8.8 |
| | 150 | 2.0 | 3.0 | 4.0 | 5.0 | 6.0 | 7.0 | 8.0 |
| | 200 | 1.9 | 2.8 | 3.8 | 4.7 | 5.7 | 6.6 | 7.5 |
| | 250 | 1.8 | 2.7 | 3.6 | 4.6 | 5.5 | 6.4 | 7.3 |
| | 300 | 1.8 | 2.7 | 3.6 | 4.5 | 5.3 | 6.2 | 7.1 |
| 25 | 40 | 3.5 | 5.3 | 7.1 | 8.8 | 10.6 | 12.4 | 14.1 |
| | 50 | 3.1 | 4.7 | 6.3 | 7.9 | 9.4 | 11.0 | 12.6 |
| | 60 | 2.9 | 4.3 | 5.8 | 7.2 | 8.6 | 10.1 | 11.5 |
| | 70 | 2.7 | 4.0 | 5.4 | 6.7 | 8.1 | 9.4 | 10.8 |
| | 80 | 2.6 | 3.8 | 5.1 | 6.4 | 7.7 | 8.9 | 10.2 |
| | 90 | 2.4 | 3.7 | 4.9 | 6.1 | 7.3 | 8.6 | 9.8 |
| | 100 | 2.4 | 3.5 | 4.7 | 5.9 | 7.1 | 8.2 | 9.4 |
| | 150 | 2.1 | 3.1 | 4.2 | 5.2 | 6.3 | 7.3 | 8.4 |

| Bar size | (mm) | 100 | 150 | 200 | 250 | 300 | 350 | 400 |
|---|---|---|---|---|---|---|---|---|
| | | | | Stress $f$ (N/mm$^2$) | | | | |
| | 200 | 2.0 | 2.9 | 3.9 | 4.9 | 5.9 | 6.9 | 7.9 |
| | 250 | 1.9 | 2.8 | 3.8 | 4.7 | 5.7 | 6.6 | 7.5 |
| | 300 | 1.8 | 2.7 | 3.7 | 4.6 | 5.5 | 6.4 | 7.3 |
| 32 | 40 | 4.1 | 6.1 | 8.2 | 10.2 | 12.3 | 14.3 | 16.3 |
| | 50 | 3.6 | 5.4 | 7.2 | 9.0 | 10.7 | 12.5 | 14.3 |
| | 60 | 3.2 | 4.9 | 6.5 | 8.1 | 9.7 | 11.4 | 13.0 |
| | 70 | 3.0 | 4.5 | 6.0 | 7.5 | 9.0 | 10.5 | 12.0 |
| | 80 | 2.8 | 4.2 | 5.7 | 7.1 | 8.5 | 9.9 | 11.3 |
| | 90 | 2.7 | 4.0 | 5.4 | 6.7 | 8.1 | 9.4 | 10.8 |
| | 100 | 2.6 | 3.9 | 5.2 | 6.4 | 7.7 | 9.0 | 10.3 |
| | 150 | 2.2 | 3.4 | 4.5 | 5.6 | 6.7 | 7.8 | 9.0 |
| | 200 | 2.1 | 3.1 | 4.1 | 5.2 | 6.2 | 7.3 | 8.3 |
| | 250 | 2.0 | 3.0 | 3.9 | 4.9 | 5.9 | 6.9 | 7.9 |
| | 300 | 1.9 | 2.9 | 3.8 | 4.8 | 5.7 | 6.7 | 7.6 |

Concrete Grade 30
Radii of bends in bar sizes ($r = K\phi$).

| Bar size | (mm) | 100 | 150 | 200 | 250 | 300 | 350 | 400 |
|---|---|---|---|---|---|---|---|---|
| | | | | Stress $f$ (N/mm$^2$) | | | | |
| 16 | 40 | 2.4 | 3.5 | 4.7 | 5.9 | 7.1 | 8.2 | 9.4 |
| | 50 | 2.1 | 3.2 | 4.3 | 5.4 | 6.4 | 7.5 | 8.6 |
| | 60 | 2.0 | 3.0 | 4.0 | 5.0 | 6.0 | 7.0 | 8.0 |
| | 70 | 1.9 | 2.9 | 3.8 | 4.8 | 5.7 | 6.7 | 7.6 |
| | 80 | 1.8 | 2.7 | 3.7 | 4.6 | 5.5 | 6.4 | 7.3 |
| | 90 | 1.8 | 2.7 | 3.5 | 4.4 | 5.3 | 6.2 | 7.1 |
| | 100 | 1.7 | 2.6 | 3.5 | 4.3 | 5.2 | 6.0 | 6.9 |
| | 150 | 1.6 | 2.4 | 3.2 | 4.0 | 4.8 | 5.6 | 6.4 |
| | 200 | 1.5 | 2.3 | 3.0 | 3.8 | 4.6 | 5.3 | 6.1 |
| | 250 | 1.5 | 2.2 | 3.0 | 3.7 | 4.4 | 5.2 | 5.9 |
| | 300 | 1.4 | 2.2 | 2.9 | 3.6 | 4.3 | 5.1 | 5.8 |
| 20 | 40 | 2.6 | 3.9 | 5.2 | 6.5 | 7.9 | 9.2 | 10.5 |
| | 50 | 2.4 | 3.5 | 4.7 | 5.9 | 7.1 | 8.2 | 9.4 |
| | 60 | 2.2 | 3.3 | 4.4 | 5.5 | 6.5 | 7.6 | 8.7 |
| | 70 | 2.1 | 3.1 | 4.1 | 5.1 | 6.2 | 7.2 | 8.2 |
| | 80 | 2.0 | 2.9 | 3.9 | 4.9 | 5.9 | 6.9 | 7.9 |
| | 90 | 1.9 | 2.8 | 3.8 | 4.7 | 5.7 | 6.6 | 7.6 |
| | 100 | 1.8 | 2.7 | 3.7 | 4.6 | 5.5 | 6.4 | 7.3 |
| | 150 | 1.7 | 2.5 | 3.3 | 4.1 | 5.0 | 5.8 | 6.6 |
| | 200 | 1.6 | 2.4 | 3.1 | 3.9 | 4.7 | 5.5 | 6.3 |
| | 250 | 1.5 | 2.3 | 3.0 | 3.8 | 4.6 | 5.3 | 6.1 |
| | 300 | 1.5 | 2.2 | 3.0 | 3.7 | 4.5 | 5.2 | 5.9 |

| Bar size | (mm) | 100 | 150 | Stress $f$ (N/mm$^2$) 200 | 250 | 300 | 350 | 400 |
|---|---|---|---|---|---|---|---|---|
| 25 | 40 | 2.9 | 4.4 | 5.9 | 7.4 | 8.8 | 10.3 | 11.8 |
| | 50 | 2.6 | 3.9 | 5.2 | 6.5 | 7.9 | 9.2 | 10.5 |
| | 60 | 2.4 | 3.6 | 4.8 | 6.0 | 7.2 | 8.4 | 9.6 |
| | 70 | 2.2 | 3.4 | 4.5 | 5.6 | 6.7 | 7.9 | 9.0 |
| | 80 | 2.1 | 3.2 | 4.3 | 5.3 | 6.4 | 7.4 | 8.5 |
| | 90 | 2.0 | 3.1 | 4.1 | 5.1 | 6.1 | 7.1 | 8.1 |
| | 100 | 2.0 | 2.9 | 3.9 | 4.9 | 5.9 | 6.9 | 7.9 |
| | 150 | 1.7 | 2.6 | 3.5 | 4.4 | 5.2 | 6.1 | 7.0 |
| | 200 | 1.6 | 2.5 | 3.3 | 4.1 | 4.9 | 5.7 | 6.5 |
| | 250 | 1.6 | 2.4 | 3.1 | 3.9 | 4.7 | 5.5 | 6.3 |
| | 300 | 1.5 | 2.3 | 3.1 | 3.8 | 4.6 | 5.3 | 6.1 |
| 32 | 40 | 3.4 | 5.1 | 6.8 | 8.5 | 10.2 | 11.9 | 13.6 |
| | 50 | 3.0 | 4.5 | 6.0 | 7.5 | 9.0 | 10.4 | 11.9 |
| | 60 | 2.7 | 4.1 | 5.4 | 6.8 | 8.1 | 9.5 | 10.8 |
| | 70 | 2.5 | 3.8 | 5.0 | 6.3 | 7.5 | 8.8 | 10.0 |
| | 80 | 2.4 | 3.5 | 4.7 | 5.9 | 7.1 | 8.2 | 9.4 |
| | 90 | 2.2 | 3.4 | 4.5 | 5.6 | 6.7 | 7.8 | 9.0 |
| | 100 | 2.1 | 3.2 | 4.3 | 5.4 | 6.4 | 7.5 | 8.6 |
| | 150 | 1.9 | 2.8 | 3.7 | 4.7 | 5.6 | 6.5 | 7.5 |
| | 200 | 1.7 | 2.6 | 3.5 | 4.3 | 5.2 | 6.0 | 6.9 |
| | 250 | 1.6 | 2.5 | 3.3 | 4.1 | 4.9 | 5.8 | 6.6 |
| | 300 | 1.6 | 2.4 | 3.2 | 4.0 | 4.8 | 5.6 | 6.4 |

Concrete Grade 35
Radii of bends in bar sizes ($r = K\phi$).

| Bar size | (mm) | 100 | 150 | Stress $f$ (N/mm$^2$) 200 | 250 | 300 | 350 | 400 |
|---|---|---|---|---|---|---|---|---|
| 16 | 40 | 2.0 | 3.0 | 4.0 | 5.0 | 6.1 | 7.1 | 8.1 |
| | 50 | 1.8 | 2.8 | 3.7 | 4.6 | 5.5 | 6.4 | 7.4 |
| | 60 | 1.7 | 2.6 | 3.4 | 4.3 | 5.2 | 6.0 | 6.9 |
| | 70 | 1.6 | 2.5 | 3.3 | 4.1 | 4.9 | 5.7 | 6.5 |
| | 80 | 1.6 | 2.4 | 3.1 | 3.9 | 4.7 | 5.5 | 6.3 |
| | 90 | 1.5 | 2.3 | 3.0 | 3.8 | 4.6 | 5.3 | 6.1 |
| | 100 | 1.5 | 2.2 | 3.0 | 3.7 | 4.4 | 5.2 | 5.9 |
| | 150 | 1.4 | 2.0 | 2.7 | 3.4 | 4.1 | 4.8 | 5.4 |
| | 200 | 1.3 | 2.0 | 2.6 | 3.3 | 3.9 | 4.6 | 5.2 |
| | 250 | 1.3 | 1.9 | 2.5 | 3.2 | 3.8 | 4.4 | 5.1 |
| | 300 | 1.2 | 1.9 | 2.5 | 3.1 | 3.7 | 4.3 | 5.0 |
| 20 | 40 | 2.2 | 3.4 | 4.5 | 5.6 | 6.7 | 7.9 | 9.0 |
| | 50 | 2.0 | 3.0 | 4.0 | 5.0 | 6.1 | 7.1 | 8.1 |
| | 60 | 1.9 | 2.8 | 3.7 | 4.7 | 5.6 | 6.5 | 7.5 |

| Bar size | (mm) | 100 | 150 | 200 | 250 | 300 | 350 | 400 |
|---|---|---|---|---|---|---|---|---|
| | | | | Stress $f$ (N/mm$^2$) | | | | |
| | 70 | 1.8 | 2.6 | 3.5 | 4.4 | 5.3 | 6.2 | 7.1 |
| | 80 | 1.7 | 2.5 | 3.4 | 4.2 | 5.0 | 5.9 | 6.7 |
| | 90 | 1.6 | 2.4 | 3.2 | 4.1 | 4.9 | 5.7 | 6.5 |
| | 100 | 1.6 | 2.4 | 3.1 | 3.9 | 4.7 | 5.5 | 6.3 |
| | 150 | 1.4 | 2.1 | 2.8 | 3.6 | 4.3 | 5.0 | 5.7 |
| | 200 | 1.3 | 2.0 | 2.7 | 3.4 | 4.0 | 4.7 | 5.4 |
| | 250 | 1.3 | 2.0 | 2.6 | 3.3 | 3.9 | 4.6 | 5.2 |
| | 300 | 1.3 | 1.9 | 2.5 | 3.2 | 3.8 | 4.5 | 5.1 |
| 25 | 40 | 2.5 | 3.8 | 5.0 | 6.3 | 7.6 | 8.8 | 10.1 |
| | 50 | 2.2 | 3.4 | 4.5 | 5.6 | 6.7 | 7.9 | 9.0 |
| | 60 | 2.1 | 3.1 | 4.1 | 5.1 | 6.2 | 7.2 | 8.2 |
| | 70 | 1.9 | 2.9 | 3.8 | 4.8 | 5.8 | 6.7 | 7.7 |
| | 80 | 1.8 | 2.7 | 3.6 | 4.6 | 5.5 | 6.4 | 7.3 |
| | 90 | 1.7 | 2.6 | 3.5 | 4.4 | 5.2 | 6.1 | 7.0 |
| | 100 | 1.7 | 2.5 | 3.4 | 4.2 | 5.0 | 5.9 | 6.7 |
| | 150 | 1.5 | 2.2 | 3.0 | 3.7 | 4.5 | 5.2 | 6.0 |
| | 200 | 1.4 | 2.1 | 2.8 | 3.5 | 4.2 | 4.9 | 5.6 |
| | 250 | 1.3 | 2.0 | 2.7 | 3.4 | 4.0 | 4.7 | 5.4 |
| | 300 | 1.3 | 2.0 | 2.6 | 3.3 | 3.9 | 4.6 | 5.2 |
| 32 | 40 | 2.9 | 4.4 | 5.8 | 7.3 | 8.8 | 10.2 | 11.7 |
| | 50 | 2.6 | 3.8 | 5.1 | 6.4 | 7.7 | 9.0 | 10.2 |
| | 60 | 2.3 | 3.5 | 4.6 | 5.8 | 7.0 | 8.1 | 9.3 |
| | 70 | 2.1 | 3.2 | 4.3 | 5.4 | 6.4 | 7.5 | 8.6 |
| | 80 | 2.0 | 3.0 | 4.0 | 5.0 | 6.1 | 7.1 | 8.1 |
| | 90 | 1.9 | 2.9 | 3.8 | 4.8 | 5.8 | 6.7 | 7.7 |
| | 100 | 1.8 | 2.8 | 3.7 | 4.6 | 5.5 | 6.4 | 7.4 |
| | 150 | 1.6 | 2.4 | 3.2 | 4.0 | 4.8 | 5.6 | 6.4 |
| | 200 | 1.5 | 2.2 | 3.0 | 3.7 | 4.4 | 5.2 | 5.9 |
| | 250 | 1.4 | 2.1 | 2.8 | 3.5 | 4.2 | 4.9 | 5.6 |
| | 300 | 1.4 | 2.0 | 2.7 | 3.4 | 4.1 | 4.8 | 5.4 |

Concrete Grade 40
Radii of bends in bar sizes ($r = K\phi$).

| Bar size | (mm) | 100 | 150 | 200 | 250 | 300 | 350 | 400 |
|---|---|---|---|---|---|---|---|---|
| | | | | Stress $f$ (N/mm$^2$) | | | | |
| 16 | 40 | 1.8 | 2.7 | 3.5 | 4.4 | 5.3 | 6.2 | 7.1 |
| | 50 | 1.6 | 2.4 | 3.2 | 4.0 | 4.8 | 5.6 | 6.4 |
| | 60 | 1.5 | 2.3 | 3.0 | 3.8 | 4.5 | 5.3 | 6.0 |
| | 70 | 1.4 | 2.1 | 2.9 | 3.6 | 4.3 | 5.0 | 5.7 |
| | 80 | 1.4 | 2.1 | 2.7 | 3.4 | 4.1 | 4.8 | 5.5 |
| | 90 | 1.3 | 2.0 | 2.7 | 3.3 | 4.0 | 4.7 | 5.3 |

| Bar size | (mm) | Stress $f$ (N/mm$^2$) 100 | 150 | 200 | 250 | 300 | 350 | 400 |
|---|---|---|---|---|---|---|---|---|
| | 100 | 1.3 | 1.9 | 2.6 | 3.2 | 3.9 | 4.5 | 5.2 |
| | 150 | 1.2 | 1.8 | 2.4 | 3.0 | 3.6 | 4.2 | 4.8 |
| | 200 | 1.1 | 1.7 | 2.3 | 2.8 | 3.4 | 4.0 | 4.6 |
| | 250 | 1.1 | 1.7 | 2.2 | 2.8 | 3.3 | 3.9 | 4.4 |
| | 300 | 1.1 | 1.6 | 2.2 | 2.7 | 3.2 | 3.8 | 4.3 |
| 20 | 40 | 2.0 | 2.9 | 3.9 | 4.9 | 5.9 | 6.9 | 7.9 |
| | 50 | 1.8 | 2.7 | 3.5 | 4.4 | 5.3 | 6.2 | 7.1 |
| | 60 | 1.6 | 2.5 | 3.3 | 4.1 | 4.9 | 5.7 | 6.5 |
| | 70 | 1.5 | 2.3 | 3.1 | 3.9 | 4.6 | 5.4 | 6.2 |
| | 80 | 1.5 | 2.2 | 2.9 | 3.7 | 4.4 | 5.2 | 5.9 |
| | 90 | 1.4 | 2.1 | 2.8 | 3.5 | 4.3 | 5.0 | 5.7 |
| | 100 | 1.4 | 2.1 | 2.7 | 3.4 | 4.1 | 4.8 | 5.5 |
| | 150 | 1.2 | 1.9 | 2.5 | 3.1 | 3.7 | 4.4 | 5.0 |
| | 200 | 1.2 | 1.8 | 2.4 | 2.9 | 3.5 | 4.1 | 4.7 |
| | 250 | 1.1 | 1.7 | 2.3 | 2.8 | 3.4 | 4.0 | 4.6 |
| | 300 | 1.1 | 1.7 | 2.2 | 2.8 | 3.3 | 3.9 | 4.5 |
| 25 | 40 | 2.2 | 3.3 | 4.4 | 5.5 | 6.6 | 7.7 | 8.8 |
| | 50 | 2.0 | 2.9 | 3.9 | 4.9 | 5.9 | 6.9 | 7.9 |
| | 60 | 1.8 | 2.7 | 3.6 | 4.5 | 5.4 | 6.3 | 7.2 |
| | 70 | 1.7 | 2.5 | 3.4 | 4.2 | 5.0 | 5.9 | 6.7 |
| | 80 | 1.6 | 2.4 | 3.2 | 4.0 | 4.8 | 5.6 | 6.4 |
| | 90 | 1.5 | 2.3 | 3.1 | 3.8 | 4.6 | 5.3 | 6.1 |
| | 100 | 1.5 | 2.2 | 2.9 | 3.7 | 4.4 | 5.2 | 5.9 |
| | 150 | 1.3 | 2.0 | 2.6 | 3.3 | 3.9 | 4.6 | 5.2 |
| | 200 | 1.2 | 1.8 | 2.5 | 3.1 | 3.7 | 4.3 | 4.9 |
| | 250 | 1.2 | 1.8 | 2.4 | 2.9 | 3.5 | 4.1 | 4.7 |
| | 300 | 1.1 | 1.7 | 2.3 | 2.9 | 3.4 | 4.0 | 4.6 |
| 32 | 40 | 2.6 | 3.8 | 5.1 | 6.4 | 7.7 | 8.9 | 10.2 |
| | 50 | 2.2 | 3.4 | 4.5 | 5.6 | 6.7 | 7.8 | 9.0 |
| | 60 | 2.0 | 3.0 | 4.1 | 5.1 | 6.1 | 7.1 | 8.1 |
| | 70 | 1.9 | 2.8 | 3.8 | 4.7 | 5.6 | 6.6 | 7.5 |
| | 80 | 1.8 | 2.7 | 3.5 | 4.4 | 5.3 | 6.2 | 7.1 |
| | 90 | 1.7 | 2.5 | 3.4 | 4.2 | 5.0 | 5.9 | 6.7 |
| | 100 | 1.6 | 2.4 | 3.2 | 4.0 | 4.8 | 5.6 | 6.4 |
| | 150 | 1.4 | 2.1 | 2.8 | 3.5 | 4.2 | 4.9 | 5.6 |
| | 200 | 1.3 | 1.9 | 2.6 | 3.2 | 3.9 | 4.5 | 5.2 |
| | 250 | 1.2 | 1.8 | 2.5 | 3.1 | 3.7 | 4.3 | 4.9 |
| | 300 | 1.2 | 1.8 | 2.4 | 3.0 | 3.6 | 4.2 | 4.8 |

# Appendix 4

# TABLES OF SHEAR RESISTANCE FOR LINKS

**Table 1. Values of $A_{sv}/s_v$**

$$A_{sv}/s_v = b_v(v - v_c)/0.87f_{yv}.$$

Calculate right-hand side of equation, then read from table a suitable size, $\phi$ and spacing, $s_v$.

| | $\phi$ | | | | |
|---|---|---|---|---|---|
| $s_v$ | 6 | 8 | 10 | 12 | 16 |
| 50 | 1.13 | 2.01 | 3.14 | 4.52 | 8.04 |
| 75 | 0.75 | 1.34 | 2.09 | 3.02 | 5.36 |
| 100 | 0.57 | 1.01 | 1.57 | 2.26 | 4.02 |
| 125 | 0.45 | 0.80 | 1.26 | 1.81 | 3.22 |
| 150 | 0.38 | 0.67 | 1.05 | 1.51 | 2.68 |
| 175 | 0.32 | 0.57 | 0.90 | 1.29 | 2.30 |
| 200 | 0.28 | 0.50 | 0.79 | 1.13 | 2.01 |
| 225 | 0.25 | 0.45 | 0.70 | 1.01 | 1.79 |
| 250 | 0.23 | 0.40 | 0.63 | 0.90 | 1.61 |
| 275 | 0.21 | 0.37 | 0.57 | 0.82 | 1.46 |
| 300 | 0.19 | 0.34 | 0.52 | 0.75 | 1.34 |
| 325 | 0.17 | 0.31 | 0.48 | 0.70 | 1.24 |
| 350 | 0.16 | 0.29 | 0.45 | 0.65 | 1.15 |
| 375 | 0.15 | 0.27 | 0.42 | 0.60 | 1.07 |
| 400 | 0.14 | 0.25 | 0.39 | 0.57 | 1.01 |

**Table 2. Values of $0.87f_{yv}A_{sv}/s_v$**

$$b_v(v-v_c)=0.87f_{yv}A_{sv}/s_v.$$

Calculate left-hand side of equation, then read from table a suitable size, $\phi$ and spacing, $s_v$.

(a) $f_{yv}=250$ N/mm$^2$

| | $\phi$ | | | | |
|---|---|---|---|---|---|
| $s_v$ | 6 | 8 | 10 | 12 | 16 |
| 50 | 246 | 437 | 683 | 984 | 1749 |
| 75 | 164 | 292 | 456 | 656 | 1166 |
| 100 | 123 | 219 | 342 | 492 | 875 |
| 125 | 98 | 175 | 273 | 394 | 700 |
| 150 | 82 | 146 | 228 | 328 | 583 |
| 175 | 70 | 125 | 195 | 281 | 500 |
| 200 | 62 | 109 | 171 | 246 | 437 |
| 225 | 55 | 97 | 152 | 219 | 389 |
| 250 | 49 | 87 | 137 | 197 | 350 |
| 275 | 45 | 80 | 124 | 179 | 318 |
| 300 | 41 | 73 | 114 | 164 | 292 |
| 325 | 38 | 67 | 105 | 151 | 269 |
| 350 | 35 | 62 | 98 | 141 | 250 |
| 375 | 33 | 58 | 91 | 131 | 233 |
| 400 | 31 | 55 | 85 | 123 | 219 |

(b) $f_{yv}=460$ N/mm$^2$

| | $\phi$ | | | | |
|---|---|---|---|---|---|
| $s_v$ | 6 | 8 | 10 | 12 | 16 |
| 50 | 453 | 805 | 1257 | 1810 | 3219 |
| 75 | 302 | 536 | 839 | 1207 | 2146 |
| 100 | 227 | 402 | 629 | 905 | 1609 |
| 125 | 181 | 322 | 503 | 724 | 1287 |
| 150 | 151 | 268 | 419 | 603 | 1073 |
| 175 | 129 | 230 | 359 | 517 | 920 |
| 200 | 113 | 201 | 314 | 453 | 805 |
| 225 | 101 | 179 | 279 | 402 | 715 |
| 250 | 91 | 161 | 251 | 362 | 644 |
| 275 | 82 | 146 | 229 | 329 | 584 |
| 300 | 76 | 134 | 210 | 302 | 536 |
| 325 | 70 | 124 | 193 | 279 | 495 |
| 350 | 65 | 115 | 180 | 259 | 460 |
| 375 | 60 | 107 | 168 | 241 | 429 |
| 400 | 57 | 101 | 157 | 226 | 402 |

# INDEX

60023